MONOGRAPHS ON
STATISTICS AND APPLIED PROBABILITY

General Editors

D.R. Cox, D.V. Hinkley, N. Reid, D.B. Rubin and B.W. Silverman

28 Sequential Methods in Statistics, 3rd edition *G.B. Wetherill* (1986)

29 Tensor Methods in Statistics *P. McCullagh* (1987)

30 Transformation and Weighting in Regression *R.J. Carroll and D. Ruppert* (1988)

31 Asymptotic Techniques for use in Statistics *O.E. Barndorff-Nielsen and D.R. Cox* (1989)

32 Analysis of Binary Data, 2nd edition *D.R. Cox and E.J. Snell* (1989)

33 Analysis of Infectious Disease Data *N.G. Becker* (1989)

34 Design and Analysis of Cross-Over Trials *B. Jones and M.G. Kenward* (1989)

35 Empirical Bayes Method, 2nd edition *J.S. Maritz and T. Lwin* (1989)

36 Symmetric Multivariate and Related Distributions *K.-T. Fang, S. Kotz and K. Ng* (1989)

37 Generalized Linear Models, 2nd edition *P. McCullagh and J.A. Nelder* (1989)

38 Cyclic Designs *J.A. John* (1987)

39 Analog Estimation Methods in Econometrics *C.F. Manski* (1988)

40 Subset Selection in Regression *A.J. Miller* (1990)

41 Analysis of Repeated Measures *M. Crowder and D.J. Hand* (1990)

42 Statistical Reasoning with Imprecise Probabilities *P. Walley* (1990)

43 Generalized Additive Models *T.J. Hastie and R.J. Tibshirani* (1990)

44 Inspection Errors for Attributes in Quality Control *N.L. Johnson, S. Kotz and X. Wu* (1991)

45 The Analysis of Contingency Tables, 2nd edition *B.S. Everitt* (1992)

46 The Analysis of Quantal Response Data *B.J.T. Morgan* (1992)

47 Longitudinal Data with Serial Correlation: A state-space Approach *R.H. Jones*

48 Differential Geometry and Statistics *Michael K. Murray and John W. Rice*

49 Markov Models and Optimization *M.H.A. Davis*

50 Networks and Chaos—Statistical and Probabilistic Aspects *O.E. Barndorff-Nielsen, J.L. Jensen and W.S. Kendall* (1993)

51 Number Theoretic Methods in Statistics *K.-T. Fang and W. Yuan* (1993)

52 Inference and Asymptotics *D.R. Cox and O.E. Barndorff-Nielsen* (1994)

53 Practical Risk Theory for Actuaries *C.D. Daykin, T. Pentikainen and M. Pesonen* (1993)

(Full details concerning this series are available from the Publishers.)

Markov Models and Optimization

M.H.A. DAVIS
Centre for Process Systems Engineering and Department of Electrical and Electronic Engineering, Imperial College of Science, Technology and Medicine, London

CHAPMAN & HALL
London · Glasgow · New York · Tokyo · Melbourne · Madras

Published by Chapman & Hall, 2–6 Boundary Row, London SE1 8HN

Chapman & Hall, 2–6 Boundary Row, London SE1 8HN, UK

Blackie Academic & Professional, Wester Cleddens Road, Bishopbriggs, Glasgow G64 2NZ, UK

Chapman & Hall Inc., 29 West 35th Street, New York NY10001, USA

Chapman & Hall Japan, Thomson Publishing Japan, Hirakawacho Nemoto Building, 6F, 1-7-11 Hirakawa-cho, Chiyoda-ku, Tokyo 102, Japan

Chapman & Hall Australia, Thomas Nelson Australia, 102 Dodds Street, South Melbourne, Victoria 3205, Australia

Chapman & Hall India, R. Seshadri, 32 Second Main Road, CIT East, Madras 600 035, India

First edition 1993

Typeset in 10/12 Times by Thomson Press (India) Ltd, New Delhi
Printed in Great Britain by St. Edmundsbury Press, Bury St. Edmunds, Suffolk

ISBN 0 412 31410 X

A catalogue record for this book is available from the British Library

Library of Congress Cataloging-in-Publishing data
Davis, M.H.A.
Markov models and optimization / M.H.A. Davis. – 1st ed.
p. cm. – (Monographs on statistics and applied probability)
Includes bibliographical references and indexes.
ISBN 0–412–31410–X
1. Markov processes. 2. Mathematical optimization. 3. Control theory. I. Title. II. Series.
QA274.7.D34 1993 92–39557
003′.85 – dc20 CIP

∞ Printed on permanent acid-free text paper, manufactured in accordance with the proposed ANSI/NISO Z 39.48-199X and ANSI Z 39.48-1984

TO JESSICA
for consenting to a ménage à trois *with Elonex*

Contents

Preface

This book is about evaluating and optimizing the performance of continuous-time dynamical systems under uncertainty (more specifically, those systems in which the basic source of uncertainty is a sequence of random occurrences taking place at deterministic or random times). This covers an enormous variety of applications in engineering systems, operations research, management science, economics and applied probability; a few examples are queueing systems (the random occurrences being arrival of customers or completion of service), investment planning (changes of interest rate or demand level), stochastic scheduling (completion of jobs or failure of machines), naval target tracking (changes of course by a manoeuvring target), insurance analysis (occurrence of claims, changes in premium rates), and optimal exploitation of resources such as fisheries, forestry or oil (amounts of resources found, random factors associated with harvesting or production, changes in market prices). All these examples – and there are many more – are *dynamic* in that actions are taken over time and actions taken now have repercussions in the future, and *stochastic* in that they involve uncertainty of the sort just described.

In this book a class of stochastic models called *piecewise-deterministic Markov processes* (referred to throughout as PDPs) is proposed as a general framework for studying problems of this kind. In the first half of the book (Chapters 1–3) the PDP is introduced, its properties are studied and methods are described for evaluating the probabilities of events and expectations and moments of random variables in stochastic system models. The second half of the book (Chapters 4 and 5) concerns optimization, i.e. stochastic control theory for PDP models. It differs radically, however, from other treatments of stochastic control, and the following remarks are intended to set the subject in context.

Stochastic control theory has developed in two main and rather separate streams, for discrete- and continuous-time models, respectively. The discrete-time theory, some of which is outlined briefly in section 47 of this book, is mainly a development of the ideas behind dynamic programming. In addition to its use for problems originally formulated in a discrete time setting, this theory is useful as a computational technique for continuous-time models, replacing the original model with a discrete-time approximation. Continuous-time stochastic control theory has, up to now, been concerned largely with systems represented by Ito stochastic differential equations (SDEs). This is the natural generalization of deterministic optimal control theory, as developed by L.S. Pontryagin (1962) and others, which concerns control of differential systems, i.e. systems represented by ordinary differential equations. In the stochastic case, the SDE model can be thought of as a deterministic differential system subject to 'white noise' perturbations. The theory for this case is extensive, elegant and largely complete. It capitalizes on the intimate connection between SDEs and second-order partial differential equations (PDEs) and the fact that the properties of the latter equations are so well understood.

It is, however, hard to deny that the SDE model has received more than its fair share of attention. The theory is in excellent shape, but the budget of genuine applications is, it must be admitted, disappointingly small. Exceptions to this sweeping statement are the LQG (linear system/quadratic cost/Gaussian disturbances) model introduced by R.E. Kalman (1960), which can be thought of as a general 'small-signal' model and has won a permanent place in engineering design; the use of diffusion models as 'heavy-traffic' approximations in queueing systems and other discrete models; and – in a very different vein – the much more recent applications of stochastic analysis to option pricing and other problems in financial economics, where the SDE model seems to play a very natural role. There are, of course, other applications, in at least some of which one gets the uneasy feeling that the stochastic modelling has been forced unnaturally into an SDE mould merely because the theory is available. There are also a very large number of applied problems, such as those mentioned above, for which the SDE model is completely inapplicable. Such problems have typically been tackled in piecemeal ways, using special techniques applicable to specific cases; yet many of them clearly have a lot in common. The piecewise-deterministic Markov process is intended to fill this gap. It has two advantages:

(a) it covers all examples of the sort described above, and in fact virtually all continuous-time stochastic models arising in applications *except* diffusions, and (b) it is technically simpler than the SDE model, in that the latter involves the theory of Brownian motion and Ito integration, the mastering of which is a not inconsiderable overhead. It has an apparent disadvantage, namely that the role played by familiar PDEs in the SDE theory is here played by certain systems of integro-differential equations, for which there is not (and cannot be) a unified theory such as exists for PDEs. At one time this was thought to be a serious obstacle to the adoption of PDPs as a general methodology. However, as we will show in Chapter 3, the integro-differential equations in question are readily solved by a simple iterative method that is highly parallelizable and hence amenable to efficient computation. Thus if one takes the reasonable view that a problem in applied probability is 'solved' when one has printed out the answer, the situation of PDP models is actually rather favourable.

The book proper begins with Chapter 2. Chapter 1 contains a summary of notions and results in analysis, probability and stochastic processes that are required subsequently. It is not a self-contained introduction to these topics – which are covered fully in many excellent texts – but more in the nature of a whistle-stop tour: readers already familiar with the sights can refresh their memories, while those to whom the terrain is new can plan a more extended tour later. Chapter 2 begins with a discussion of some of the examples mentioned above, with a view to showing how equivalent Markov models can be constructed by the inclusion of so-called *supplementary variables*. The importance of Markov models is that they are the only ones for which systematic methods of calculation of distributions and average values are available. The examples lead to the general definition of the piecewise-deterministic Markov process, which covers them all as special cases. The remainder of the chapter is devoted to establishing the Markov properties of the PDP, namely the strong Markov property, a characterization of the extended generator, and a demonstration of the fact that PDPs are *Borel right processes*. Chapter 3 develops the *systematic methods of calculation*. These are based on the differential formula for PDPs – a formula that plays the role of the 'change of variables' formula of ordinary calculus, or the Ito formula of Ito calculus. The significance of the differential formula is that it provides the link between PDPs and the integro-differential equations characterizing expected values of certain functionals of the process.

The iterative method by which these equations may be solved is then given, together with some examples. Section 34 concerns the important topic of *stationary distributions*. It is shown that there is a relation between stationary distributions for a PDP and stationary distributions of an imbedded Markov chain; this is helpful in studying the former since the latter is the more tractable object.

The remaining two chapters deal with control theory: Chapter 4 deals with *continuous control*, where the decision-maker applies control action at all time moments, and Chapter 5 with *impulse control*, where the decision-maker takes occasional interventions. The main objective in Chapter 4 is to show that the so-called *value function* of dynamic programming satisfies a generalized version of the *Bellman equation*, a nonlinear analogue of the integro-differential equations of Chapter 3 which characterizes *minimum* expected values over certain classes of control strategies. The optimal strategy itself is easily determined once the value function is known. In Chapter 5, two characterizations of the value function for impulse control are given, one based on an iterative procedure, and the other on what is known as a *quasi-variational inequality*. The results in these chapters give analogues for PDPs of the principal results in stochastic control theory for stochastic differential equations, and indeed they look much the same, with the integro-differential operators associated with PDPs replacing the second-order differential operators associated with SDEs. However, the detailed interpretation and methods of proof are of course substantially different and are given here in self-contained form.

Not included in this book are systematic development of large-scale computational methods or detailed analysis of specific applied problems; needless to say, both of these form an essential part of the subject. To fill this gap, the author and Michael Dempster intend to edit a second volume containing case studies by different authors covering a spectrum of applications taken all the way from model formulation to numerical results. The objective of the present volume is to give an account of the main results of PDP theory in a compact and self-contained manner. The greatest possible level of generality has not always been aimed at: the aim has been to give, where possible, relatively straightforward arguments leading to the most useful results rather than to obtain the most general results possible. In the interests of completeness, some more technical results have been included, for example the martingale representation property for stochastic jump processes (given in the Appendix) and the fact that PDPs are *Borel right processes* (section 27). Readers entirely new to the subject can afford to skip over these things at first. Basic prerequisites for reading this book are knowledge of stochastic processes and Markov processes as contained in courses in applied probability, and some familiarity with axiomatic probability theory.

This project really began in about 1980 with some work on water resource management at IIASA (International Institute for Applied Systems Analysis, Laxenburg, Austria) leading to a problem in optimal capacity expansion which fell outside the scope of any existing theory of stochastic control (it is discussed in Chapter 2), and an IIASA Working Paper by Domokos Vermes (1980) developing control theory for so-called piecewise-linear processes, which are a special case of PDPs. The PDP concept originated from discussions I had with Vermes and Dempster, based on these problems. Vermes subsequently wrote the first paper on optimal control of PDPs whereas Dempster and Juan–Juan Ye (1992) put this subject in more or less the form in which it now appears in Chapter 4. The theory of optimal stopping and impulse control, as it appears in Chapter 5, was largely developed by Ulrich Gugerli (1986) and Oswaldo Costa (see Costa and Davis (1988, 1989)). Other authors, in particular Suzanne Lenhart (1989) and Dariusz Gątarek (1990, 1991, 1992), have given independent treatments of these problems which have influenced the presentation here. In a different vein, Angelos Dassios and Paul Embrechts (1989) have pioneered applications of PDPs in risk theory and insurance, and their influence is very evident in Chapter 3. Each chapter ends with a Notes and References section in which I have attempted, no doubt unsuccessfully, to mention all related work.

This book was largely written in Spring 1991 while I was a Visiting Professor at the Mathematics Institute of the University of Oslo. There I had the exceptional opportunity to present most of the material in a seminar course to a small but select (and long-suffering) audience who provided a lot of useful feedback. I am particularly grateful to Bernt Øksendal who went to a great deal of trouble in making all the arrangements for this visit, including provision of the best skiing conditions in Nordmarka for several winters past. At Imperial College, London, thanks are due to Martin Clark for innumerable conversations, to Richard Vinter for, among other things, teaching me all I know about relaxed controls, and to Doris Abeysekera, who not only word-processed most of the manuscript (using the excellent $\mathbb{EXP}$ system) but also ably held the fort in London while I was away in Oslo. I am very grateful to Juan–Juan Ye and to Manfred Schäl for their careful reading of the manuscript, which unearthed an egregious number of misprints as well as some embarrassing mistakes. Still more errors were brought to light by Matthieu Biron, Takis Fotopoulos, Antonio Smania and Michalis Zervos in an informal seminar at Imperial College. I can only hope that not too many remain. I am grateful to the Research Section of the Royal Statistical Society for the opportunity to launch the Piecewise-Deterministic Process at an Ordinary Meeting of the Society and to the SERC for research funding. Finally, I very much appreciate the continuing interest of the Series Editor, Sir David Cox, in this project and the helpful collaboration of the editorial staff at Chapman & Hall.

Mark H.A. Davis
London

CHAPTER 1

Analysis, probability and stochastic processes

J.L. Doob (1953) complains in his classic treatise *Stochastic Processes* that 'any rigorous treatment of probability is threatened with the charge that it is absurdly overloaded with extraneous mathematics'. Time has, however, proved such charges to be ill-founded and, as Doob foresaw, the axiomatic formulation of probability based on measure theory – originally introduced by Kolmogorov (1933) – has become an indispensable framework for the study of stochastic processes. In particular, this book is largely about Markov processes, and since the definition of the Markov property is a statement about conditional expectations, we cannot do without an adequate formulation of this concept, which is one of the many things the measure-theoretic framework provides.

Axiomatic probability theory also has considerable intuitive content, particularly in the context of decision making under uncertainty, a major topic of this book: families of sigma-fields represent information flow; stopping times are closely associated with decision strategies; martingale properties characterize optimal decisions; and so on. There is far more to it than an arid concern for mathematical rectitude (important as the latter is). The purpose of this introductory chapter is to summarize those notions and results in probability and stochastic process theory that are needed in the remainder of the book. Some effort has been made to keep these prerequisites to a minimum, and only the basic facts about probability spaces, integration, continuous-time stochastic processes and Markov processes are needed. This chapter is not a self-contained introduction to these topics, which are covered in detail in several excellent textbooks, some of which are listed in section 15 at the end of this chapter. This chapter gives a quick summary: it can be used as a reference for specific results and as an outline map to orient the reader in consulting the textbooks for a more detailed treatment.

There is an intimate connection between probability and analysis, and we start in section 11 by discussing briefly some topics in real and functional analysis. The next two sections cover probability and stochastic processes, while the final section deals with Markov processes, with the emphasis on the so-called *differential generator* of a Markov process which plays such a major role in the remainder of the book.

11 Analysis

11.1 Continuity and compactness

The most fundamental notion in mathematical analysis is *continuity*, the study of which is the province of *topology*. A *topological space* is a set X together with a collection $\mathscr{T}$ of subsets called *open sets* that have the following properties:

1. $X \in \mathscr{T}$ and $\varnothing \in \mathscr{T}$;
2. If A, $B \in \mathscr{T}$ then $A \cap B \in \mathscr{T}$;
3. If $A_\alpha \in \mathscr{T}$ for all $\alpha \in J$, where J is an arbitrary index set, then $\bigcup_{\alpha \in J} A_\alpha \in \mathscr{T}$.

In 1, the symbol $\varnothing$ denotes the empty set. It is immediately obvious from 2, that if $A_1, A_2, \ldots, A_n \in \mathscr{T}$ then $\bigcap_{i=1}^n A_i \in \mathscr{T}$. Thus, briefly put, 2 and 3 state that $\mathscr{T}$ is closed under finite intersections and arbitrary unions. A set A is *closed* if its complement $A^c = X \backslash A = \{x \in X : x \notin A\}$ is open[1]. The *closure* $\bar{A}$ of A is the intersection of all closed sets containing A (this is a closed set, in view of 3 above), i.e. the smallest closed set containing A. The fundamental example is the real line $\mathbb{R} =]-\infty, +\infty[$. An *open interval* in $\mathbb{R}$ is a set of the form $]a, b[= \{x \in \mathbb{R}: a < x < b\}$ while a *closed interval* is $[a, b] = \{x \in \mathbb{R}: a \leqslant x \leqslant b\}$. The *usual topology* in $\mathbb{R}$ is defined as follows: a set A is open if, and only if, for every $x \in A$ there exists $\varepsilon > 0$ such that $]x - \varepsilon, x + \varepsilon[\subset A$. It follows easily that A is open if, and only if, it is a countable union of open intervals, and that closed intervals are indeed closed sets.

We now turn to the notion of *convergence* in topological spaces. For $x \in X$, a *neighbourhood* of x is a set A such that $x \in O \subset A$ for some open set O.

[1] $\{x: P\}$ is read as 'the set of x such that P'.

(11.1) *Definition.* A sequence $(x_n, n = 1,2,\ldots)$ in X converges to $x \in X$ if for any neighbourhood A of x there is a number m such that $x_n \in A$ for all $n > m$. We say, 'x_n is eventually in every neighbourhood of x', and write $x_n \to x$. A point $z \in X$ is a *cluster point* of the sequence (x_n) if $x_n \in A$ for infinitely many indices n whenever A is a neighbourhood of z. (Thus every limit point is a cluster point.)

If X and Y are sets, a *function* f from X to Y (written $F: X \to Y$) is simply a rule that associates a unique element $f(x) \in Y$ with each $x \in X$. For subsets $A \subset X, B \subset Y$ we write $f(A) = \{f(x): x \in A\}$ and $f^{-1}(B) = \{x \in X: f(x) \in B\}$. The function f is *onto* if $f(X) = Y$ and *one-to-one* if $f^{-1}(\{y\})$ is a single-point set for each $y \in Y$. In this case $f^{-1}: Y \to X$ is also a function. If $(X, \mathscr{T})$ and $(Y, \mathscr{S})$ are topological spaces then $f: X \to Y$ *is continuous* if $f^{-1}(\mathscr{S}) \subset \mathscr{T}$, i.e. $B \in \mathscr{S}$ implies $f^{-1}(B) \in \mathscr{T}$.

(11.2) *Theorem.* If $f:(X, \mathscr{T}) \to (Y, \mathscr{S})$ is continuous then $x_n \to x$ implies $f(x_n) \to f(x)$.

Proof. Let B be a neighbourhood of $f(x)$ and $f(x) \in O \subset B$ where $O \in \mathscr{S}$. Because f is continuous, $f^{-1}(O) \in \mathscr{T}$ and hence $f^{-1}(B)$ is a neighbourhood of x. Thus $x_n \in f^{-1}(B)$ for large n, which is equivalent to $f(x_n) \in B$. □

A set $K \subset X$ is *compact* if every open cover of K has a finite sub-cover, i.e. whenever $K \subset \bigcup_{\alpha \in J} O_\alpha$, where J is an arbitrary index set and O_α is open for each $\alpha \in J$, then $K \subset \bigcup_{i=1}^n O_{\alpha_i}$ for some $\alpha_1, \ldots, \alpha_n \in J$. Sets in Euclidean d-space $\mathbb{R}^d = \{(x_1, \ldots, x_d): x_i \in \mathbb{R}, i = 1, \ldots, d\}$ are compact if and only if they are closed and bounded, but this is not true of every topological space.

(11.3) *Theorem.* Let K be a compact set.

1. Every infinite sequence in K has a cluster point.
2. If $f: K \to \mathbb{R}$ is continuous then f attains its supremum.
3. If A is a closed subset of K then A is compact.

Proof.

1. Let $x_n \in K, n = 1,2,\ldots$. To show that (x_n) has a cluster point is equivalent to showing that $\bigcap_n F_n \neq \varnothing$, where $F_n = \overline{\{x_k: k > n\}}$. If in fact $\bigcap_n F_n = \varnothing$ then $\bigcup_n (K \backslash F_n) = K$. This is an open covering,

and hence $\bigcup_{n=1}^{N}(K\backslash F_n)=K$ for some N. However, $x_{N+1}\in K$, but $x_{N+1}\in\bigcup_{n=1}^{N}(K\backslash F_n)$, which is a contradiction.

2. It is easy to show that $f(K)$ is compact in $\mathbb{R}$ if f is continuous, so that $f(K)$ is closed and bounded. Hence $y^*:=\sup\{y:y\in f(K)\}$ is finite and in $f(K)$, so there exists $x\in K$ such that $f(x)=y^*$.
3. This follows from the fact that if $\{O_\alpha:\alpha\in J\}$ is an open cover of A then $\{O_\alpha:\alpha\in J\}$, together with A^c, is an open cover of K. □

A function $f:(X,\mathcal{T})\to(Y,\mathcal{S})$ is a *homeomorphism* if f is one-to-one and onto and both f and f^{-1} are continuous. Topological spaces $(X,\mathcal{T}),(Y,\mathcal{S})$ are *homeomorphic* if a homeomorphism exists between them. Homeomorphic spaces are topologically identical: for example, a sequence $(x_n)\subset X$ converges if and only if the sequence $(y_n=f(x_n))\subset Y$ converges, and a set K is compact in X if and only if $f(K)$ is compact in Y.

11.2 Metric and linear spaces

In most applications the topology on a space X is determined by a distance function or *metric*. A function $d:X\times X\to\mathbb{R}$ is a metric if:

1. $d(x,y)=d(y,x)\geqslant 0$ for all $x,y\in X$;
2. $d(x,y)=0$ if and only if $x=y$;
3. (Triangle inequality) $d(x,y)\leqslant d(x,z)+d(z,y)$ for all $x,y,z\in X$.

For $\varepsilon>0$ the *ε-ball* around $x\in X$ is the set $B_\varepsilon(x)=\{y:d(x,y)<\varepsilon\}$. The *metric topology* on X is defined as follows: a set O is open if and only if $x\in O$ implies $B_\varepsilon(x)\subset O$ for some $\varepsilon>0$. Then a sequence (x_n) converges to x if $d(x_n,x)\to 0$ as $x\to\infty$. Note that the usual topology on $\mathbb{R}$ is the metric topology corresponding to the metric $d(x,y)=|x-y|$. A topological space $(X,\mathcal{T})$ is *metrizable* if there exists a metric d such that $\mathcal{T}$ is the metric topology; this d is never unique. If K is a compact subset of a metrizable space then any infinite sequence in K has a convergent subsequence; this follows easily from Theorem 11.3(1).

A sequence (x_n) in a metric space (X,d) is a *Cauchy sequence* if $d(x_n,x_m)\to 0$ as $n,m\to\infty$. If $x_n\to x$ then (x_n) is Cauchy, by the triangle inequality. (X,d) is *complete* if the converse holds, namely for every Cauchy sequence (x_n) there exists $x\in X$ such that $x_n\to x$. (X,d) is *separable* if there is a countable dense subset, i.e. a sequence (x_n) such that for any $x\in X$ and $\varepsilon>0$ there exists m such that $d(x,x_m)<\varepsilon$.

The real line $\mathbb{R}$ is both complete and separable; the set $\mathbb{Q}$ of rational numbers is countable and dense in $\mathbb{R}$.

A set X is a *linear space* if operations of addition and scalar multiplication are defined on X, i.e.:

1. To every pair $x, y \in X$ corresponds an element $x+y$ such that $x+y=y+x$ and $x+(y+z)=(x+y)+z$; there is a unique element $0 \in X$ such that $x+0=x$ for each x, and for each $x \in X$ there is an element $-x \in X$ such that $x+(-x)=0$.
2. To every pair $(a, x) \in \mathbb{R} \times X$ corresponds $ax \in X$ such that $1x=x$, $a(bx)=(ab)x$ and $a(x+y)=ax+ay$, $(a+b)x=ax+bx$ for $a, b \in \mathbb{R}$, $x, y \in X$. It follows that $0x=0$.

X is a *normed linear space* if it is a linear space and there is a function $\|\cdot\|: X \to \mathbb{R}_+$, the *norm*, such that $\|x+y\| \leqslant \|x\| + \|y\|$, $\|ax\| = |a| \, \|x\|$ and $\|x\| > 0$ if $x \neq 0$. The norm defines a metric $d(x, y) = \|x-y\|$ and hence a metric topology. If X is complete in this metric then it is a *Banach space*. The simplest examples of Banach spaces are the Euclidean spaces $\mathbb{R}^d = \{(x_1, x_2, \ldots, x_n): x_i \in \mathbb{R}, i=1, \ldots, n\}$ with the usual Euclidean norm, but there are more interesting examples, including the following classes of *function spaces* which will often be used in the sequel.

(11.4) *Continuous functions*

For $T > 0$ we denote by $C[0, T]$ the set of continuous real-valued functions on $[0, T]$. This is a linear space with $(f_1 + f_2)(t) = f_1(t) + f_2(t)$, and the *uniform norm* is $\|f\| = \max_{0 \leqslant t \leqslant T} |f(t)|$ (the maximum is attained, in view of Theorem (11.3)(b)). $C[0, T]$ is a Banach space, one countable dense set consisting of (polygonal) functions that are linear on subintervals $[(i-1)T/k, iT/k]$, $i=1, \ldots, k$, $k=1, 2, \ldots$, and take rational values at the points iT/k.

(11.5) *RCLL functions*

A function $f:[0, T] \to \mathbb{R}$ is RCLL (right-continuous with left-hand limits) if $\lim_{s \downarrow t} f(s) = f(t)$ for all $t \in [0, T[$ and $f(t-) := \lim_{s \uparrow t} f(s)$ exists for $t \in]0, T]$. The set of all such functions is denoted $D[0, T]$. The uniform norm as above is not very useful in this case because, for example, if $f(t) = I_{(t \geqslant a)}$ for some $a \in]0, T[$ and $f_n(t) = I_{(t \geqslant a + 1/n)}$, then $\|f_n - f\| = 1$, so that the sequence (f_n) fails to converge to f, as it plainly 'ought' to do. The *Skorokhod metric* d_s in $D[0, T]$ is defined as follows: let Λ be the set of strictly increasing continuous functions λ mapping $[0, T]$ on to itself; thus $\lambda(0)=0, \lambda(T)=T$. Then $d_s(f,g)$ is the infimum of $\varepsilon>0$ such that there exists $\lambda\in\Lambda$ with $\sup_{t\in[0,T]} |\lambda(t)-t| \leqslant \varepsilon$ and $\sup_{t\in[0,T]} |f(\lambda(t))-g(t)| \leqslant \varepsilon$. The map λ amounts to a 'distortion of the time scale', and the reader can check that $d_s(f_n, f)\to 0$ for the example above. Under this metric the space $D[0, T]$ is not complete; however, it is possible (Billingsley, 1968, section 14) to define an equivalent metric (i.e. one generating the same topology) under which $D[0, T]$ is also complete.

11.3 Differentiability

We need to discuss some differentiability properties of functions on the real line. These involve the Lebesgue integral, and the reader can refer ahead to section 12 for the properties of this integral.

A function $f:\mathbb{R}\to\mathbb{R}$ is *differentiable* at $t\in\mathbb{R}$ if there exists a number, denoted $f'(t)$, such that given $\varepsilon>0$ there is a $\delta>0$ for which

$$\left|\frac{1}{s}(f(t+s)-f(t))-f'(t)\right|<\varepsilon, \quad \text{if } 0<|s|<\delta.$$

f is *differentiable* if it is differentiable at every $t\in\mathbb{R}$. f is *absolutely continuous* if the following holds: for every finite interval $[a, b]$ and $\varepsilon>0$ there is a $\delta>0$ such that

$$\sum_{i=1}^{n}(b_i-a_i)<\delta \qquad \text{implies } \sum_{i=1}^{n}|f(b_i)-f(a_i)|<\varepsilon$$

for any disjoint intervals $]a_i, b_i[$, $i=1,\ldots,n$ contained in $[a, b]$.

(11.6) *Theorem*

1. If $f:\mathbb{R}\to\mathbb{R}$ is absolutely continuous it is differentiable almost everywhere (i.e. at all points $t\in\mathbb{R}$ except a set having Lebesgue measure zero).
2. f is an indefinite integral, i.e. there exists a measurable function $f':\mathbb{R}\to\mathbb{R}$ such that, for all $a<b$, $f(b)-f(a)=\int_a^b f'(t)\mathrm{d}t$, if and only if f is absolutely continuous.

Absolute continuity is not an easy property to check directly from the definition. Usually easier to verify is the somewhat stronger condition of *Lipschitz continuity*, which is widely used in this book. A function $f:\mathbb{R}\to\mathbb{R}$ is Lipschitz continuous on $A\subset\mathbb{R}$ if there exists a constant L such that

$$(11.7)\qquad |f(t)-f(s)|\leqslant L|t-s| \qquad \text{for all } t,s\in A.$$

f is *globally Lipschitz* if (1.1) holds with $A=\mathbb{R}$ and *locally Lipschitz* if it is Lipschitz on any bounded interval $[a,b]$, (or, equivalently, any compact set). f is *Lipschitz near* $x\in\mathbb{R}$ if it is Lipschitz on $[x-\varepsilon,x+\varepsilon]$ for some $\varepsilon>0$. Then 1. f is locally Lipschitz if and only if 2. it is Lipschitz near every $x\in\mathbb{R}$. Indeed, 1. clearly implies 2. while for the converse note that $\{(x-\varepsilon_x,x+\varepsilon_x:x\in[a,b]\}$ is an open cover of $[a,b]$, where ε_x is such that (11.7) holds for some $L=L(x)$ with $A=[x-\varepsilon_x,x+\varepsilon_x]$. Since $[a,b]$ is compact there is a finite sub-cover $\{(x-\varepsilon_{x_i},x+\varepsilon_{x_i}):\ i=1,\dots,n\}$ and (11.7) then holds with $A=[a,b]$ and $L=\max\limits_{1\leqslant i\leqslant n}L(x_i)$.

A locally Lipschitz function is clearly absolutely continuous, and we therefore have the following.

(11.8) *Proposition.* A locally Lipschitz continuous function is differentiable almost everywhere and is equal to the integral of its derivative.

The definition of Lipschitz continuity extends in an obvious way to functions taking values in $\mathbb{R}^d$.

Let $\{f_\alpha,\alpha\in J\}$ be a collection of $\mathbb{R}_+$-valued functions indexed by an arbitrary set J, such that each function f_α is Lipschitz continuous. Then

$$(11.9)\qquad f(t)=\inf_{\alpha\in J}f_\alpha(t)$$

exists but is not necessarily Lipschitz. For example, let $J=\mathbb{N}$ and $f_\alpha(t)=1$ for $t\leqslant 0$, $f_\alpha(t)=1-\alpha t$ for $0\leqslant t\leqslant 1/\alpha$ and $f_\alpha(t)=0$ for $t\geqslant 1/\alpha$; then each f_α is globally Lipschitz but $f(t)=I_{(t\leqslant 0)}$ which is not even continuous. This cannot happen, however, if all the f_α have the same Lipschitz constant.

(11.10) *Theorem.* Suppose that for each $\alpha\in J$, $f_\alpha:[a,b]\to\mathbb{R}_+$ is Lipschitz continuous with constant L not depending on α. Then f given by (11.9) is Lipschitz with the same constant.

Proof. For t, $s\in[a,b]$

$$\begin{aligned} f(t)-f(s) &= \inf_{\alpha\in J} f_\alpha(t) - \inf_{\beta\in J} f_\beta(s) \\ &\leqslant f_\alpha(t) - \inf_{\beta\in J} f_\beta(s) \end{aligned}$$

for any $\alpha\in J$. Fix $\varepsilon>0$ and choose α such that $f_\alpha(s)\leqslant \inf_{\beta\in J} f_\beta(s)+\varepsilon$.

Then

$$\begin{aligned} f(t)-f(s) &\leqslant f_\alpha(t)-f_\alpha(s)+\varepsilon \\ &\leqslant |f_\alpha(t)-f_\alpha(s)|+\varepsilon\leqslant L|t-s|+\varepsilon. \end{aligned}$$

Thus $f(t)-f(s)\leqslant L|t-s|$, and applying the same argument with s,t interchanged we also have $f(s)-f(t)\leqslant L|t-s|$. The result follows. □

12 Probability theory

The *basic space* Ω in axiomatic probability theory contains points ω representing possible realizations of some random phenomenon, and *events* are subsets of Ω to which a probability can be assigned. It turns out that the natural collection of events for assignment of probabilities is a *σ-field*. A σ-field $\mathfrak{F}$ in Ω is a class of subsets such that:

1. $F\in\mathfrak{F}$ implies $F^c(=\Omega\backslash F)\in\mathfrak{F}$; and
2. if $F_i\in\mathfrak{F}$ $(i=1,2,\ldots)$ then $\bigcup_i F_i\in\mathfrak{F}$ (i.e. $\mathfrak{F}$ is closed under countable unions).

In view of 1. $\mathfrak{F}$ is also closed under countable intersections. If $(\mathfrak{F}_\alpha, \alpha\in J)$ is an indexed family of σ-fields with arbitrary index set J then $\mathfrak{G}=\bigcap_\alpha \mathfrak{F}_\alpha=\{F\in\mathfrak{F}: F\in\mathfrak{F}_\alpha \text{ for all } \alpha\in J\}$ is also a σ-field. It follows from this that for any collection $\mathfrak{C}$ of subsets of Ω there is a unique smallest σ-field containing $\mathfrak{C}$, namely the intersection of all σ-fields containing $\mathfrak{C}$ (and there is always at least one such σ-field, namely the σ-field of all subsets of Ω). This σ-field is denoted $\sigma(\mathfrak{C})$ and called the σ-field *generated by* $\mathfrak{C}$.

If $(X,\mathscr{T})$ is a topological space then the *Borel σ-field* $\mathfrak{B}(X)$ in X is the σ-field generated by the open sets $\mathscr{T}$. The Borel sets $\mathfrak{B}=\mathfrak{B}(\mathbb{R})$ of the real line $\mathbb{R}$ are the members of the Borel σ-field corresponding to the usual topology. A primary justification for this whole theory is that there are subsets of $\mathbb{R}$ that are *not* Borel sets.

A pair $(\Omega, \mathfrak{F})$, where Ω is a set and $\mathfrak{F}$ a σ-field of subsets of Ω, is called a *measurable space.* If $(Y, \mathfrak{Y})$ is another measurable space and $f:\Omega \to Y$ a function, then the σ-field in Ω *generated by* f is $\mathfrak{S}_f = f^{-1}(\mathfrak{Y}) = \{f^{-1}(Y): Y \in \mathfrak{Y}\}$. It is easily checked that $\mathfrak{S}_f$ is indeed a σ-field. The following theorem plays an important role in elucidating how σ-fields and functions interrelate.

(12.1) *Theorem.* Let Ω, Y, f be as above and suppose that $g:\Omega \to \mathbb{R}$ is $\mathfrak{S}_f$-measurable, i.e. $g^{-1}(\mathfrak{B}) \subset \mathfrak{S}_f$. Then there exists a function $h: Y \to \mathbb{R}$ such that $g = h \circ f$ (i.e. $g(\omega) = h(f(\omega))$, $\omega \in \Omega$).

Thus g is $\mathfrak{S}_f$-measurable if and only if g is a *function of* f. This functional dependence is often the property we are interested in, but mathematically it is more easily handled when couched in terms of measurability.

A *probability measure* on a measurable space $(\Omega, \mathfrak{F})$ is a countably additive set function of total mass one; more formally, it is a map $P:\mathfrak{F} \to [0, 1]$ such that $P\Omega = 1$ and if $F_1, F_2, \ldots,$ are disjoint sets in $\mathfrak{F}$ then[2] $P(\bigcup_{i=1}^{\infty} F_i) = \sum_{i=1}^{\infty} PF_i$; it follows in particular that $P\varnothing = 0$. Of course, there may be other sets $N \in \mathfrak{F}$ such that $PN = 0$; these are *sets of probability zero* or *null sets.* There is a minor inconvenience in the theory at this point: since a null set N is interpreted as an event that occurs with probability zero, clearly any subset of N ought to be a null set; but there is no guarantee that such sets are measurable (i.e. in $\mathfrak{F}$), so no probability may be attached to them. *Completion* gets around this. Let $\mathfrak{F}^P$ be the collection of subsets of Ω expressible in the form $F = F_1 \cup F_2$, where $F_1 \in \mathfrak{F}$ and $F_2 \subset F_3 \in \mathfrak{F}$ with $PF_3 = 0$, and define $PF = PF_1$. Then $\mathfrak{F}^P$ is a σ-field. $\mathfrak{F}^P \supset \mathfrak{F}$, P is unambiguously defined on $\mathfrak{F}^P$, and it is indeed true that if $A \subset N$ (where N is a null set of $\mathfrak{F}$) then $A \in \mathfrak{F}^P$ and $PA = 0$. $\mathfrak{F}^P$ is called the *P-completion of* $\mathfrak{F}$.

Defining a probability measure involves the apparently arduous task of specifying its value for all $F \in \mathfrak{F}$; it is however a key fact that specifying it on a much smaller class of sets may imply its value on all the remaining sets. A class $\mathfrak{G}$ of sets is a *field* if:

1. $G \in \mathfrak{G}$ implies $G^c \in \mathfrak{G}$, and
2. $G_1, G_2 \in \mathfrak{G}$ implies $G_1 \cup G_2 \in \mathfrak{G}$.

The *Carathéodory extension theorem* states that if P is a countably additive set function on $\mathfrak{G}$ with $P\Omega = 1$ then there is a unique

[2] PF or $P(F)$ is a convenient notation for the probability of a set $F \in \mathfrak{F}$.

probability measure $\bar{P}$ on $\sigma(\mathfrak{G})$ such that $\bar{P}F = PF$ for all $F \in \mathfrak{G}$. The classic example is the construction of Lebesgue measure on the unit interval $[0, 1]$: we take $\mathfrak{G}$ to be the class of sets G that can be written $G =]a_1, b_1] \cup]a_2, b_2] \cup \cdots \cup]a_n, b_n]$ with $0 < a_1 < b_1 < a_2 < \cdots < b_n$, and define $l(G) = \sum_{i=1}^{n} (b_i - a_i)$. It can be shown – and this is not obvious – that l is countably additive on $\mathfrak{G}$; hence it extends uniquely to a measure l on $\mathfrak{B}[0, 1]$ (the Borel sets of $[0, 1]$), since $\mathfrak{B}[0, 1] = \sigma(\mathfrak{G})$. The l-completion of $\mathfrak{B}[0, 1]$ gives the Lebesgue measurable sets of $[0, 1]$. The same construction applies on $\mathbb{R}$, but then of course l is not a probability measure. To construct a probability measure we use a *distribution function*, which is a function $F : \mathbb{R} \to [0, 1]$ such that F is monotone increasing and right-continuous, with

$$F(-\infty) := \lim_{t \to -\infty} F(t) = 0 \quad \text{and} \quad F(\infty) := \lim_{t \to +\infty} F(t) = 1.$$

For a set G represented as above (but not restricting a_i, b_i to lie in $[0, 1]$) we define $PG = \sum_1^n (F(b_i) - F(a_i))$. Then it can be shown that P is countably additive on $\mathfrak{G}$ and $P\mathbb{R} = 1$, so that F determines a probability measure $\bar{P}$ on $(\mathbb{R}, \mathfrak{B})$ with the property – in particular – that $\bar{P}(]-\infty, a]) = F(a)$ for $a \in \mathbb{R}$. Conversely, if P is a given probability measure on $(\mathbb{R}, \mathfrak{B})$ then $F(a) := P(]-\infty, a])$ is a distribution function. Thus there is a one-to-one correspondence between distribution functions and probability measures on $(\mathbb{R}, \mathfrak{B})$. It is often convenient to specify a distribution instead by using the *survivor function* $H(a) := P(]a, \infty[)$; this is right-continuous and decreasing with $H(-\infty) = 1, H(\infty) = 0$.

The triple $(\Omega, \mathfrak{F}, P)$, where Ω is set, $\mathfrak{F}$ is a σ-field of subsets of Ω, and P a probability measure on $(\Omega, \mathfrak{F})$, is called a *probability space.* Sets $A \in \mathfrak{F}$ are called *events,* and we can read the statement '$\omega \in A$' as 'A happens'. There is a simple but very powerful result concerning sequences of events which is known as the Borel–Cantelli lemma. If $A_n \in \mathfrak{F}$ for $n = 1, 2, \ldots$, we define

$$\limsup A_n = \bigcap_{n=1}^{\infty} \bigcup_{k=n}^{\infty} A_k, \qquad \liminf A_n = \bigcup_{n=1}^{\infty} \bigcap_{k=n}^{\infty} A_k. \tag{1.2}$$

Then $\omega \in \limsup A_n$ if and only if A_n happens infinitely often, and $\omega \in \liminf A_n$ if and only if A_n happens for all but finitely many n. It follows from the laws of set theory, and is easily verified from this informal description, that $(\liminf A_n)^c = \limsup A_n^c$.

(12.2) *Lemma* (Borel–Cantelli). Suppose that $A_n \in \mathfrak{F}$ for $n = 1, 2, \ldots$, and that $\sum_{n=1}^{\infty} PA_n < \infty$. Then $P(\limsup A_n) = 0$.

Proof. Define $B_n = \bigcup_{k=n}^{\infty} A_k$; then B_n is a decreasing sequence of sets and $P(\limsup A_n) = \lim P(B_n)$. But $P(B_n) \leqslant \sum_{k=n}^{\infty} PA_k$ and this converges to 0 if $\sum_{n=1}^{\infty} PA < \infty$. □

A *random variable* (r.v.) is a measurable function $X:(\Omega, \mathscr{F}) \to (\mathbb{R}, \mathfrak{B})$. It is customary to suppress the ω in the notation for r.v.s: for example we write $X \geqslant 0$ when, strictly speaking, we mean $X(\omega) \geqslant 0$ for all $\omega \in \Omega$, and $(X \geqslant 0)$ for $\{\omega: X(\omega) \geqslant 0\}$. A particular example of an r.v. is the *indicator function* I_A of a set $A \in \mathfrak{F}$, defined by

$$I_A(\omega) = \begin{cases} 1 & \omega \in A \\ 0 & \omega \notin A. \end{cases}$$

A random variable X is *simple* if it takes the form

$$X(\omega) = \sum_{i=1}^{n} a_i I_{A_i}(\omega)$$

for some constants a_i and sets $A_i \in \mathfrak{F}$. The *expectation* or *mean* of a simple r.v. is denoted $\mathbb{E}X$ and defined by

$$\mathbb{E}X = \sum_{i=1}^{n} a_i PA_i.$$

If X is a non-negative (but not necessarily simple) r.v., i.e. $X \geqslant 0$, we define

$$\mathbb{E}X = \sup\{\mathbb{E}Y : Y \text{ is simple and } Y \leqslant X\};$$

this is always well defined but may be equal to $+\infty$. An arbitrary r.v. can always be written $X = X^+ - X^-$, where $X^+ = XI_{(X \geqslant 0)}$ and $X^- = -XI_{(X<0)}$. Then $X^+, X^- \geqslant 0$ and $|X| = X^+ + X^-$. If $\mathbb{E}|X| < \infty$ then X is *integrable* and we define $\mathbb{E}X = \mathbb{E}X^+ - \mathbb{E}X^-$. In particular, the expectation always exists if X is bounded: $|X| \leqslant K$ for some non-random constant K.

We say a property holds *almost surely* (a.s.) on a probability space $(\Omega, \mathfrak{F}, P)$ if it holds for every ω except for a set N such that $PN = 0$. If X, Y are two r.v.s such that $X = Y$ a.s. then X is integrable if and only if Y is integrable, and then $\mathbb{E}X = \mathbb{E}Y$. It is clear from the definition that expectation is a linear and monotone operator, i.e. if X, Y are integrable r.v.s and $a, b \in \mathbb{R}$ then $\mathbb{E}[aX + bY] = a\mathbb{E}X + b\mathbb{E}Y$, and if $X \geqslant Y$ a.s. then $\mathbb{E}X \geqslant \mathbb{E}Y$. Expectations are often written using the conventional notation for integrals:

$$\mathbb{E}[XI_A] = \int_A X(\omega)\mathrm{d}P(\omega) \qquad \left(\text{or} \int_A X\mathrm{d}P\right).$$

An important example is the *unit interval probability space*, where $\Omega = [0, 1]$, $\mathfrak{F}$ = Borel sets and $P = l$ (Lebesgue measure). Then we write the expectation in any of the forms

$$\mathbb{E}[X] = \int_{[0,1]} X \mathrm{d}l = \int_0^1 X(x)\mathrm{d}x,$$

and this is the *Lebesgue integral.*

The main theorems of integration theory follow.

(12.3) *Theorem*

1. (Monotone convergence) if X_n, X are non-negative r.v.s such that $X_n \uparrow X$ then $\mathbb{E}X_n \uparrow \mathbb{E}X$.
2. (Dominated convergence) If (X_n), X, Y are r.v.s such that Y is integrable, $|X_n| \leqslant Y$ for all n and $X_n \to X$ a.s. then X is integrable and $\mathbb{E}X_n \to \mathbb{E}X$. (A special case is *bounded convergence* in which $Y \equiv K$, a constant.)
3. (Fatou Lemma) If (X_n), Y are r.v.s such that Y is integrable and $X_n \geqslant Y$, then

$$\mathbb{E}(\liminf_{n\to\infty} X_n) \leqslant \liminf_{n\to\infty} \mathbb{E}X_n.$$

It follows easily from the definition of expectation that if X is an integrable r.v. then, given $\varepsilon > 0$, there exists $\delta > 0$ such that, for $A \in \mathfrak{F}$, $PA < \delta$ implies $\mathbb{E}[|X|I_A] < \varepsilon$. We also have the *Markov inequality*

$$P(|X| > c) \leqslant \frac{1}{c}\mathbb{E}[|X|], \qquad c \in \mathbb{R}_+,$$

which follows from the monotone property of the expectation and the fact that $h(X) \leqslant |X|$, where h is the function $h(x) = cI_{(|x|>c)}$. From these two facts we see that for any integrable r.v. X,

$$\lim_{c\to\infty} \mathbb{E}[|X|I_{(|X|>c)}] = 0.$$

A collection $\{X_\alpha : \alpha \in J\}$ of r.v.s, where J is an arbitrary index set, is *uniformly integrable* if this convergence is uniform in α, i.e.

$$\lim_{c\to\infty} \sup_{\alpha\in J} \mathbb{E}[|X_\alpha| I_{(|X_\alpha|>c)}] = 0.$$

If $(\Omega, \mathfrak{F}, P)$ is a probability space and $(Y, \mathfrak{Y})$ another measurable space then any measurable function $\phi: \Omega \to Y$ determines a probability measure P^ϕ on $(Y, \mathfrak{Y})$ by the recipe $P^\phi A = P(\phi^{-1}(A)) = P\{\omega \in \Omega : \phi(\omega) \in A\}$. Then $(Y, \mathfrak{Y}, P^\phi)$ is a probability space. If $g: Y \to \mathbb{R}$ is a measurable function, we can think of g as a random variable on $(Y, \mathfrak{Y}, P^\phi)$ or the composition $g \circ \phi$ as a random variable on Ω. It is important that the expectation is the same either way:

$$\int_\Omega g \circ \phi(\omega) \mathrm{d}P(\omega) = \int_Y g(y) \mathrm{d}P^\phi(y). \tag{12.4}$$

(1.3)

This formula is valid for all P^ϕ-integrable functions g.

If $X_1, \ldots, X_d$ are random variables on Ω then $\phi(\omega) = (X_1(\omega), \ldots, X_d(\omega))$ is a measurable map from Ω to $\mathbb{R}^d$; the function $F: \mathbb{R}^d \to [0, 1]$ defined by

$$\begin{aligned} F(a) &= P(X_1 \leqslant a_1, \ldots, X_d \leqslant a_d) \\ &= P^\phi(]-\infty, a_1] \times \ldots \times]-\infty, a_d]), \qquad a \in \mathbb{R}^d \end{aligned}$$

is the *joint distribution function* of $(X_1, \ldots, X_d)$, and specifying $F(.)$ is enough to determine P^ϕ. In particular, when $d = 1$, F is the distribution function of X_1 and we write (12.4) as

$$\mathbb{E}g(X_1) = \int_\mathbb{R} g(y) \mathrm{d}P^{X_1}(y) =: \int_\mathbb{R} g(y) \mathrm{d}F(y).$$

The integral on the right is known as the *Lebesgue–Stieltjes integral* of the function g with respect to the distribution function F. If F is absolutely continuous then its derivative $f(y) = (\mathrm{d}/\mathrm{d}y)F(y)$ is the *density function* of X_1 and the expectation is given in terms of an ordinary Lebesgue integral as

$$\mathbb{E}g(X_1) = \int_\mathbb{R} g(y) f(y) \mathrm{d}y.$$

In the case of the unit interval probability space $([0, 1], \mathfrak{B}, l)$ the identity function $U(x) = x$ is a random variable with the *uniform* $[0, 1]$ *distribution*, denoted by $U[0, 1]$. The corresponding distribution and density functions are[3] $F(y) = (y \vee 0) \wedge 1$ and $f(y) = I_{[0,1]}(y)$. In this case, for $g: [0, 1] \to \mathbb{R}$ we have simply

$$\mathbb{E}g(U) = \int_0^1 g(y) \mathrm{d}y.$$

[3]Notation: $a \vee b = \max\{a, b\}$, $a \wedge b = \min\{a, b\}$.

Convergence of sequences $(X_n, n = 1,2,\dots)$ of random variables on a probability space $(\Omega, \mathfrak{F}, P)$ can be defined in various different ways. Almost sure convergence, denoted $X_n \to X$ a.s., has already been introduced. $X_n \to X$ *in probability* if $P(|X_n - X| > \varepsilon) \to 0$ as $n \to \infty$ for every $\varepsilon > 0$. For $1 \leqslant p < \infty$, we say that $X \in L_p$ if $\mathbb{E}|X|^p < \infty$, and $X_n \to X$ *in* L_p if $\mathbb{E}|X_n|^p < \infty$ for all n and $\mathbb{E}|X_n - X|^p \to 0$ (this implies that $X \in L_p$ also). L_1 is just the set of integrable r.v.s. The relations between these modes of convergence follow.

(12.5) *Theorem.* Let $(X_n), X$ be random variables on a given probability space $(\Omega, \mathfrak{F}, P)$. Then:

1. $X_n \to X$ a.s. implies $X_n \to X$ in probability.
2. $X_n \to X$ in L_p for some $p \geqslant 1$ implies that $X_n \to X$ in probability.
3. If $X_n \to X$ in L_p then $X_n \to X$ in L_r for $1 \leqslant r \leqslant p$.
4. $X_n \to X$ in probability implies there exists a subsequence X_{n_k} such that $X_{n_k} \to X$ a.s. as $k \to \infty$.
5. If $X_n \to X$ a.s. then X is integrable and $X_n \to X$ in L_1 if and only if the X_n are uniformly integrable.

The most common case of L_p convergence is $p = 2$ when it is sometimes called *quadratic mean convergence*. Note from 3. that if $X \in L_2$, then $X \in L_1$ and $(X - \mathbb{E}X) \in L_2$. The quantity $\mathbb{E}(X - \mathbb{E}X)^2$ is the *variance* of X. The proof of 4. is a simple and typical application of the Borel–Cantelli lemma. Let a_n, b_n be sequences of positive numbers such that $a_n, b_n \downarrow 0$ and $\sum_n b_n < \infty$. If $X_n \to X$ in probability, then for each n there is an integer $k(n)$ such that $P(|X_k - X| > a_n) < b_n$ for all $k \geqslant k(n)$; hence $\sum_n P(|X_{k(n)} - X| > a_n) < \infty$ and by the Borel–Cantelli lemma $P[\limsup(|X_{k(n)} - X| > a_n)] = 0$. Thus with probability 1, $|X_{k(n)} - X| \leqslant a_n$ for all but finitely many n, which implies that $X_{k(n)} \to X$ a.s. as $n \to \infty$.

Let $(\Omega, \mathfrak{F}, P)$ be a probability space. Events $A, B \in \mathfrak{F}$ are *independent* if $P(A \cap B) = P(A)P(B)$. If $\mathfrak{G}, \mathfrak{H}$ are two sub-σ-fields of $\mathfrak{F}$, then $\mathfrak{G}$ and $\mathfrak{H}$ are independent if G, H are independent for all $G \in \mathfrak{G}$, $H \in \mathfrak{H}$. Random variables X_1, X_2 are independent if $\mathfrak{G}_1, \mathfrak{G}_2$ are independent, where $\mathfrak{G}_i = X_i^{-1}(\mathfrak{B})$, $i = 1, 2$. Thus $P(X_1 \in A_1$ and $X_2 \in A_2) = P(X_1 \in A_1)P(X_2 \in A_2)$ for all A_1, $A_2 \in \mathfrak{B}$.

(12.6) *Theorem.* If $X_1, X_2 \in L_1$ and X_1, X_2 are independent, then $X_1 X_2 \in L_1$ and $\mathbb{E}[X_1 X_2] = \mathbb{E}X_1 \mathbb{E}X_2$.

A common way of creating independent σ-fields and random variables is by the formation of *product spaces*. If $(\Omega_1, \mathfrak{F}_1, P_1)$, $(\Omega_2, \mathfrak{F}_2, P_2)$ are probability spaces, let $\mathfrak{C}$ be the field in $\Omega = \Omega_1 \times \Omega_2$ consisting of finite disjoint unions of sets of the form $A = A_1 \times A_2$, where $A_1 \in \mathfrak{F}_1$, $A_2 \in \mathfrak{F}_2$, and $\mathfrak{F} = \mathfrak{F}_1 \otimes \mathfrak{F}_2$ is the smallest σ-field in Ω containing $\mathfrak{C}$. For $A = A_1 \times A_2$ define $P(A) = P_1(A_1)P_2(A_2)$ and extend this definition to $\mathfrak{C}$ by additivity over disjoint subsets. By the Carathéodory extension theorem there is a unique extension of the measure P to $\mathfrak{F}$; this is the *product measure*, denoted $P = P_1 \otimes P_2$. Now let $\mathfrak{F}^1 = \{F \in \mathfrak{F}: F = F_1 \times \Omega_2, F_1 \in \mathfrak{F}_1\}$ and $\mathfrak{F}^2 = \{F \in \mathfrak{F}: F = \Omega_1 \times F_2, F_2 \in \mathfrak{F}_2\}$; it is easily seen that these are independent. For $i = 1, 2$, a random variable X_i defined on Ω_i defines a r.v. X^i on Ω by the formula $X^i(\omega) = X_i(\omega_i)$ for $\omega = (\omega_1, \omega_2) \in \Omega$. Then X^1, X^2 are independent. The same construction applies to products of $d \geqslant 2$ probability spaces; in particular, if $(\Omega_i, \mathscr{F}_i, P_i)$, $i = 1, \ldots, d$ are copies of the unit interval probability space then $P = P_1 \otimes \cdots \otimes P_d$ is just a d-dimensional Lebesgue measure on $\Omega = [0, 1]^d$ and the coordinate functions $U_i(x_1, \ldots, x_d) = x_i$, $i = 1, \ldots, d$ are independent $U[0, 1]$-distributed random variables. A sequence (X_n) of random variables is i.i.d. (independent and identically distributed) if the elements are independent with a common distribution function. A fundamental result about i.i.d. sequences follows.

(12.7) *Theorem* (Strong law of large numbers). Let $X_1, X_2, \ldots,$ be an i.i.d. sequence of random variables such that X_1 is integrable, and denote $\mu = \mathbb{E}X_1$. Then

$$\lim_{n \to \infty} \frac{1}{n} \sum_{k=1}^{n} X_k = \mu \qquad \text{a.s.}$$

This is a major result of probability theory, and a primary justification of the axiomatic approach, since it shows that the expectation coincides with the 'average of a long sequence of independent trials', as intuition and experience suggest that it should.

If $X \in L_1$ and $\mathfrak{G}$ is a sub-σ-field of $\mathfrak{F}$, the *conditional expectation* of X given $\mathfrak{G}$ is any random variable Z such that

(12.8) Z is $\mathfrak{G}$-measurable and $\mathbb{E}(XI_G) = \mathbb{E}(ZI_G)$ for all $G \in \mathfrak{G}$.

Any two Zs satisfying (12.8) agree almost surely, and we write $\mathbb{E}(X \mid \mathfrak{G}) = Z$, where Z is *any* r.v. satisfying (12.8). It is important to remember that $\mathbb{E}(X|\mathfrak{G})$ is uniquely defined only up to almost sure equivalence. If $X \in L_1$ and Y is any r.v., the conditional expectation of X given Y is

$$\mathbb{E}(X \mid Y) = \mathbb{E}(X \mid \sigma(Y)).$$

By Theorem (12.1) there is a function $h:\mathbb{R} \to \mathbb{R}$ such that $\mathbb{E}(X \mid Y)(\omega) = h(Y(\omega))$. Again, this function h is defined only to a.s. equivalence.

(12.9) *Theorem* (properties of conditional expectation). Suppose $X \in L_1$ and $\mathfrak{G}, \mathfrak{H}$ are sub-σ-fields of $\mathfrak{F}$. Then:

1. $\mathbb{E}(X|\mathfrak{G}) = X$ if X is $\mathfrak{G}$-measurable.
2. $\mathbb{E}(X|\mathfrak{G}) = \mathbb{E}X$ if X is independent of $\mathfrak{G}$.
3. $\mathbb{E}(XY|\mathfrak{G}) = \mathbb{E}(X|\mathfrak{G})Y$ if Y is bounded and $\mathfrak{G}$-measurable.
4. $\mathbb{E}[\mathbb{E}(X|\mathfrak{H})|\mathfrak{G}] = \mathbb{E}(X|\mathfrak{G})$, if $\mathfrak{G} \subset \mathfrak{H}$.

A very important special case of the last statement is when $\mathfrak{G} = (\Omega, \varnothing\}$ (the *trivial* σ-field). An r.v. is measurable with respect to this σ-field only if it is a.s. constant, and $\mathbb{E}(X|\{\varnothing, \Omega\}) = \mathbb{E}X$. Hence applying 4. with $\mathfrak{G} = (\Omega, \varnothing\}$ gives

(12.10) $$\mathbb{E}(\mathbb{E}(X|\mathfrak{G})) = \mathbb{E}X.$$

A further important property is that

(12.11) $$X \geqslant Y \text{ a.s. implies } \mathbb{E}(X|\mathfrak{G}) \geqslant \mathbb{E}(Y|\mathfrak{G}) \qquad \text{a.s.}$$

Indeed, it follows immediately from the definition of conditional expectation that $\mathbb{E}(X|\mathfrak{G}) \geqslant 0$ a.s. if $X \geqslant 0$ a.s. and (12.11) is then clear.

13 Stochastic processes

13.1 Processes, filtrations and stopping times

A probability space $(\Omega, \mathfrak{F}, P)$ is fixed throughout this section and all random variables are defined on this probability space. A *stochastic process* is a family $X = (X_t, t \in J)$ of random variables with some index set $J \subset \mathbb{R}$. If $J = \{1, 2, \ldots, d\}$ then X is just a random vector (an $\mathbb{R}^d$-valued random variable). If $J = \mathbb{Z}_+ = \{0, 1, 2, \ldots\}$ then X is a *discrete-time* process, while if $J = [0, T]$ for some $T > 0$ or $J = \mathbb{R}_+$, then X is a *continuous-time* process. These are the only cases that will be considered, with the main accent on the continuous-time case. The *finite-dimensional distributions* of a process is the collection of joint distribution functions $F_{t_1,\ldots,t_n}(a_1,\ldots,a_n)$ of the random variables $(X_{t_1},\ldots,X_{t_n})$ for all n, and $t_i \in J$ with $t_1 < t_2 < \cdots < t_n$. Obviously, these are *consistent*, in the sense that

$$F_{t_1,\ldots,t_{j-1},t_{j+1},\ldots,t_n}(a_1,\ldots,a_{j-1},a_{j+1},\ldots,a_n)$$
$$= F_{t_1,\ldots,t_n}(a_1,\ldots,a_{j-1},\infty,a_{j+1},\ldots,a_n).$$

A fundamental theorem of Kolmogorov states that given any consistent family of finite-dimensional distributions there is a probability space and a stochastic process defined on this probability space whose finite-dimensional distributions coincide with the given ones.

Consider the continuous-time case $J = \mathbb{R}_+$. There are actually three different ways of thinking of a stochastic process, each of them useful in different contexts (we will write $X_t(\omega), X(t,\omega)$ interchangeably):

1. a family of random variables $(X_t, t \in \mathbb{R}_+)$;
2. a function on the product space: $X: \mathbb{R}_+ \times \Omega \to \mathbb{R}$; and
3. a function-space valued random variable, i.e. a measurable map $\tilde{X}: \Omega \to \Xi$ such that $\tilde{X}(\omega)(t) = X(t,\omega)$ for each (t,ω).

In 3, Ξ is some set of real-valued functions such as $C(\mathbb{R}_+)$ or $D(\mathbb{R}_+)$. The function $t \to X(t,\omega)$ for fixed ω is called a *sample function* of the process, and it is generally a highly non-trivial question whether given a set of finite-dimensional distributions it is possible to construct the corresponding process in such a way that all (or almost all) its sample functions belong to a particular space Ξ.

One problem with definition 1 is that it says nothing about the properties of $X(t,\omega)$ as a function of t. X is *measurable* if the function $(t,\omega) \to X(t,\omega)$ of definition 2 is measurable on the product space $(\mathbb{R} \times \Omega, \mathfrak{B} \otimes \mathfrak{F})$. If X is measurable then for all ω the function $t \to X(t,\omega)$ is measurable, and random variables such as

$$\int_0^t g(X(s,\omega))\mathrm{d}s$$

(where g is a bounded measurable function) are well defined. Two processes X, X' are *modifications* of each other if $P(X_t = X'_t) = 1$ for all $t \in \mathbb{R}_+$ (note that X, X' then have the same finite-dimensional distributions) and are *indistinguishable* if the sample functions $t \to X(t,\omega), t \to X'(t,\omega)$ are identical for $\omega \in G$ with $PG = 1$. The latter is much the stronger requirement: it is easy to construct examples of processes X, X' which are modifications of each other but whose sample functions are not identical for any $\omega \in \Omega$.

Let $\mathring{\mathfrak{F}}_t^X$ be the smallest σ-field in $\mathfrak{F}$ with respect to which all the random variables $(X_s, s \in [0,t])$ are measurable (this is always well defined; section 1.2). By definition, $\mathring{\mathfrak{F}}_{t_1}^X \subset \mathring{\mathfrak{F}}_{t_2}^X$ if $t_1 \leqslant t_2$, i.e. $(\mathring{\mathfrak{F}}_t^X, t \in \mathbb{R}_+)$ is an increasing family of sub-σ-fields of $\mathfrak{F}$. It is called the *natural filtration* of X. For each $t \in \mathbb{R}_+$ $(\Omega, \mathring{\mathfrak{F}}_t^X, P)$ is a probability space and we could obtain a complete probability space by adjoining to $\mathring{\mathfrak{F}}_t^X$ all subsets of null sets, as described in section 12. However, the normal procedure is to use the $(\mathfrak{F}, P)$-completion of $\mathring{\mathfrak{F}}_t^X$: this is the σ-field $\mathfrak{F}_t^X$ of subsets A of Ω expressible in the form $A = A_1 \cup A_2$, where $A_1 \in \mathring{\mathfrak{F}}_t^X$ and $A_2 \subset A_3$, where $A_3 \in \mathfrak{F}$ and $PA_3 = 0$. This has the advantage that the class of null sets is the same for every $t \in \mathbb{R}_+$. It will usually be understood that natural filtrations are completed in this way.

(13.1) *Example*

As a concrete example, and one that will be useful later, take $\Omega = \mathbb{R}_+$, $\mathfrak{F} = \mathfrak{B}(\mathbb{R}_+)$, P to be the measure determined by a given survivor function F on $\mathbb{R}_+$ and T the identity function $T(\omega) = \omega$. Then T is simply a non-negative random variable with survivor function F. Now let $X_t = I_{(t \geqslant T)}$; thus $X_t = 1$ if T 'has happened' at some time before or at t, and $X_t = 0$ otherwise. The reader can check that $\mathring{\mathfrak{F}}_t^X$ consists of all Borel sets of $[0,t]$ together with the single set $]t,\infty[$; $\mathfrak{F}_t^X$ is the completion described above. By Theorem (12.1) any $\mathfrak{F}$-measurable random variable Y is expressible in the form $Y(\omega) = h(T(\omega))$ for some measurable function h. We claim that

$$\mathbb{E}[Y \mid \mathfrak{F}_t^X] = X_t Y - (1 - X_t)\frac{1}{F(t)}\int_{]t,\infty[} h(s)\mathrm{d}F(s),$$

where the coefficient of $(1 - X_t)$ takes an arbitrary value (say, 0) when $t \geqslant \inf\{s : F(s) = 0\}$. Indeed, denoting the expression on the right by Z, it is easily checked that Z is $\mathfrak{F}_t^X$-measurable and that the defining property $\mathbb{E}[YI_G] = \mathbb{E}[ZI_G]$ holds for $G \in \mathfrak{F}_t^X$ (take $G = A \cup N$ and $G =]t,\infty[\cup N$, where $A \in \mathfrak{B}([0,t])$ and N is a null set.)

This example points the way to the interpretation of $\mathfrak{F}_t^X$, namely it represents what is known at time t by observing the process $(X_s, s \leqslant t)$. In the example, if $X_t = 1$ then we know the value of T

and hence the conditional expectation of $Y = h(T)$ is equal to $h(T)$, whereas if $X_t = 0$ we know that T occurs at some time in the future, and the conditional expectation is the integral of h with respect to the conditional distribution $-\mathrm{d}F/F$. We can also write the conditional expectation as a function of the sample function of the observed process, namely as

$$Z = X_t h(\inf\{s \in \mathbb{Q}_+ : X_s = 1\}) + (1 - X_t)g(t),$$

where $\mathbb{Q}_+$ is the set of non-negative rational numbers (we use $\mathbb{Q}_+$ rather than $\mathbb{R}_+$ to ensure that Z is a measurable function from $D[0,t]$ to $\mathbb{R}$) and $g(t) = -\int_{]t,\infty[} h(s)\mathrm{d}F(s)/F(t)$. This exemplifies a general result which is actually a special case of Theorem (12.1).

(13.2) *Theorem.* Let $(X_t, t \in \mathbb{R}_+)$ be a measurable process almost all of whose sample functions belong to the space $D(\mathbb{R}_+)$ of RCLL functions on $\mathbb{R}_+$. Let Y be an $\mathfrak{F}_t^X$-measurable random variable, for some $t \geqslant 0$. Then there is a measurable function $g_t: D[0,t] \to \mathbb{R}$ such that $Y = g_t(X_s, 0 \leqslant s \leqslant t)$ a.s.

Thus any $\mathfrak{F}_t^X$-measurable random variable is a function of the observed process up to time t.

A *filtered probability space* $(\Omega, \mathfrak{F}, (\mathfrak{F}_t)_{t \in \mathbb{R}_+}, P)$ is a probability space $(\Omega, \mathfrak{F}, P)$ together with a filtration $(\mathfrak{F}_t, t \in \mathbb{R}_+)$, a filtration being any family of sub-σ-fields such that $\mathfrak{F}_s \subset \mathfrak{F}_t$ for $s \leqslant t$. In applications[4] $\mathfrak{F}_t$ will invariably be the natural filtration of some collection of processes, but for many purposes it is not necessary to mention these processes explicitly. If we denote by $\mathfrak{F}_\infty$ the smallest σ-field containing $\mathfrak{F}_t$ for all $t \in \mathbb{R}_t$, then it will usually be assumed that $\mathfrak{F} = \mathfrak{F}_\infty$ and that $\mathfrak{F}_t$ is completed with all null sets of $\mathfrak{F}$ as described earlier.

A process X is *adapted* to a filtration $\mathfrak{F}_t$ if, for each $t \in \mathbb{R}_+$, X_t is an $\mathfrak{F}_t$-measurable random variable. X is *progressively measurable* if for each $t \in \mathbb{R}_+$ the process restricted to $[0,t]$, i.e. the function $X^t: [0,t] \times \Omega \to \mathbb{R}$ with $X^t(s,\omega) = X(s,\omega)$ is measurable with respect to the σ-field $\mathfrak{B}[0,t] \otimes \mathfrak{F}_t$ in $[0,t] \times \Omega$. Thus a progressively measurable process is measurable and adapted. The converse is not generally true, but any measurable adapted process with right-continuous sample paths is progressively measurable; this is the most useful case for our purposes.

[4]We will write just $\mathfrak{F}_t$ instead of $(\mathfrak{F}_t, t \in \mathbb{R}_+)$ where no confusion seems likely.

A *stopping time* on a filtered probability space is a random variable taking values in $\mathbb{R}_+ \cup \{+\infty\}$ such that $(T \leqslant t) \in \mathfrak{F}_t$ for all $t \in \mathbb{R}_+$. Thus, given $\mathfrak{F}_t$ one knows whether T has happened by time t or not. Clearly any constant time t_0 is a stopping time, and it is easy to show that if S, T are stopping times then so are $S \wedge T, S \vee T$ and $S + T$. If (T_n) is a sequence of stopping times then $\sup_n T_n$ is a stopping time; $\inf_n T_n$ is not generally a stopping time, but it is one if the filtration $\mathfrak{F}_t$ is right-continuous, i.e. for each $t \in \mathbb{R}_+$, $\bigcap_{\varepsilon > 0} \mathfrak{F}_{t+\varepsilon} = \mathfrak{F}_t$.

If T is an $\mathfrak{F}_t$-stopping time, $\mathfrak{F}_T$ denotes the σ-field of events prior to T, defined by

$$\mathfrak{F}_T = \{A \in \mathfrak{F} : A \cap (T \leqslant t) \in \mathfrak{F}_t \text{ for all } t \in \mathbb{R}_+\}.$$

One can check that $\mathfrak{F}_T$ is indeed a σ-field, that T is $\mathfrak{F}_T$-measurable, and that if T, S are stopping times then $\mathfrak{F}_{T \wedge S} = \mathfrak{F}_T \cap \mathfrak{F}_S$ and the events $(T < S)$, $(T \leqslant S)$ belong to $\mathfrak{F}_{T \wedge S}$; if $S \leqslant T$ a.s. then $\mathfrak{F}_S \subset \mathfrak{F}_T$.

If T is a stopping time and X a stochastic process then X_T denotes the function $\omega \to X_{T(\omega)}(\omega)$, defined on $(T < \infty)$. $X_T I_{(T<\infty)}$ is a random variable if X is measurable, and is $\mathfrak{F}_T$-measurable if X is progressively measurable. (This fact is the main reason for introducing the concept of progressive measurability.) In the latter case the *stopped process* X^T defined by $X_t^T = X_{T \wedge t}$ is also progressively measurable.

13.2 Martingales

A process $M = (M_t, t \in \mathbb{R}_+)$ is a *martingale* of a filtration $\mathfrak{F}_t$ if M is adapted to $\mathfrak{F}_t$, the random variable M_t is integrable for each $t \in \mathbb{R}_+$ and for each $s, t \in \mathbb{R}_+$ with $s \leqslant t$,

$$\mathbb{E}[M_t | \mathfrak{F}_s] = M_s \qquad \text{a.s.} \tag{13.3}$$

It is a *supermartingale* if $\mathbb{E}[M_t | \mathfrak{F}_s] \leqslant M_s$ a.s. and a *submartingale* if $\mathbb{E}[M_t | \mathfrak{F}_t] \geqslant M_s$. Note in particular, using (12.10), that a martingale has constant expectation, i.e. $\mathbb{E}M_t = \mathbb{E}M_0$ for all $t \in \mathbb{R}_+$, whereas a sub(super)-martingale has monotone increasing (decreasing) expectation. Also note that $-M = (-M_t, t \in \mathbb{R}_+)$ is a supermartingale if M is a submartingale, and that M is a martingale if and only if it is both a super- and a submartingale. The following is a basic result.

(13.4) *Theorem.* Suppose M is a submartingale of $\mathfrak{F}_t$, where $\mathfrak{F}_t$ is right-continuous and completed in the way described earlier. Then

M has a right-continuous modification if and only if the function $t \to \mathbb{E}M_t$ is right-continuous. This modification can be chosen to be RCLL and adapted to $\mathfrak{F}_t$.

The significance of this result is that if the filtration satisfies the two conditions mentioned (and virtually all the filtrations encountered in this book do) and $t \to \mathbb{E}M_t$ is right-continuous, we can assume that M_t has RCLL paths; in particular, M is progressively measurable, and $M_T I_{(T<\infty)}$ is $\mathfrak{F}_T$-measurable for any stopping time T.

An *increasing process* is a stochastic process A such that with probability one the sample path $t \to A(t, \omega)$ is a non-decreasing function. Note that if A_t is integrable for each t then A is a submartingale; this follows from (1.6). Similarly $X_t = M_t + A_t$ is a submartingale if M is a martingale and A is an increasing process.

The two major results of martingale theory that we shall need are the *submartingale convergence theorem* and the *optional sampling theorem*. The first of these is as follows. A stochastic process X_t is uniformly integrable if the collection of random variables $\{X_t : t \in \mathbb{R}_+\}$ is uniformly integrable in the sense described in section 12.

(13.5) *Theorem* (submartingale convergence). Let X be a right-continuous submartingale of a filtration $\mathfrak{F}_t$ and suppose that $\sup_{t \in \mathbb{R}_+} \mathbb{E}[X_t \vee 0] < \infty$. Then $X_\infty(\omega) = \lim_{t \to \infty} X_t(\omega)$ exists for almost all $\omega \in \Omega$ and X_∞ is an integrable random variable.

As easy consequences, we have the following.

(13.6) *Corollary*

1. Let X be a right-continuous non-negative supermartingale. Then $X_\infty = \lim_{t \to \infty} X_t$ exists a.s. and $X_t \geqslant \mathbb{E}[X_\infty | \mathfrak{F}_t]$ a.s. for each $t \in \mathbb{R}_+$.
2. If in Theorem (13.5) X is a uniformly integrable martingale then $X_t = \mathbb{E}[X_\infty | \mathfrak{F}_t]$.

In these cases we sometimes say that X_∞ *closes* the supermartingale in that the supermartingale inequality $X_s \geqslant \mathbb{E}[X_t | \mathfrak{F}_s]$ or martingale equality $X_s = \mathbb{E}[X_t | \mathfrak{F}_s]$ is also valid when $t = \infty$.

The traditional picture of a martingale M is that it represents the evolution of a player's fortune in successive plays of a fair game.

If M_s is the fortune now at time s, then 'fair' means that, on average, the fortune at some later time t, after more plays, should be neither more nor less than it is now, and this is precisely the martingale equality. Similarly, super- and submartingales represent unfavourable and favourable games, respectively. The optional sampling theorem addresses the question of whether, in a fair game, it is possible to tilt the odds in one's favour by an ingenious choice of playing strategy, or more precisely, whether the martingale equality $M_S = \mathbb{E}[M_T | \mathfrak{F}_S]$ holds when S, T are stopping times with $S \leqslant T$ a.s., rather than fixed times. The answer is: maybe. For example if M_t is a martingale with continuous sample paths and $M_0 = 0$ a.s., define $T = \inf\{t : M_t = 1\}$ and assume that $T < \infty$ a.s. Then $M_T = 1$ a.s., so clearly $\mathbb{E}[M_T | \mathfrak{F}_0] = 1 \neq M_0 = 0$. Thus some restrictions on the stopping times considered are necessary to exclude cases of this sort. The basic result is as follows.

(13.7) *Theorem* (optional sampling). Let X_t be a right-continuous supermartingale of a filtration $\mathfrak{F}_t$ and suppose there exists an integrable random variable Y such that $X_t \geqslant \mathbb{E}[Y | \mathfrak{F}_t]$ for all $t \in \mathbb{R}_+$. Let S, T be $\mathfrak{F}_t$ stopping times such that $S \leqslant T$ a.s. and define $X_T = Y$ when $T = \infty$, $X_S = Y$ when $S = \infty$. Then the random variables X_S, X_T are integrable, and

(13.8) $$X_S \geqslant \mathbb{E}[X_T | \mathfrak{F}_S] \quad \text{a.s.}$$

Inequality (13.8) holds in particular if S, T are bound by a constant: $S, T \leqslant a$.

The last statement follows from the fact that the process $X'_t = X_{t \wedge a}$ is also supermartingale and (13.8) is unchanged if X is replaced by X'; but for X' the condition of the theorem is satisfied by $Y = X_a$, by the supermartingale property. If X_t is a uniformly integrable martingale then the conditions of Theorem 1.14 are met for both X_t and $-X_t$ with $Y = X_\infty$, from Corollary (13.6) part 1. We thus obtain Corollary (13.9).

(13.9) *Corollary*. If X_t is a uniformly integrable martingale then $X_S = \mathbb{E}[X_T | \mathfrak{F}_S]$ a.s. for any stopping times S, T with $S \leqslant T$.

Finally, a process (X_t) is a *local martingale* if there exists an increasing sequence of stopping times T_n such that $T_n \uparrow \infty$ a.s. and the process $M^n_t := M_{t \wedge T_n}$ is a uniformly integrable martingale for each n.

14 Markov processes

14.1 The Markov and strong Markov properties

Thus far, a stochastic process on a filtered probability space $(\Omega, \mathfrak{F}, (\mathfrak{F}_t), P)$ has always been real valued. One can, however, consider a process X taking values in some other measurable space $(E, \mathscr{E})$ as a collection $X = (X_t, t \in \mathbb{R}_+)$ of measurable functions $X_t: \Omega \to E$. If $B(E)$ denotes the set of bounded measurable functions on $f: E \to \mathbb{R}$ then $Y_t^f := f(X_t)$ will be a real-valued stochastic process for any $f \in B(E)$. We will always assume that E is a Borel space (i.e. E is a topological space which is homeomorphic to a Borel subset of a complete separable metric space (Bertsekas and Shreve, 1978, Definition 7.7, p. 118)) and $\mathscr{E}$ is the Borel σ-field. In particular this implies that singleton sets $\{x\}$ for $x \in E$ are closed and hence in $\mathscr{E}$, X is a *Markov process* if for any times t, s with $t \geqslant s$ and any $f \in B(E)$

$$\mathbb{E}[f(X_t) \mid \mathfrak{F}_s] = \mathbb{E}[f(X_t) \mid X_s]. \tag{14.1}$$

This says that the only information relevant to evaluating the behaviour of the process beyond time s is the value of the current state, X_s. It implies in particular that X is adapted to $\mathfrak{F}_t$. E is called the *state space* of X, and points $x \in E$ are *states*. A function $p(s, x, t, A)$ defined for $s, t \in \mathbb{R}_+, s \leqslant t, x \in E$ and $A \in \mathscr{E}$ is the *transition function* for X if it has the following properties:

1. For fixed s, t, x the function $A \to p(s, x, t, A)$ is a probability measure on $(E, \mathscr{E})$.
2. For fixed s, t, A the function $x \to p(s, x, t, A)$ is $\mathscr{E}$-measurable.
3. For all (s, x, A), $p(s, x, s, A) = I_A(x)$.
4. For any $s \leqslant t, x \in E$ and $A \in \mathscr{E}$

$$\mathbb{E}[I_A(X_t) \mid \mathfrak{F}_s] = p(s, X_s, t, A) \qquad \text{a.s.}$$

Parts 1 and 2 are technical conditions; parts 3 and 4 indicate that $p(s, x, t, A)$ is the probability that, if $X_s = x$, X_t will lie in the set A. In view of 1, the right-hand side of (14.1) is then given by

$$\mathbb{E}[f(X_t) \mid X_s] = \int_E f(y) p(s, X_s, t, \mathrm{d}y) \qquad \text{a.s.} \tag{14.2}$$

Now consider three times $r \leqslant s \leqslant t$. From property 4 of Theorem

(12.9) of conditional expectation

$$\mathbb{E}[f(X_t)|\mathfrak{F}_r] = \mathbb{E}[\mathbb{E}[f(X_t)|\mathfrak{F}_s]|\mathfrak{F}_r] \quad \text{a.s.} \tag{14.3}$$

From (14.1) and (14.2) the inner conditional expectation on the right is given by

$$\int_E f(y)p(s, X_s, t, \mathrm{d}y)$$

and using (14.1), (14.2) again the right-hand side of (14.3) is equal almost surely to

$$\int_E \int_E f(y)p(s, z, t, \mathrm{d}y)p(r, X_r, s, \mathrm{d}z).$$

Thus, taking $f = I_A$ for some $A \in \mathscr{E}$, (14.3) is (nearly) the same thing as saying that

$$p(r, x, t, A) = \int_E p(s, z, t, A)p(r, x, s, \mathrm{d}z). \tag{14.4}$$

This relation is known as the *Chapman–Kolmogorov equation.* Now suppose that p is a transition function satisfying (14.4) for all $x \in E$, $A \in \mathscr{E}$ and $r \leqslant t$, and that μ is a probability measure on $(E, \mathscr{E})$. For times $0 < t_1 < t_2 < \cdots < t_n$ define a probability $P_{0,t_1,\ldots,t_n}$ on E^{n+1} by

$$\begin{aligned} P_{0,t_1,\ldots,t_n}&(A_0 \times A_1 \times \cdots \times A_n) \\ &= \int_{A_0} \int_{A_1} \cdots \int_{A_{n-2}} \int_{A_{n-1}} \\ &\times p(t_{n-1}, x_{n-1}, t_n, A_n)p(t_{n-2}, x_{n-2}, t_{n-1}, \mathrm{d}x_{n-1}) \cdots \\ &\times p(0, x_0, t_1, \mathrm{d}x_1)\mu(\mathrm{d}x_0). \end{aligned} \tag{14.5}$$

One can check, using (14.4), that this defines a consistent family of finite-dimensional distributions, and hence by Kolmogorov's theorem there is a probability space $(\Omega, \mathscr{F}, P)$ and a process X such that $P_{0,t_1,\ldots,t_n}$ is the family of finite-dimensional distributions for X. Then X is a Markov process with p as its transition function, and μ is the distribution of the initial state x_0. (We can take $\mathfrak{F}_t = \mathfrak{F}_t^X$, the natural filtration.) It is possible to construct examples of Markov processes that do not have transition functions, but the normal procedure is to start with a transition function (or something equivalent to it) and construct the process as outlined above. Note that only one ingredient other than the transition function is required, namely the *initial distribution* μ.

A transition function is time invariant if for all $r, s, t, \in \mathbb{R}_+$ with $s \leqslant t$, $x \in E$ and $A \in \mathscr{E}$ we have

$$p(s+r, x, t+r, A) = p(s, x, t, A).$$

Taking $s = 0$ and $t = u - r$ with $u \geqslant r$ this becomes

$$p(r, x, u, A) = p(0, x, u - r, A),$$

showing that the transition between times r and u depends only on the time *difference* $u - r$. We define a three-parameter function, which is also (by slight abuse of notation) denoted p, by

$$p(t, x, A) = p(0, x, t, A).$$

In this new notation the Chapman–Kolmogorov equation (14.4) becomes

$$p(r+s, x, A) = \int_E p(s, z, A)p(r, x, \mathrm{d}z), \tag{14.6}$$

for $r, s \in \mathbb{R}_+, x \in \mathscr{E}$. For the remainder of this section we will discuss only Markov processes with time-invariant transition functions. It will be seen in the applications to piecewise-deterministic processes in Chapters 2 and 3 that a simple redefinition of the process transforms a process with a time-varying transition function into another with a time-invariant transition function, so there is from our point of view no extra generality in discussing the time-varying case.

Suppose we are given a transition function p and we define the finite-dimensional distributions as in (14.5) taking $\mu = \delta_x$, where δ_x denotes the *Dirac measure* at $x \in E$ defined by $\delta_x(\{x\}) = 1, \delta_x(A) = 0$ if $x \notin A$ (recall that $\{x\} \in \mathscr{E}$). We can then construct a Markov process X_t having p as its transition function and $P(X_0 = x) = 1$. It is however possible that the same construction taking $\mu = \delta_{x'}$ for some $x' \neq x$ would lead to a process defined on some other probability space. A *Markov family* is a collection $(\Omega, \mathfrak{F}, (\mathfrak{F}_t), (X_t), (P_x, x \in E))$, where $(\Omega, \mathfrak{F})$ is a measurable space, $(\mathfrak{F}_t)$ a filtration, $(X_t, t \in \mathbb{R}_+)$ a family of E-valued random variables such that X_t is $\mathfrak{F}_t$-measurable for each t, and, for each $x \in E$, P_x is a probability measure on $(\Omega, \mathfrak{F})$ such that (X_t) is a Markov process on $(\Omega, \mathfrak{F}, P_x)$ with transition function p and initial distribution δ_x, i.e. $P_x[X_0 = x] = 1$. We will write $\mathbb{E}_x$ for the expectation with respect to the measure P_x. Note that the measure P_x and transition function p are related by

(14.7) $$P_x[X_t \in A] = p(t, x, A) \qquad (t, x, A) \in \mathbb{R}_+ \times E \times \mathscr{E}.$$

In a Markov family, only the measure P_x depends on the initial point $x \in E$; all the other ingredients are the same for every x. This provides yet another way of expressing the Markov property: because the transition function is the same for every P_x we easily see that for $f \in B(E)$, $x \in E$ and $s, t \geqslant 0$ we have

$$\mathbb{E}_x[f(X_{s+t}) | \mathfrak{F}_s] = \mathbb{E}_z[f(X_t)] |_{z = X_s}.$$

Thus the behaviour of the process beyond time s is just that of another process started at X_s.

In applications Markov families are normally constructed on *canonical spaces*. The most common such space, certainly for the purposes of this book, is the space $\Omega = D_E[0, \infty[$ of right-continuous E-valued functions on $\mathbb{R}_+$ with left-hand limits. We let x_t denote the coordinate function $x_t(\omega) = \omega(t)$ for $\omega \in D_E[0, \infty[$, $\mathfrak{F}_t$ be the natural filtration $\mathfrak{F}_t = \sigma\{x_s, s \leqslant t\}$ and[5] $\mathfrak{F} = \bigvee_{t \leqslant 0} \mathfrak{F}_t$. If we start with a transition function p and $x \in E$ then we can construct, as outlined above, a Markov process X_t on *some* probability space, which we will now denote $(\Omega', \mathscr{F}', P')$, with the property that $P'[X_0 = x] = 1$. Suppose that this process has the property that, on a set $\Omega'' \in \mathfrak{F}'$ of P'-probability 1, the sample function $\omega' \to X_t(\omega')$ lies in $D_E[0, \infty[$. (Of course, this places a non-trivial restriction on the class of transition functions we can consider.) Then we can define a map $\phi: \Omega' \to D_E[0, \infty[$ by $\phi(\omega')(t) = X_t(\omega')$ for $\omega' \in \Omega''$, $\phi(\omega') = \omega_0$ for $\omega' \in \Omega' \backslash \Omega''$, where ω_0 is an arbitrary element of $D_E[0, \infty[$. The corresponding image probability is denoted $P_x := P'\phi^{-1}$. It is easy to see that $(\Omega, \mathfrak{F}, (\mathfrak{F}_t), (x_t), P_x)$ is also a Markov process with transition function p and initial measure δ_x, i.e. we have *realized* the process on the canonical space $(\Omega, \mathfrak{F}, (\mathfrak{F}_t))$. Since this construction can be carried out for any $x \in E$, we have in fact created a Markov family (x_t, P_x) on $(\Omega, \mathfrak{F}, (\mathfrak{F}_t))$ in this way. From now on we will generally consider Markov families rather than Markov processes, and use the notation (x_t, P_x) for a Markov family whether or not it is constructed on some canonical space.

[5] $\bigvee_{t \geqslant 0} \mathfrak{F}_t = \sigma(\mathfrak{F}_t, t \in \mathbb{R}_+)$, the smallest σ-field containing all $\mathfrak{F}_t$.

A strong Markov process (or family) is one for which the Markov property holds at stopping times of the σ-field $\mathfrak{F}_t$, not just at fixed times as hitherto considered. More precisely, a family $(\Omega, \mathfrak{F}, (\mathfrak{F}_t), (x_t), P_x)$ is a strong Markov family if (x_t) is progressively measurable with respect to $\mathfrak{F}_t$ and for every $x \in E$, $A \in \mathscr{E}$, $\mathfrak{F}_t$-stopping time T and $t \geqslant 0$

(14.8) $$P_x[x_{T+t} \in A \mid \mathfrak{F}_T] I_{T<\infty} = p(t, x_T, A) I_{T<\infty}.$$

(Recall that p and P_x are related by (14.7).) This says that the probabilistic evolution of the process after the stopping time T is just that of another process 'restarted' at T, or, more colloquially, the process 'restarts at stopping times'. The following standard example shows that it is possible for a process to be Markov without being strong Markov.

(14.9) *Example.* Take $E = \mathbb{R}_+$. The measure P_+ is such that if $x > 0$ then $x_t = x + t$, while if $x = 0$ then $x_t = 0$ for $t \leqslant S$ and $x_t = t - S$ for $t \geqslant S$, where S is a random variable with $P_0[S > t] = \mathrm{e}^{-t}$. This is clearly a Markov family with respect to the natural filtration of (x_t) since knowing x_t enables one to reconstruct the whole sample path $\{x_s, 0 \leqslant s \leqslant t\}$. The transition function is

$$p(t, x, A) = \begin{cases} I_A(x + t), & x > 0, \\ I_A(0)\mathrm{e}^{-t} + \int_0^t I_A(t-s)\mathrm{e}^{-s}\mathrm{d}s, & x = 0. \end{cases}$$

S is a stopping time of the *completed* natural filtration $(\mathfrak{F}_t)$ with respect to P_0 since $(S \leqslant t) = (S < t) \cup (S = t) = (x_t > 0) \cup (S = t)$, and $(x_t > 0)$ is an element of $\mathfrak{F}_t$ while $P_x(S = t) = 0$ for $t > 0$. Now consider (1.16) with $T = S$, $t = 1$, $x = 0$ and $A = \mathbb{R}_+ \backslash \{0\}$. Then $p(t, x_s, A) = 1 - \mathrm{e}^{-1}$ while $P_0[x_{S+1} > 0 \mid \mathfrak{F}_S] = 1$. The strong Markov property thus fails.

14.2 Semigroups and generators

Let $(\Omega, \mathfrak{F}, (\mathfrak{F}_t), (x_t), P_x)$ be a Markov family on a state space E. As before, we denote by $B(E)$ the set of all bounded measurable functions $f: E \to \mathbb{R}$. This is a Banach space under the norm

$$\| f \| = \sup_{x \in E} | f(x) |$$

with the linear space structure $(f_1 + f_2)(x) = f_1(x) + f_2(x)$, etc. For $t \in \mathbb{R}_+$ define an operator $P_t : B(E) \to B(E)$ by

$$P_t f(x) = \mathbb{E}_x f(x_t).$$

Because $|\mathbb{E}_x f(x_t)| \leqslant \mathbb{E}_x[\|f\|] = \|f\|$, P_t is a *contraction*, i.e. $\|P_t f\| \leqslant \|f\|$, and the Chapman–Kolmogorov equation (14.6) is equivalent to the following *semigroup property* of P_t: for $t, s \in \mathbb{R}_+$,

$$P_t P_s = P_{t+s},$$

or, in more detail, for $f \in B(E)$ and $x \in E$,

$$P_t(P_s f)(x) = P_{t+s} f(x).$$

Associated with the semigroup P_t is its *strong generator* which, loosely speaking, is the derivative of P_t at $t = 0$. Let $\mathscr{D}(\hat{\mathfrak{A}}) \subset B(E)$ be the set of functions f for which the following limit exists

$$\lim_{t \downarrow 0} \frac{1}{t}(P_t f - f)$$

and denote this limit $\hat{\mathfrak{A}} f$. The limit refers to convergence in the norm $\|\cdot\|$, i.e. for $f \in \mathscr{D}(\hat{\mathfrak{A}})$ we have

$$\lim_{t \downarrow 0} \left\| \frac{1}{t}(p_t f - f) - \hat{\mathfrak{A}} f \right\| = 0.$$

It is important to appreciate that the domain $\mathscr{D}(\hat{\mathfrak{A}})$ may not be (and usually is not) all of $B(E)$. It is quite possible to have two different Markov processes x_t^1 and x_t^2 whose generators $\hat{\mathfrak{A}}_1$ and $\hat{\mathfrak{A}}_2$ are the same operator but which act on different classes of functions (i.e. $\mathscr{D}(\hat{\mathfrak{A}}_1) \cap \mathscr{D}(\hat{\mathfrak{A}}_2) = \varnothing$); many examples of this will be found in Chapter 2. Thus specifying its domain $\mathscr{D}(\hat{\mathfrak{A}})$ is an essential part of specifying the operator $\hat{\mathfrak{A}}$.

Let B_0 be the subset of $B(E)$ consisting of those functions f for which $\lim_{t \downarrow 0} \|P_t f - f\| = 0$. The semigroup P_t is said to be *strongly continuous* on B_0. B_0 is clearly a linear subspace of $B(E)$; it is also closed, since if $f_n \to f$, $f_n \in B_0$, then

$$\limsup_{t \downarrow 0} \|P_t f - f\| \leqslant \limsup_{t \downarrow 0} \|P_t f - P_t f_n\| + \lim_{t \downarrow 0} \|P_t f_n - f_n\| + \|f_n - f\|.$$

The second term on the right is zero, and

$$\|P_t f - P_t f_n\| = \|P_t(f - f_n)\| \leqslant \|f - f_n\|,$$

by the contraction property. Hence $f \in B_0$. Since B_0 is a closed linear subspace of $B(E)$ it is itself a Banach space. For any $t \in \mathbb{R}_+$ the operator P_t maps B_0 into B_0; indeed, for $t, s \in \mathbb{R}_+$ we have for $f \in B_0$

$$\|P_{t+s}f - P_t f\| = \|P_t(P_s f - f)\| \leqslant \|P_s f - f\|$$

and hence $P_s(P_t f) \to P_t f$ as $s \downarrow 0$; i.e. $P_t f \in B_0$. We can define derivatives and Riemann integrals of continuous functions $g: \mathbb{R}_+ \to B_0$ in the same way as for real-valued functions, namely

$$\frac{\mathrm{d}}{\mathrm{d}t} g(t) = \lim_{\delta \to 0} \frac{1}{\delta} (g(t + \delta) - g(t))$$

$$\int_a^b g(t)\mathrm{d}t = \lim_{n \to \infty} \frac{b-a}{n} \sum_{k=0}^{n-1} g\left(a + \frac{k}{n}(b-a)\right).$$

We now have the following result.

(14.10) *Proposition*

1. If $f \in B_0$ and $t \in \mathbb{R}_+$ then $\int_0^t P_s f \,\mathrm{d}s \in \mathscr{D}(\hat{\mathfrak{A}})$ and

$$\hat{\mathfrak{A}}\left(\int_0^t P_s f \,\mathrm{d}s\right) = P_t f - f.$$

2. If $f \in \mathscr{D}(\hat{\mathfrak{A}})$ and $t \in \mathbb{R}_+$, then $P_t f \in \mathscr{D}(\hat{\mathfrak{A}})$ and

$$\frac{\mathrm{d}}{\mathrm{d}t} P_t f = \hat{\mathfrak{A}} P_t f = P_t \hat{\mathfrak{A}} f.$$

3. If $f \in \mathscr{D}(\hat{\mathfrak{A}})$ and $t \in \mathbb{R}_+$, then

$$P_t f - f = \int_0^t \hat{\mathfrak{A}} P_s f \,\mathrm{d}s = \int_0^t P_s \hat{\mathfrak{A}} f \,\mathrm{d}s.$$

Proof

1. For $\delta > 0$, and with I denoting the identity mapping $If = f$, we have

$$\frac{1}{\delta}(P_\delta - I)\int_0^t P_s f\,\mathrm{d}s = \frac{1}{\delta}\int_0^t (P_{s+\delta}f - P_s f)\mathrm{d}s$$

$$= \frac{1}{\delta}\int_t^{t+\delta} P_s f\,\mathrm{d}s - \frac{1}{\delta}\int_0^\delta P_s f\,\mathrm{d}s.$$

As $\delta \downarrow 0$, the right-hand side converges to $P_t f - f$.

2. By the semigroup property, for $\delta > 0$

$$\frac{1}{\delta}(P_\delta(P_t f) - P_t f) = P_t\left(\frac{1}{\delta}(P_\delta f - f)\right)$$

and hence $(\mathrm{d}^+/\mathrm{d}t)P_t f = P_t\hat{\mathfrak{A}}f = \hat{\mathfrak{A}}P_t f$ for $f \in \mathscr{D}(\hat{\mathfrak{A}})$, where $\mathrm{d}^+/\mathrm{d}t$ denotes the right derivative. For the left derivative we have for $t > 0$ and $0 < \delta \leqslant t$

$$\frac{1}{-\delta}(P_{t-\delta}f - P_t f) = P_{t-\delta}\left(\frac{1}{\delta}(P_\delta f - f)\right),$$

from which it follows that $\mathrm{d}^- P_t f/\mathrm{d}t = P_t\hat{\mathfrak{A}}f = \mathrm{d}^+ P_t f/\mathrm{d}t$.

3. This follows from 2 and the fact that

$$\int_a^b \frac{\mathrm{d}}{\mathrm{d}t}g(t)\mathrm{d}t = g(b) - g(a)$$

for continuously differentiable B_0-valued functions g. □

Let us re-express these results in probabilistic terms, using the definitions of P_t as $P_t f(x) = \mathbb{E}_x f(x_t)$. First, from 2, if we define $V(t,x) = \mathbb{E}_x f(x_t)$ for $f \in \mathscr{D}(\hat{\mathfrak{A}})$, then V satisfies

(14.11) $$\frac{\partial}{\partial t}V(t,x) = \hat{\mathfrak{A}}V(t,x), \qquad V(0,x) = f(x),$$

which is an abstract version of the so-called *Kolmogorov backward equation.* It shows that giving the generator $(\hat{\mathfrak{A}}, \mathscr{D}(\hat{\mathfrak{A}}))$ of the process is enough to determine the entire probability measure P_x: the semigroup P_t is determined by solving (14.11) and this determines the transition function $p(t,x,A)$, since for $f \in B(E)$

$$\int_E f(y)p(t,x,\mathrm{d}y) = P_t f(x).$$

Now the transition function determines the finite-dimensional distributions and hence the measure P_x. Of course, some technical conditions are required to turn this calculation into reality, and no-one would think of carrying it out in practice except for artificial examples, but it does show that the process is in principle completely characterized by its generator. Much more important for practical calculations is part 3 of Proposition (14.10) which we can write as

$$\mathbb{E}_x f(x_t) = f(x) + \mathbb{E}_x \int_0^t \hat{\mathfrak{A}} f(x_s) \mathrm{d}s, \qquad f \in \mathscr{D}(\hat{\mathfrak{A}}). \tag{14.12}$$

This is known as the *Dynkin formula* and virtually every computation of expected values encountered in this book is ultimately based on it, as the reader will discover in Chapter 3. The key property turns out to be the close connection between the Dynkin formula and certain martingales, which we give next.

(14.13) *Proposition.* For $f \in \mathscr{D}(\hat{\mathfrak{A}})$ we define the real-valued process $(C_t^f, t \in \mathbb{R}_+)$ by

$$C_t^f = f(x_t) - f(x_0) - \int_0^t \hat{\mathfrak{A}} f(x_s) \mathrm{d}s.$$

Then for any $x \in E$, the process $(C_t^f)_{t \geqslant 0}$ is a martingale on $(\Omega, \mathfrak{F}, (\mathfrak{F}_t), P_x)$.

Proof. C_t^f is clearly adapted to $\mathfrak{F}_t$, and for $t \geqslant s$ we have from the definitions

$$\mathbb{E}_x[C_t^f - C_s^f \mid \mathfrak{F}_s] = \mathbb{E}_x[f(x_t) \mid \mathfrak{F}_s] - f(x_s) - \mathbb{E}_x\left[\int_s^t \hat{\mathfrak{A}} f(x_u) \mathrm{d}u \,\middle|\, \mathfrak{F}_s\right]. \tag{14.14}$$

Using the time-homogeneity, the first term on the right is equal to $P_{t-s} f(x_s)$, while

$$\begin{aligned} \mathbb{E}_x\left[\int_s^t \hat{\mathfrak{A}} f(x_u) \mathrm{d}u \,\middle|\, \mathfrak{F}_s\right] &= \mathbb{E}_x\left[\int_s^t \hat{\mathfrak{A}} f(x_u) \mathrm{d}u \,\middle|\, x_s\right] \\ &= \int_s^t P_{u-s} \hat{\mathfrak{A}} f(x_s) \mathrm{d}u \\ &= \int_0^{t-s} P_r \hat{\mathfrak{A}} f(x_s) \mathrm{d}r. \end{aligned}$$

From part 3 of Proposition (14.10) we now see that the right-hand side of (14.14) is equal to zero, establishing the martingale property. □

The function $\hat{\mathfrak{A}}f$ is defined for $f \in \mathscr{D}(\hat{\mathfrak{A}})$ and then the Dynkin formula (14.12) or the martingale property of Proposition (14.13) holds. There may however be other functions f, not in $\mathscr{D}(\hat{\mathfrak{A}})$, for which something akin to (14.12) and Proposition (14.13) is still true. With this in mind, we make the following definition.

(14.15) *Definition.* Let $\mathscr{D}(\mathfrak{A})$ denote the set of measurable functions $f: E \to \mathbb{R}$ with the following property: there exists a measurable function $h: E \to \mathbb{R}$ such that the function $t \to h(x_t)$ is integrable P_x-a.s. for each $x \in E$ and the process

$$C_t^f = f(x_t) - f(x_0) - \int_0^t h(x_s)\mathrm{d}s \tag{14.16}$$

is a local martingale. Then we write $h = \mathfrak{A}f$ and call $(\mathfrak{A}, \mathscr{D}(\mathfrak{A}))$ the *extended generator* of the process x_t.

Local martingales were introduced in section 13.2. The definition is stated in terms of local martingales, rather than martingales, for the technical reason that the former are easier to characterize in applications; this will be seen in Chapter 2. Note that if $f \in \mathscr{D}(\hat{\mathfrak{A}})$ then C_t^f given by (14.16) is a martingale, and *a fortiori* a local martingale, with $h = \hat{\mathfrak{A}}f$. This shows that $\mathfrak{A}$ is – as the name implies – an extension of $\hat{\mathfrak{A}}$ in that $\mathscr{D}(\mathfrak{A}) \supset \mathscr{D}(\hat{\mathfrak{A}})$ and $\mathfrak{A}f = \hat{\mathfrak{A}}f$ for $f \in \mathscr{D}(\hat{\mathfrak{A}})$. The function $\mathfrak{A}f$ is defined uniquely: if C^f of (14.16) were a local martingale with $h = h_1$ and $h = h_2$, then the process $M_t = \int_0^t (h_1(x_s) - h_2(x_s))\mathrm{d}s$ would be a local martingale whose sample paths are continuous and of bounded variation, and it is known (Elliott, 1982, Lemma 11.39) that any such local martingale is identically equal to a constant, which is zero in this case since $M_0 = 0$. Since this is true P_x-a.s. for every $x \in E$, we must have $h_1(x) = h_2(x)$, except possibly on some set $A \in \mathscr{E}$ such that

$$\int_0^\infty I_A(x_s)\mathrm{d}s = 0 \qquad P_x\text{-a.s.} \qquad \text{for every } x \in E.$$

A set A with this property is said to have *potential zero*. The process x_t 'spends no time' in A, regardless of the starting point, so the process C_t^f of (14.16) does not depend of the values of h for $x \in A$, and h is unique up to sets of zero potential.

If C_t^f is a martingale then $\mathbb{E}_x C_t^f = C_0^f = 0$ which implies that the Dynkin formula holds in the form

$$\mathbb{E}_x f(x_t) = f(x) + \mathbb{E}_x \int_0^t \mathfrak{A} f(x_s) ds. \tag{14.17}$$

In fact $\{f \in \mathscr{D}(\mathfrak{A}) : C_t^f \text{ is a martingale}\}$ is precisely the widest class of functions for which (14.17) is true. Since, as mentioned above, the Dynkin formula is our primary result for calculating expectations, we naturally want it to hold as widely as possible; hence the interest in the extended generator concept. In particular, it is often quite difficult to characterize $\mathscr{D}(\hat{\mathfrak{A}})$, but there may be (and are, in the applications of Chapter 2) easily checked sufficient conditions for membership of $\mathscr{D}(\mathfrak{A})$.

15 Notes and references

The prerequisites for reading this book are a course in applied probability and stochastic processes as presented, for example, in the excellent books of Cox and Miller (1965) or Çinlar (1975), and some knowledge of measure-theoretic probability at the level of, say, Williams (1990). Wong and Hajek (1984) provide a readable introduction to most of these matters at the appropriate level. As mentioned, the latter half of this book is concerned with stochastic control theory. For an insightful general introduction to this subject the reader can consult Whittle (1982). More specialized texts covering discrete- and continuous-time cases, respectively, are Bertsekas and Shreve (1978) and Fleming and Rishel (1975). The book by Tijms (1986) presents a computationally oriented approach that covers some of the same ground as the present volume. Approximation of continuous-time models by discrete-time ones and 'heavy traffic' limit theorems are discussed in many places, for example Kushner (1977) and Ethier and Kurtz (1986). For the LQG problem, see Whittle (1982) or Davis (1977). Financial applications are covered by Merton (1991).

11 Dudley (1989) is a superb reference for the material in this section and section 12. The earlier text by Kingman and Taylor (1966) is also highly recommended. Royden (1988) provides comprehensive coverage of topics in analysis.

12 This material is covered in many texts on 'advanced' probability theory, for example Williams (1990) and Billingsley (1986), both of which are excellent. Wong and Hajek (1984) provide a concise summary. As mentioned at the beginning of the chapter, the systematic formulation of axiomatic probability is due to Kolmogorov (1933). The major achievement of Kolmogorov and other workers at the time, such as Ulam (1932), was to realize that no *topological* structure is required for a probability space. Constructing measures (i.e. set functions with the countable additivity property) in particular cases does generally bring in topological properties – as for example in the construction of Lebesgue measure on the real line – but once this is done the topology plays no further role in probabilistic studies of independence, convergence, etc.

13 Wong and Hajek (1984) is probably the best introductory account. An unmatched exposition is given by Meyer (1966), a book which is unfortunately long out of print. Dellacherie and Meyer (1975, 1980) give an authoritative expanded account. Brémaud (1981) can also be consulted.

14 Treatments of Markov process theory slanted in the appropriate way are surprisingly hard to find: standard textbooks tend to be either from the 'applied probability' school, concentrating too much on countable-state Markov chains, or from the 'pure mathematics' school, where anything smacking of computations – in particular any mention of generators – is aristocratically discarded. Honourable exceptions are Wentzell (1981), Wong and Hajek (1984), and – at a considerably more advanced level – Ethier and Kurtz (1986). A classic reference is Dynkin (1965). The extended generator concept is essentially due to Stroock and Varadhan (1969).

CHAPTER 2

Piecewise-deterministic Markov processes

The aim of this book is to analyse stochastic models in which the randomness appears as point events, i.e. there is a sequence of random occurrences at fixed or random times $T_1 < T_2 < T_3 \ldots$ but there is no additional component of uncertainty between these times. In this chapter we introduce a general class of processes called *piecewise-deterministic Markov processes* (PDPs) which are stochastic models having precisely this property, and sufficiently general to cover a wide variety of applied problems as special cases. Constructing the PDP model corresponding to a particular application proceeds by way of one or both of two well-known principles: the inclusion of *supplementary variables* to turn an initially non-Markov process into a Markov process, and the use of indicator variables to keep track of the various possible 'configurations' of the system under study. For example, in a queueing network each storage buffer can be full, empty or somewhere in between, and clearly the 'state' of the network – however interpreted – changes when, for instance, one buffer becomes full. In the first section of the chapter we describe several examples which illustrate the application of these principles and lead to Markovian stochastic models which will turn out to be PDPs. The discussion is at an informal level with no pretensions, at this stage, to mathematical rigour: everything can be substantiated in a more formal way later by applying the general PDP theory as developed in the remainder of the chapter.

A PDP consists of a mixture of deterministic motion and random jumps. Before giving a formal definition, some information about deterministic differential equations and vector fields, and about methods for constructing random variables with specified distributions, is given in sections 22 and 23, respectively. The general definition is given in section 24, and the remaining three sections

25–27 are devoted to establishing the main properties of PDPs in their capacity as Markov processes, namely the strong Markov property, an exact characterization of the extended generator, and the fact that PDPs are *right processes.*

21 Markov models and supplementary variables

A *Leitmotiv* of stochastic modelling is that calculations are greatly facilitated if the model is formulated as a Markov process; then general methods for computing distributions and expectations, based mainly on the Dynkin formula and the Kolmogorov backward equation, are available. For some non-Markovian models, in renewal theory for instance, special techniques have been developed, but these tend not to extend beyond the very special assumptions on which the models are based, so that any generalization will require a completely fresh approach. In this section we explore the formulation of some stochastic models involving randomness in the form of point events. The prototype such process is, of course, the Poisson process, but before discussing this let us say something about that remarkable construct, the *exponential distribution.* An $\mathbb{R}_+$-valued random variable T is exponentially distributed if its survivor function $F(t) = P[T > t]$ is given by

$$F(t) = \mathrm{e}^{-\lambda t}, \quad t \geqslant 0$$

for some constant $\lambda > 0$. The mean and standard deviation of T are both equal to $1/\lambda$. Invariably, T will represent a random time. The key property of the exponential distribution relates to the conditional distribution of the remaining time given that $T > s$; this is given by

$$(21.1) \qquad P[T > t + s \mid T > s] = \frac{F(t+s)}{F(s)} = \frac{\mathrm{e}^{-\lambda(t+s)}}{\mathrm{e}^{-\lambda s}} = \mathrm{e}^{-\lambda t}.$$

This expresses the 'memoryless' property of the distribution: the conditional distribution is just the same as the unconditional distribution of T, regardless of the elapsed time s. Another way of expressing this is in terms of the *hazard rate*, which is by definition a function $h(t)$, satisfying

$$P[T \in]s, s + \delta] \mid T > s] = h(s)\delta + o(\delta),$$

where $o(\delta)$ denotes a function such that $o(\delta)/\delta \to 0$ as $\delta \to 0$. Thus $h(s)\delta$ expresses, to first order, the probability that T occurs 'now' given that it has not occurred 'so far'. In the exponential case we have

$$P[T\in]s,s+\delta]\,|\,T>s]=\frac{1}{F(s)}(F(s)-F(s+\delta))$$
$$=1-e^{-\lambda\delta}=\lambda\delta+o(\delta).$$

The hazard rate is thus a constant, $h(s)=\lambda$. For a non-negative random variable with general density function ψ and corresponding survivor function $\Psi(s)=\int_s^\infty \psi(t)\mathrm{d}t$ the hazard rate is given by

(21.2) $$h(s)=-\frac{1}{\Psi(s)}\frac{\mathrm{d}}{\mathrm{d}s}\Psi(s)=\frac{\psi(s)}{\Psi(s)}.$$

The function h is defined for $s<c=\inf\{t:\Psi(t)=0\}$. In fact, there is a one-to-one correspondence between h and Ψ: equation (21.2) gives h in terms of Ψ, while conversely, integrating (21.2) and using the fact that $\Psi(0)=1$ gives

(21.3) $$\Psi(t)=\exp\left(-\int_0^t h(s)\mathrm{d}s\right).$$

Thus the exponential is the *only* distribution with constant hazard rate.

(21.4) *The Poisson process*

Let $S_1,S_2,\ldots$ be a sequence of independent, identically distributed (i.i.d.) random variables with exponential distribution, $P[S_1>t]=e^{-\lambda t}$ for some $\lambda>0$, and define $T_1=S_1, T_i=S_1+\cdots+S_i$ for $i\geqslant 2$ and for $t\geqslant 0$

$$N_t=\sum_{i=1}^{\infty} I_{(t\geqslant T_i)}.$$

The sample functions of N_t are thus right-continuous step functions with jumps of height 1 at each T_i; in particular they belong to the space $D[0,\infty[$. The random variable N_t has the Poisson distribution $P[N_t=n]=e^{-\lambda t}(\lambda t)^n/n!$, with mean $\mathbb{E}N_t=\lambda t$. We let $\mathfrak{F}_t$ denote the natural filtration, $\mathfrak{F}_t=\sigma\{N_s, s\leqslant t\}$. Fix $t>0$ and let $T_{k(\omega)}$ denote the last jump time before t. Now let $S_1^*=T_{k(\omega)+1}-t$ and $S_i^*=S_{k(\omega)-i}$, $i\geqslant 2$. Because of the memoryless property (21.1), the conditional distribution of S_1^* given $\mathfrak{F}_t$ (or, equivalently, given that $S_{k(\omega)+1}>t-T_k$) is exponential, $P[S_1^*>s\,|\,\mathfrak{F}_t]=e^{-\lambda s}$, and S_i^*, $i=1,2,\ldots$ are i.i.d.

It follows that $N_s^* = N_{t+s} - N_t$, $s > 0$, is a Poisson process which is independent of $\mathfrak{F}_t$, i.e. the process 'restarts' at each time t. In particular, N_t has independent increments: $(N_{t_4} - N_{t_3})$ and $(N_{t_2} - N_{t_1})$ are independent random variables for any $0 \leqslant t_1 \leqslant t_2 \leqslant t_3 \leqslant t_4$, and this implies that (N_t) is a Markov process. Indeed, if f is any bounded function and $t \geqslant s$ then

$$\mathbb{E}[f(N_t)|\mathfrak{F}_s] = \sum_{k=0}^{\infty} f(k + N_s)\frac{1}{k!}\mathrm{e}^{-\lambda(t-s)}(\lambda(t-s))^k \tag{21.5}$$

since $N_t - N_s$ is Poisson distributed with mean $\lambda(t-s)$. We can also consider the Poisson process as a Markov family on the state space $E = \mathbb{Z} = \{0, \pm 1, \pm 2, \ldots\}$ or $\mathbb{Z}_+ = \{0, 1, 2, \ldots\}$, the measure P_x being such that $N_0 = x \in E$ and $(N_t - x)$ is a Poisson process as described above. From (21.5) with $s = 0$, we see that the transition function is

$$p(t, x, \{z\}) = \begin{cases} \mathrm{e}^{-\lambda t}\dfrac{1}{(z-x)!}(\lambda t)^{z-x}, & z \geqslant x \\ 0, & z < x. \end{cases} \tag{21.6}$$

Let us now consider the generator $\mathfrak{A}$ of this family. It is easy to see what the formula for $\mathfrak{A}f$ is, although determining the exact domain $\mathscr{D}(\mathfrak{A})$ is a little more tricky. For small t we have from (21.6)

$$P_x[N_t = x] = 1 - \lambda t + o(t)$$

$$P_x[N_t = x+1] = \lambda t + o(t)$$

$$P_x[N_t \geqslant x+2] = o(t),$$

and hence, certainly if f is bounded,

$$\mathbb{E}_x f(N_t) = f(x)(1 - \lambda t + o(t)) + f(x+1)(\lambda t + o(t)) + o(t).$$

Thus $(\mathbb{E}_x f(N_t) - f(x)) = \lambda(f(x+1) - f(x))t + o(t)$, and therefore

$$\mathfrak{A}f(x) = \lim_{t \downarrow 0} \frac{1}{t}(\mathbb{E}_x f(N_t) - f(x)) = \lambda(f(x+1) - f(x)).$$

The domain $\mathscr{D}(\hat{\mathfrak{A}})$ of the *strong* generator $\hat{\mathfrak{A}}$ is the set of functions f for which the limit exists *uniformly in* x. It is not necessary to describe this set precisely here. We have seen however that for $f \in \mathscr{D}(\hat{\mathfrak{A}})$, the process

$$C_t^f = f(N_t) - f(N_0) - \lambda \int_0^t (f(N_s + 1) - f(N_s))\mathrm{d}s$$

is a martingale (of $\mathfrak{F}_t$). To see this by direct calculation, note that by the independent increment property, for $t > s$,

$$\mathbb{E}_x[C_t^f - C_s^f | \mathfrak{F}_s] = \mathbb{E}_x[C_t^f - C_s^f | N_s]$$

$$= \mathbb{E}_x\left[f(y+\tilde{N}_t) - f(y) - \lambda \int_s^t (f(y+1+\tilde{N}_r) - f(y+\tilde{N}_r))\mathrm{d}r \right]\Bigg|_{y=N_s},$$

where $\tilde{N}_t = N_t - N_s$. Now $\tilde{N}_t$ is Poisson distributed with mean $\lambda(t-s)$, so

$$\mathbb{E}_x[f(y+\tilde{N}_t) - f(y)] = \sum_{k=0}^{\infty} f(y+k)\frac{1}{k!}\mathrm{e}^{-\lambda(t-s)}(\lambda(t-s))^k - f(y) =: h(t).$$

A short calculation shows that

$$\frac{\mathrm{d}}{\mathrm{d}t}h(t) = \lambda \sum_{k=0}^{\infty} (f(y+k+1) - f(y+k))\frac{1}{k!}\mathrm{e}^{-\lambda(t-s)}(\lambda(t-s))^k,$$

and hence, since $h(s) = 0$, that

$$\mathbb{E}_x\left[\lambda \int_s^t (f(y+1+\tilde{N}_r) - f(y+\tilde{N}_r))\mathrm{d}r\right] = \int_s^t \frac{\mathrm{d}}{\mathrm{d}r}h(r)\mathrm{d}r = h(t);$$

thus $\mathbb{E}_x[C_t^f - C_s^f | \mathfrak{F}_s] = 0$. This calculation is valid for all functions f such that $\mathbb{E}_x|f(N_t)| < \infty$ for all $x \in E, t \in \mathbb{R}_+$, so all such functions are in the domain of the *extended* generator. By the optional sampling theorem the process $C_t^{f,n} := C_{t\wedge n \wedge T_n}^f$ is also a martingale, since $\sigma_n := n \wedge T_n$ is a bounded stopping time. Note however that P_x-a.s. the process $C_t^{f,n}$ only involves the values of f on the set $A = \{x, x+1, \ldots, x+n\}$, and hence $C_t^{f,n} = C_t^{\tilde{f},n}$, where $\tilde{f}(y) = f(y)I_A(y)$. It follows that the process C_t^f is a local martingale for *any* (finite-valued) function f, since $\sigma_n \uparrow \infty$ as $n \to \infty$, P^x-a.s. Thus $\mathscr{D}(\mathfrak{A})$, the domain of the extended generator, consists of all functions $f: E \to \mathbb{R}$ *with no restrictions at all.* This illustrates why it is so convenient to define the extended generator in terms of local martingales.

(21.7) *Renewal processes*

The renewal process is a close cousin of the Poisson process. It is a point process N_t defined in a similar way to the Poisson process but with the interarrival time S_i now being i.i.d. random variables with some density function ψ on $\mathbb{R}_+$ (not necessarily exponential). The process clearly 'restarts' at each 'renewal time' T_i, and this fact immediately leads to the well-known *renewal equation* satisfied by $m(t) := \mathbb{E}N_t$ (starting at $N_0 = 0$). Indeed, we have

$$m(t) = \mathbb{E}N_t = \mathbb{E}\{\mathbb{E}[N_t | T_1]\}$$
$$= \mathbb{E}\{0I_{(T_1 > t)} + (1 + m(t - T_1))I_{(T_1 \leqslant t)}\}$$

Since T_1 has density ψ this gives

$$m(t) = \int_0^t \psi(s)\mathrm{d}s + \int_0^t m(t-s)\psi(s)\mathrm{d}s, \tag{21.8}$$

which is the renewal equation. It can be solved by Laplace transform methods.

The renewal process is not a Markov process because, if T_k denotes the last jump time before some fixed time t, the residual interarrival time $T_{k+1} - t$ has a conditional distribution given $\mathfrak{F}_t$ that depends on the time $\tau = t - T_k$ since the last jump. Given this information, however, nothing else about the past of the process has any relevance to predicting the future, since future interarrival times $S_{k+1}, S_{k+2}, \ldots$, are independent of the past. We therefore create a new, two-component process $x_t = (N_t, \tau_t)$, where N_t is the integer-valued renewal process as before and $\tau_t \in \mathbb{R}_+$ is the time since the last jump. With the inclusion of this 'supplementary variable', (x_t) is a Markov process, or a Markov family if we allow for arbitrary starting points. Its state space is $E = \mathbb{Z}_+ \times \mathbb{R}_+$ (or $\mathbb{Z} \times \mathbb{R}_+$) and a typical sample function is shown in Fig. 2.1. Note that τ_t is reset to zero each time a jump occurs. Denote the hazard rate corresponding to the renewal density ψ by $\lambda(t) = \psi(t)/\Psi(t)$, where $\Psi(t) = \int_t^\infty \psi(s)\mathrm{d}s$ is the survivor function. Then if $\tau_t = s$ the probability of getting a jump in the interval $]t, t+\delta]$ is $\lambda(s)\delta + o(\delta)$. We can now do a heuristic calculation to see what the generator of (x_t) must be. If we start at $x = (n, \tau)$ then after a short time δ either there has been no jump, in which case the process is at $(n, \tau + \delta)$, or one jump, in which case it is at

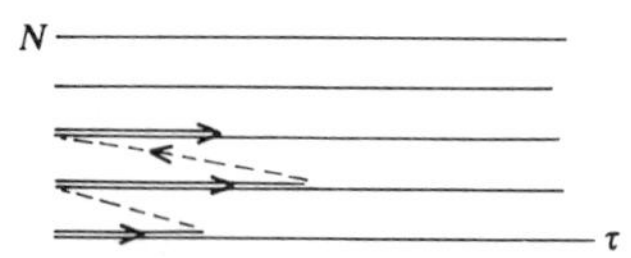

Fig. 2.1. *Markov model for the renewal process.*

$(n+1,\delta')$ for some $\delta' \in [0,\delta]$; other events have probability $o(\delta)$. Thus for a bounded function $f: E \to \mathbb{R}$ we have (roughly speaking)

$$\mathbb{E}_{(n,\tau)} f(x_\delta) = (1 - \lambda(\tau)\delta) f(n, \tau+\delta) + \lambda(\tau)\delta f(n+1,\delta') + o(\delta),$$

so that

$$\frac{1}{\delta}(\mathbb{E}_{(n,\tau)} f(x_\delta) - f(n,\tau)) = \frac{1}{\delta}(f(n,\tau+\delta) - f(n,\tau)) + \lambda(\tau)(f(n+1,\delta') - f(n,\tau+\delta)) + o(1)$$

Thus, assuming that the function $\tau \to f(n,\tau)$ is differentiable (and *a fortiori* continuous), we see that

$$\begin{aligned}\mathfrak{A} f(n,\tau) &= \lim_{\delta \downarrow 0} \frac{1}{\delta}(\mathbb{E}_{(n,\tau)} f(x_\delta) - f(n,\tau)) \\ &= \frac{\mathrm{d}}{\mathrm{d}\tau} f(n,\tau) + \lambda(\tau)(f(n+1,0) - f(n,\tau)).\end{aligned}$$

Again, for f to be in the domain of the strong generator this limit must exist uniformly in (n,τ); the domain of the extended generator is bigger but, in contrast to the previous example, membership of $\mathscr{D}(\mathfrak{A})$ does impose *some* restrictions on f. For example, it is clearly necessary that the function $\tau \to f(n,\tau)$ be differentiable if anything of the above calculation is to survive. The exact class of functions belonging to $\mathscr{D}(\mathfrak{A})$ is identified in Theorem (26.14) below.

(21.9) *Single-server queues*

Customers arrive at a queue at random times $T_1, T_2, \ldots$, and the customer arriving at time T_i requires $Y_i > 0$ units of time for processing. The total service load (measured in time units) presented up to time t is

(21.10) $$L_t = L_0 + \sum_i Y_i I_{(t \geq T_i)},$$

where $L_0 \geq 0$ is the service load existing at time 0, and the *virtual waiting time* V_t is the unique solution of the equation

$$V_t = L_t - \int_0^t I_{(V_s > 0)} \mathrm{d}s$$

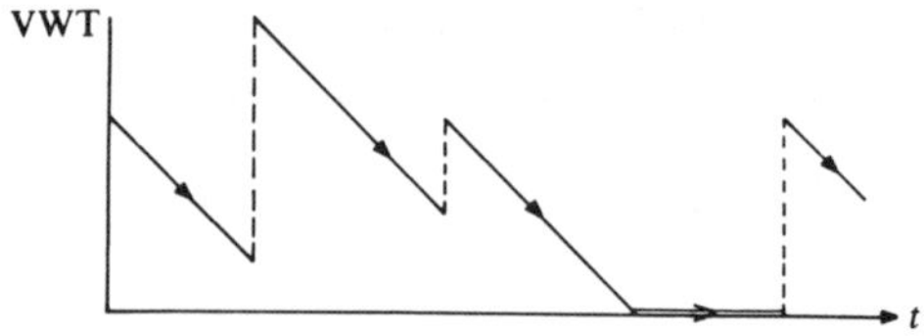

Fig. 2.2. *The virtual waiting time (VWT) for a single-server queue.*

(Fig. 2.2). In this equation the first term on the right is the total service load presented while the second is the amount of processing that has been carried out; thus V_t is the amount of unprocessed load at time t, or equivalently the time a customer arriving at t would have to wait for service to begin (hence the name). The queue has two possible *configurations*, namely *busy* and *empty*, and we will denote these by an indicator variable v with $v = 1$ when the queue is busy and $v = 0$ when it is empty.

There is a conventional classification $A/B/n$ of queueing systems where A refers to the arrival process, B to the distribution of the Y_i and n is the number of servers ($= 1$ in this discussion). We first consider the $M/G/1$ queue, where M (for 'Markov') means that the interarrival times are independent and exponential, i.e. the arrivals form a Poisson process, and G (for 'general') means that the Y_i are i.i.d. with some arbitrary distribution function F. It is also assumed that the Y_i are independent of the arrivals process. When $v = 1$, V_t decreases at unit rate between jumps; when $v = 0$ there is (because of the memoryless property of the exponential distribution) an exponential waiting time for a new arrival. We can define as state space for our Markov model the set $E = \{0, 0\} \cup (\{1\} \times]0, \infty[)$ and take $x_t = (v_t, \zeta_t)$, where v is the current configuration of the queue and $\zeta_t = V_t$ is the virtual waiting time. Then (x_t) is a Markov process which evolves as shown in Fig. 2.3: when ζ_t hits zero, x_t jumps to $(0, 0)$ and waits there until the next arrival, at which point it jumps to $(1, Y)$, where Y is the service requirement of the arriving customer. Let λ be the rate of the arrivals process. To see what the generator

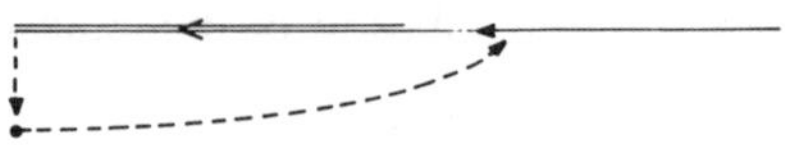

Fig. 2.3. *Markov model for the VWT process – Poisson arrivals.*

of this process must be note that, starting at $(1,\zeta)$, the state after a short time δ is, roughly, $(1,\zeta-\delta)$ with probability $(1-\lambda\delta)$, while with probability $\lambda\delta$ the process jumps to $(1,\zeta+Y-\delta)$, where Y has distribution F. Thus

$$\begin{aligned}\frac{1}{\delta}(\mathbb{E}_{(1,\zeta)}f(v_\delta,\zeta_\delta)-f(1,\zeta)) &= \frac{1}{\delta}f(1,\zeta-\delta)(1-\lambda\delta)\\ &\quad+\frac{1}{\delta}\int_0^\infty f(1,\zeta+y-\delta)F(\mathrm{d}y)\delta\\ &\quad-\frac{1}{\delta}f(1,\zeta)+o(1)\end{aligned}$$

and as $\delta\to 0$ this converges to

$$\mathfrak{A}f(1,\zeta)=-\frac{\partial}{\partial\zeta}f(1,\zeta)+\lambda\int_0^\infty(f(1,\zeta+y)-f(1,\zeta))F(\mathrm{d}y).$$

By the same argument

$$\mathfrak{A}f(0,0)=\lambda\int_0^\infty(f(1,y)-f(0,0))F(\mathrm{d}y).$$

However, it is clear that in this case the formula for $\mathfrak{A}f$ does not contain all the information about the process since in deriving it we have not used the fact that the process jumps from $(1,0)$ to $(0,0)$ when $(1,0)$ is hit. It will be seen later that this information is encoded in the specification of the domain $\mathscr{D}(\mathfrak{A})$: all $f\in\mathscr{D}(\mathfrak{A})$ must satisfy a *boundary condition* which is determined by the jump behaviour at $(1,0)$.

Let us now turn to the $GI/G/1$ queue: GI (for 'general independent') means that the interarrival times are i.i.d. with general distribution, i.e. the arrivals form a renewal process. In order to get a Markov model we now have to include the supplementary variable τ, the time since the last arrival. The state space is $E=(\{0\}\times[0,\infty[)\cup(\{1\}\times[0,\infty[\times]0,\infty[)$; see Fig. 2.4. When $v=1,\zeta=(\zeta^1,\zeta^2)$ is two dimensional with $\zeta_t^1=\tau$ and $\zeta_t^2=V_t$, the virtual waiting time. We denote $x_t=(v_t,\zeta_t)$. The sample functions are as shown in Fig. 2.4: when $v=1$, ζ_t^1 and ζ_t^2 increase and decrease respectively at unit rate. If the queue becomes empty (i.e. the process hits the line $\zeta^2=0$) then we have to continue accumulating the time since the last jump, so

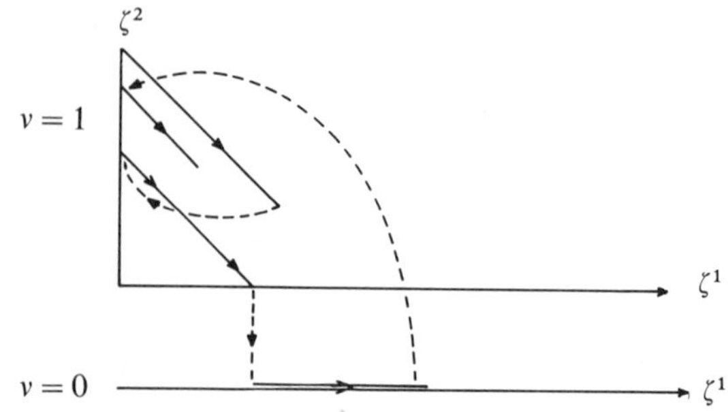

Fig. 2.4. *Markov model for the VWT process – renewal process arrivals.*

x_t jumps from $(1, \zeta^1, 0))$ to $(0, \zeta^1)$. If the next arrival occurs t time units later, bringing service requirement Y, then x_t jumps from $(0, \zeta^1 + t)$ to $(1, (0, Y))$. It is left to the reader to do the formal calculation identifying the generator $\mathfrak{A}f$ in this case; once the remainder of this chapter has been read, such formulae can be written down immediately with no calculations at all.

(21.11) *An insurance model* (Dassios and Embrechts, 1989)

This is a variant on the classical models for the analysis of insurance risk. One assumes that the cumulative claims on an insurance company form a process L_t similar to the arrival process (10) in the $M/G/1$ queue: Poisson arrivals and i.i.d. claims Y_i at the arrival times T_i. These claims are amounts of money the company has to pay out (immediately, contrary to common experience with insurance companies). Meanwhile the premium income is assumed to be a constant c per unit time. (This is the total premium income from all policy holders, so its variability is actually very small.) If the company has insufficient funds to meet a claim it can borrow money at interest rate β. When the company's debt is x the interest payment is βx per unit time and the company cannot recover from debt unless $c > \beta x$, i.e. $x < c/\beta$. A sample function of the evolution of the company's wealth x_t is shown in Fig. 2.5. x_t decreases by Y_i at claim times T_i, while between claims

$$\frac{\mathrm{d}}{\mathrm{d}t} x_t = cI_{(x_t \geqslant 0)} + (c + \beta x_t) I_{(x_t < 0)}.$$

With Poisson arrivals, (x_t) is a Markov process on the state space

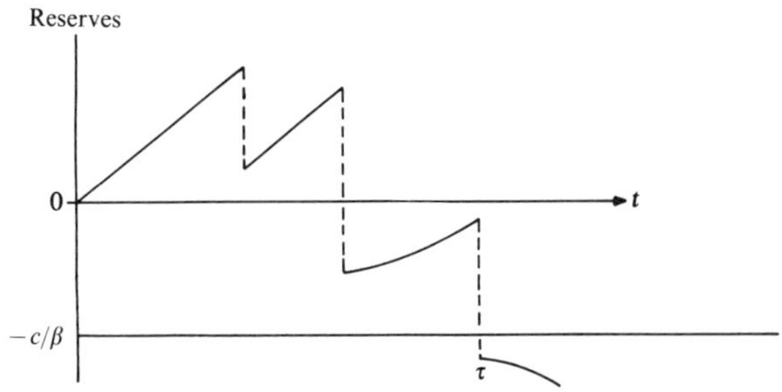

Fig. 2.5. *Free reserves process for the insurance model. The absolute ruin time is* τ.

$E = \mathbb{R}$. The stopping time $\tau_a = \inf\{t : x_t \leqslant -c/\beta\}$ is the *absolute ruin time*, and much of the analysis of this model centres around properties of τ_a: for example, whether eventual ruin is certain, $P_x[\tau_a < \infty] = 1$ or otherwise what the dependence of $P_x[\tau_a < \infty]$ on the initial level of reserves x is.

(21.12) *Permanent Health Insurance* (*PHI*) (Waters *et al.*, 1991)

This is another application from the world of insurance. PHI is a scheme whereby an insurance company contracts to maintain the insured person's income in the event of that person being sick (i.e. unable to carry out his or her normal employment) and not being paid by his/her employer. Customarily, employers will continue to pay employees while they are sick for a certain period, which might be a week or might be six months. Thus claims on the company only arise when periods of sickness exceed this so-called *elimination period*. The policy lasts from the time it is taken out until the insured person's retiring age. The question of course is what premium should be charged for a policy of this sort.

As far as actuaries are concerned, there are only three sorts of people: healthy, sick and dead. Transitions between these states are represented by Fig. 2.6. A healthy person falls sick at a random time with age-dependent hazard rate $\rho(t)$ or dies at a time with hazard rate $\mu(t)$. A sick person recovers after a sickness period at a time with hazard rate $\sigma(t, y)$ depending on age and the duration y of the current sickness period, or dies with hazard rate $\pi(t, y)$. A dead person stays dead. The functions μ, π, ρ, σ are estimated from actuarial data.

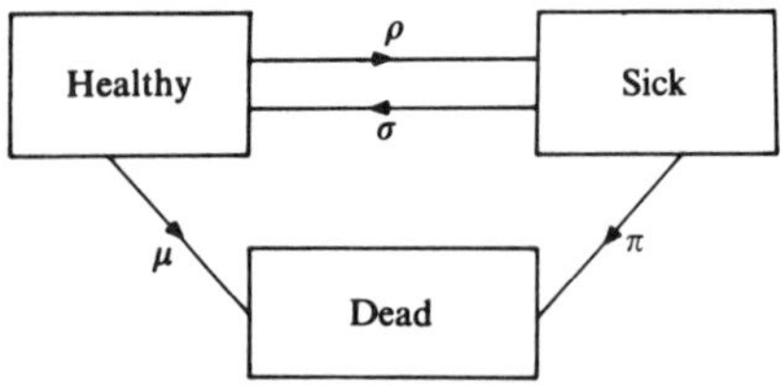

Fig. 2.6. *'State transitions' for health insurance.*

For the PHI application we have to distinguish whether a sick person is claiming or not, i.e. whether the duration of the current sickness exceeds the elimination period. It is convenient for this purpose to build a Markov model with a state space E consisting of four components labelled $v = 1$ (healthy), $v = 2$ (sick and not claiming), $v = 3$ (sick and claiming) and $v = 4$ (dead). E is shown in Fig. 2.7 along with a possible sample function. The horizontal variable is $t \in [0, t_f]$, the time since the policy was taken out. Thus $t_f = t_r - t_0$ where t_r is the insured person's retirement age and t_0 his/her age on taking out the policy; he/she is assumed to be healthy at t_0. The total hazard rate when $v = 1$ is therefore $\lambda_1(t) = \rho(t_0 + t) + \mu(t_0 + t)$ and a jump at time t goes to $v = 4$ with probability $\mu/(\rho + \mu)$ or to $(2,(t, 0))$ (i.e. the point $(t, 0)$ in the two-component) with probability $\rho/(\rho + \mu)$. When $v = 2$ the vertical component y is the

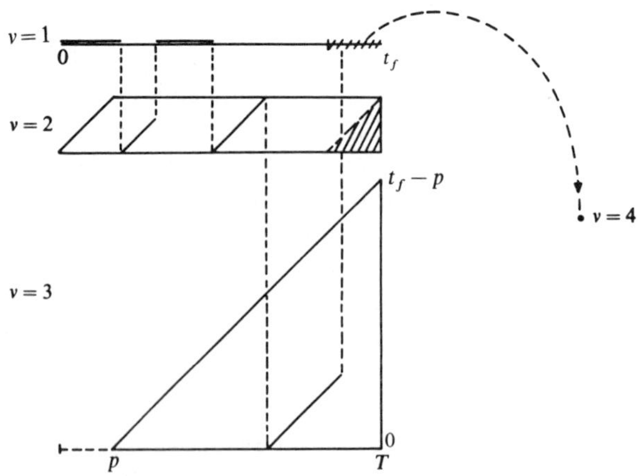

Fig. 2.7. *Markov model for the PHI process.*

duration of sickness, taking values in $[0, p]$ where p is the length of the elimination period. The trajectory moves northeast with a jump to $v = 1$ or 4 with total hazard rate $\lambda_2(t, y) = \sigma(t_0 + t, y) + \pi(t_0 + t, y)$. If the trajectory hits the barrier $y = p$ at time t there is an immediate transition to $(3, (t, 0))$. In $v = 3$ the vertical component y is the duration of the current claim, i.e. $y = y_1 - p$, where y_1 is the duration of the current sickness, and the hazard rate is $\lambda_3(t, y) = \sigma(t_0 + t, p + y) + \pi(t_0 + t, p + y)$. At time t_f the policy finishes and has no value. Denoting $x_t = (v_t, \zeta_t)$, the company's expected payout per unit of contracted income support is

$$\mathbb{E}_{(1,0)} \int_0^{t_f} I_{(v_t = 3)} \mathrm{d}t.$$

We will evaluate this in Chapter 3.

The next two examples concern optimization applications, where instead of a single stochastic process we have a family of processes indexed by some form of control or decision strategy. The problem is then to find a strategy that yields optimal performance by the yardstick of some quantitative performance measure. General theory for such problems, applicable to the examples here, is developed in Chapters 4 and 5.

(21.13) *A capacity expansion model* (David, Dampster, Sethi and Vermes, 1987)

Capacity expansion is the process of providing additional facilities to meet a rising demand for their services. The facilities we have in mind are such things as water resources, electricity supply or large-scale production plant. The common feature of these examples is that capacity expansion is undertaken only occasionally, that there is significant lead time, i.e. delay between the time a decision to install extra capacity is taken and the time at which that extra capacity comes on stream, and that there is significant randomness in the evolution of future demand, even within the lead time for an extra capacity unit. The following model is about the simplest that captures all these features in a tractable way.

We suppose that the demand process $d(t)$ is a Poisson process with rate $\mu > 0$; thus demand increments are supposed to arrive in unit discrete steps at random times. Depending on the application, this may be the way they actually do arrive, or, more generally, $d(t)$

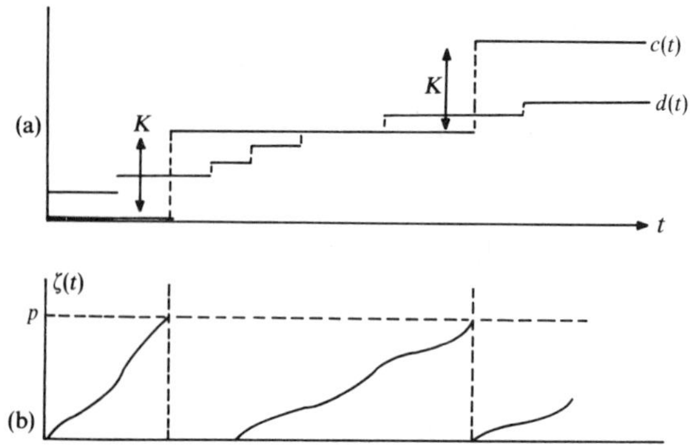

Fig. 2.8. *The capacity expansion problem: (a) supply and demand processes; (b) cumulative investment process.*

can be thought of as a discrete approximation to a general increasing curve of demand. We could, by including supplementary variables, model $d(t)$ as a renewal process or other more complicated form of point process, but for simplicity only the Poisson model will be considered here. Capacity is assumed to be provided by building a series of identical production facilities, each of which provides K units of capacity. Each production facility costs £p, and investment can be channelled into the current building project at any rate up to a maximum of £κ per week. We simply assume that the project is complete and comes on stream when the cumulative investment in it reaches £p. Thus there is a minimum lead time of p/κ to provide a new facility. Figure 2.8(a) shows typical sample functions of demand $d(t)$ and capacity $c(t)$; the latter increases by K units each time a new facility is completed, and we assume that when this happens, further investment is chanelled immediately into the next project in the series. If $u(t)$ (£/week) is the rate of investment at time t and there is a constant interest rate δ then the capital value at time zero of the total investment over an infinite time horizon is

$$\int_0^\infty e^{-\delta t}u(t)\mathrm{d}t.$$

The objective is to choose the rate of investment in such a way that capacity matches demand as closely as possible. Penalties are paid at a rate $h(d(t)-c(t))$ per week, where $h:\mathbb{Z}\to\mathbb{R}_+$ is a given function, $h(z)$ being a *shortage* penalty when $z>0$ and an *overcapacity* penalty when $z < 0$. We assume that $h(0) = 0$ and that both $h(z)$ and $h(-z)$ are increasing in $z > 0$. Denoting $v(t) = d(t) - c(t)$, the capital value of the total penalties paid is

$$\int_0^\infty e^{-\delta t} h(v(t))\,dt.$$

The *control problem* is to choose an investment strategy $u(t)$ so as to minimize the expected total capital cost

$$\mathbb{E}\int_0^\infty e^{-\delta t}(u(t) + h(v(t)))\,dt.$$

Obviously, $u(t)$ must be non-anticipative, i.e. must not depend on knowledge of future demand increments. Thus *admissible controls* are processes $u(t)$ taking values in $[0, \kappa]$ and adapted to the natural filtration $\mathfrak{F}_t^d$ of $d(t)$ (the only source of randomness in the problem). Let $\zeta(t)$ denote the cumulative investment in the current project; between completions this satisfies

$$\frac{d}{dt}\zeta(t) = u(t). \tag{21.14}$$

A project is completed when $\zeta(\cdot)$ reaches level p, at which point it is reset to zero (Fig. 2.8(b)). As the state for this system we take the process $x_t = (v_t, \zeta_t)$ evolving in the state space $E = \mathbb{Z} \times [0, p]$ as shown in Fig. 2.9. The shortage level v_t jumps up by one rung of the ladder each time a unit of demand arrives; when ζ_t hits the barrier $\zeta = p$ the process jumps from (v, p) to $(v - K, 0)$. Between jumps, ζ_t evolves according to (21.14).

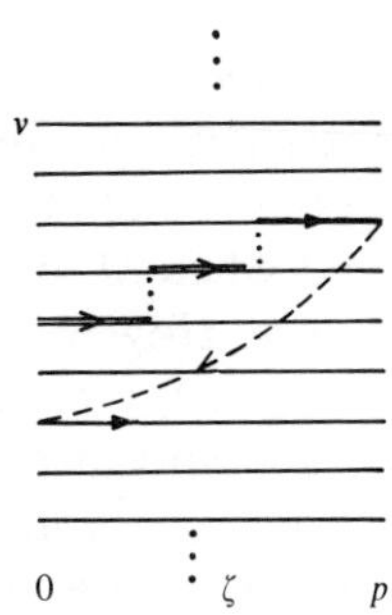

Fig. 2.9. *Markov model for the capacity expansion problem.*

As will be seen in the discussion of this problem in section 42.13, optimal strategies are *bang–bang*: one should always invest at maximum rate ($u = \kappa$) or not at all ($u = 0$). There is a barrier $\zeta = f(v)$ in the state space such that, optimally, $u = \kappa$ when $\zeta > f(v)$ and $u = 0$ when $\zeta \leqslant f(v)$. The main computational task in this problem is to determine the barrier function f. The problem can be generalized in various ways, for example by allowing for facilities with different capacities $K_1, \ldots, K_n$ and costs $p_1, \ldots, p_n$. The state space then consists of n 'ladders' indexed by the capacity of the facility currently under construction and, in addition to the investment rate decision there is also a 'boundary' control: we have to decide, on completion of each project, what sort of facility should be built next. General problems of this sort are studied in Chapter 4.

(21.15) *A repair shop model* (Costa and Davis, 1989; Costa, 1991)

This is an operations research example involving decision-making of a different kind, namely *control by intervention*. In the previous example control action $u(t)$ was exercised at every time moment t. Here the process evolves 'freely' between a sequence of times τ_1, $\tau_2, \ldots$, at which the decision-maker 'intervenes', shifting the state of the process immediately to some new point in the state space from which motion continues freely until the next intervention time. A *strategy* consists of a choice of stopping (i.e. non-anticipative random) times $\tau_1, \tau_2, \ldots$, and of the 'shifts' that take place at these times. Problems of this kind are studied in a general setting in Chapter 5, and the following is typical application.

A production plant has two machines M_1 and M_2 which operate independently. Machine $M_i (i = 1, 2)$ produces goods worth m_i per unit time while it is working. It breaks down with age-dependent hazard rate $h_i(t)$ and takes a fixed time t_i to repair. It must be taken out of service when its age reaches s_i but may be taken out of service for maintenance at any time before that. Repair work costs a_i per unit time, with a fixed cost p_i paid when M_i breaks down or $q_i < p_i$ when it is taken out for maintenance. However, the repair shop can only repair one machine at a time: if M_i breaks down while M_j is under repair then repair work on it cannot begin until the work on M_j is complete. There are thus five *configurations* which we label by an indicator variable v: M_1 and M_2 both working ($v = 1$); M_1 under repair and M_2 working and, conversely ($v = 2, 3$); M_1 under repair

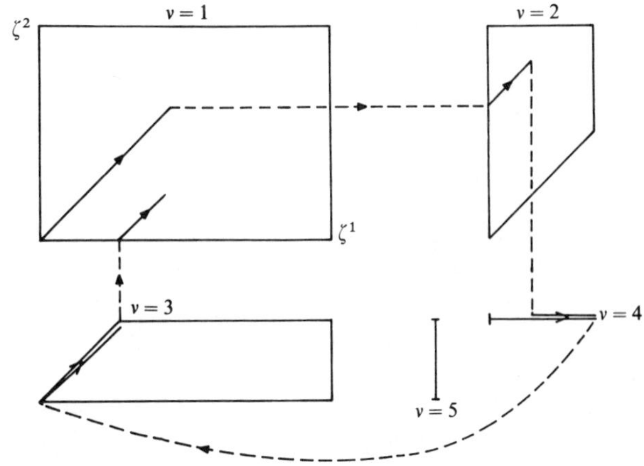

Fig. 2.10. *Markov model for the maintenance scheduling problem.*

and M_2 awaiting repair, and conversely ($\nu = 4, 5$). The state space for a Markov model is shown in Fig. 2.10.

Because the hazard rates are age dependent we need to record the age of each working machine (or more precisely, the time since the last repair on that machine was completed – repaired machines count as new) and the progress of each machine through the repair shop. We denote the state of the system by $x_t = (\nu_t, \zeta_t)$, where $\zeta_t \in \mathbb{R}^2$ for $\nu_t = 1, 2, 3$ and $\zeta_t \in \mathbb{R}^1$ for $\nu_t = 4, 5$. A typical sample function is also shown in Fig. 2.10 starting with M_1, M_2 both new. At time t, M_1 breaks down and the state jumps from $x_{t-} = (1, (t, t))$ to $x_t = (2, (0, t))$. We leave it to the reader to interpret what happens subsequently. The expected total discounted value of production minus repair costs in the absence of any interventions is

$$(21.16) \quad \mathbb{E}\left\{\int_0^\infty e^{-\delta t}\{(m_1 + m_2)I_{(\nu_t = 1)} + (m_2 - a_1)I_{(\nu_t = 2)} + (m_1 - a_2)I_{(\nu_t = 3)}\}\,dt - \sum_{j=1}^\infty e^{-\delta T_j}\{p_1 I_{(x_{T_j} \in A_1)} + p_2 I_{(x_{T_j} \in A_2)}\}\right\},$$

where $A_1 = \{x \in E : \nu = 2 \text{ and } \zeta^2 > 0 \text{ or } \nu = 5\}$ and $A_2 = \{x \in E : \nu = 3$ and $\zeta^1 > 0$ or $\nu = 4\}$; these are the possible states after a breakdown of M_1 and M_2, respectively. An intervention strategy in this context is a region $D = \{\zeta : \zeta^2 < f(\zeta^1)\}$ in the $\nu = 1$ component with the boundary $\partial D = \{\zeta : \zeta^2 = f(\zeta^1)\}$ divided into disjoint sets ∂D_1 and ∂D_2.

The process evolves as before inside D, while if D_i is hit then machine i is taken out of service for maintenance, i.e. the process jumps to the $v = i + 1$ component, $i = 1, 2$. The expected value of production is now given by (21.16) but with the term involving p_i modified to

$$(p_i + (q_i - p_i)I_{(x_{T_j -} \in \partial D_i)})$$

so that cost q_i is paid when machine i is taken out of service but p_i when it breaks down. The control problem is to determine what region D and boundary partition $\partial D_1, \partial D_2$ maximizes this value. This kind of problem, which is usually known as *impulse control*, is studied in Chapter 5.

This concludes our discussion of examples. The remainder of the chapter is devoted to developing the general piecewise-deterministic process framework in which to place them.

22 Ordinary differential equations and vector fields

Let $g: \mathbb{R}^d \to \mathbb{R}^d$ be a (globally) Lipschitz continuous function, i.e. there exists a constant K such that $|g(x) - g(y)| \leqslant K|x - y|$ for $x, y \in \mathbb{R}^d$. Then the ordinary differential equation (ODE)

(22.1) $$\frac{\mathrm{d}}{\mathrm{d}t} x(t) = g(x(t)), x(0) = x \in \mathbb{R}^d$$

has a unique solution, defined for all $t \in \mathbb{R}$. We will denote this solution $\phi(t, x)$ to emphasize the dependence on the initial point x. It follows immediately from the uniqueness of the solution that ϕ has the following two properties

(22.2). The map $\phi_t : x \mapsto \phi(t, x)$ is one-to-one and onto; indeed, $\phi^{-1}(t, x) = \phi(-t, x)$ for any $x \in \mathbb{R}^d$.

(22.3). The family $(\phi_t)_{t \in \mathbb{R}}$ is a group, i.e. for any $t, s \in \mathbb{R}$, $\phi_{t+s} = \phi_t \circ \phi_s$, or more explicitly, $\phi(t + s, x) = \phi(t, \phi(s, x))$ for all $x \in \mathbb{R}^d$.

Let $f: \mathbb{R}^d \to \mathbb{R}$ be a C^1 (continuously differentiable) function. Then with $x(t) = \phi(t, x)$,

$$\frac{\mathrm{d}}{\mathrm{d}t} f(x(t)) = \sum_i \frac{\partial f}{\partial x^i}(x(t)) g^i(x(t)),$$

where g^i is the ith component of g. Let $\mathfrak{X}$ denote the first-order differential operator

$$\mathfrak{X}f(x) = \sum_i \frac{\partial f}{\partial x^i}(x) g^i(x).$$

Then $x(t)$ satisfies (22.1) if and only if it satisfies

$$\frac{\mathrm{d}}{\mathrm{d}t} f(x(t)) = \mathfrak{X}f(x(t)) \quad \text{for all } f \in C^\infty(\mathbb{R}^d). \tag{22.4}$$

We have already seen that the solution of (22.1) satisfies (22.4); conversely, if $x(t)$ satisfies (22.4) it satisfies (22.1) (just apply (22.4) with $f(x) = x^i$, $i = 1, \ldots, d$) which shows that (22.4) determines $x(t)$ uniquely. Equation (22.4) is the so-called *coordinate-free* form of the differential equation, the operator $\mathfrak{X}$ is known as a *vector field*, and $\phi(t, x)$ is the *flow* of $\mathfrak{X}$. The coordinate-free formulation enables us to study differential equations on non-flat manifolds as well as in Euclidean space $\mathbb{R}^d$ (see, for example, Boothby, 1984), but vector fields appears naturally in the theory in any case, as will be seen below.

Let us now generalize to the case where g satisfies a local – rather than global – Lipschitz condition. This means that for any compact set $D \subset \mathbb{R}^d$ there is a constant K_D such that

$$|g(x) - g(y)| \leqslant K_D |x - y|, \quad x, y \in D.$$

Under this condition the solution $x(t)$ to (22.1) (or (22.4)) starting at $x \in D$ is uniquely determined for $t \leqslant t_D$, where t_D is the time at which the solution exits from D. It may happen, however, that t_D fails to converge to ∞ as $D \uparrow \mathbb{R}_d$, in which case there is said to be an *explosion* at $t_\infty = \lim t_D$, and $|x(t)| \to \infty$ as $t \uparrow t_\infty$. The standard example in $\mathbb{R}$ is the vector field $g(x) = x^2$ which is locally but not globally Lipschitz. The solution of the ODE $\mathrm{d}x(t)/\mathrm{d}t = x^2(t), x(0) = x$ is $x(t) = x/(1 - tx)$ which explodes at $t_\infty = 1/x$. In this book we will always assume that vector fields are at least locally Lipschitz and that there are no explosions ($t_\infty = \infty$ from any starting point). It is not difficult to cxtend the theory to allow for explosions, but this does not seem important for applications.

It is instructive to think of solutions of ODEs as Markov processes. Assuming that g is locally Lipschitz and $t_\infty \equiv \infty$ the 'trajectory' of the process is just $x_t = \phi(t, x)$ and the 'semigroup' is $P_t f(x) = f(\phi(t, x))$.

Thus for $f \in C^1$ we have from (22.4)

$$P_t f(x) - f(x) = \int_0^t \mathfrak{X} f(x_s) \mathrm{d}s \tag{22.5}$$

so that

$$C_t^f := f(x_t) - f(x) - \int_0^t \mathfrak{X} f(x_s) \mathrm{d}s := 0, \tag{22.6}$$

which is certainly a martingale. Thus $\mathfrak{X}$ is the extended generator of the deterministic process. Its domain is, however, much larger than C^1. Indeed, $\mathscr{D}(\mathfrak{X}) = \{f: E \to \mathbb{R}: f$ is measurable and for all $x \in E$ the function $t \to f(\phi(t,x))$ is absolutely continuous$\}$. If $f \in \mathscr{D}(\mathfrak{X})$ then $t \to f(\phi(t,x))$ is differentiable almost everywhere and is equal to the integral of its derivative (see Theorem (11.6)). If we denote the derivative $\mathfrak{X} f(x_t)$ then (22.4) holds for almost all t, and (22.5) – or equivalently (22.6) – holds. This is the sense in which we will interpret '$\mathfrak{X} f$' in later chapters. It is worth pointing out that to belong to $\mathscr{D}(\mathfrak{X})$ a function f only has to be directionally differentiable along the flow of $\mathfrak{X}$; it does not even need to be continuous in other directions. For example, if $E = \mathbb{R}^2$ and $\mathfrak{X} = \partial/\partial x^1$ then any function of the form $f(x) = f_1(x) I_{(x^2 \leqslant a)} + f_2(x) I_{(x^2 > a)}$, where $f_1, f_2 \in C^1(\mathbb{R}^2)$, is in $\mathscr{D}(\mathfrak{X})$ even though this function is possibly discontinuous along the (flow) line $x^2 = a$.

A related matter, which will be important in later chapters, is solution of linear scalar ODEs with general coefficients, i.e. equations of the form

$$\frac{\mathrm{d}}{\mathrm{d}t} x(t) = a(t)x(t) + b(t), \quad x(0) = x_0. \tag{22.7}$$

Of course, $x(t)$ cannot be continuously differentiable if a, b are not continuous. What we mean by (22.7) is that $x(t)$ satisfies the corresponding integral equation

$$x(t) = x_0 + \int_0^t a(s)x(s)\mathrm{d}s + \int_0^t b(s)\mathrm{d}s. \tag{22.8}$$

If $x(t)$ satisfies this then it is absolutely continuous and its derivative is almost everywhere equal to $a(t)x(t) + b(t)$. This is the correct interpretation for (22.7).

(22.9) *Proposition.* Suppose $a, b: \mathbb{R}_+ \to \mathbb{R}$ are integrable functions on any interval $[0, T]$. Then equation (2.14) has the unique solution

$$(22.10)\quad x(t) = \exp\left(\int_0^t a(s)\mathrm{d}s\right)x_0 + \int_0^t \exp\left(\int_s^t a(u)\mathrm{d}u\right)b(s)\mathrm{d}s.$$

Proof. Define $x(t)$ by (22.10); the second integral is well-defined since the exponential term is a continuous function of s, and hence bounded on $[0, t]$. A simple calculation using the Fubini theorem shows that $x(t)$ satisfies (22.8), so that (22.10) is a solution. To show uniqueness, suppose that $x_1(t)$ and $x_2(t)$ both satisfy (22.8), and define $z(t) = x_1(t) - x_2(t)$. Then $z(t)$ satisfies

$$z(t) = \int_0^t a(s)z(s)\mathrm{d}s,$$

and $z(t) = 0$ follows from the Gronwall–Bellman lemma (Fleming and Rishel, 1975, Appendix A).

23 Simulation

The simplest example of an uncountable probability space is the *unit interval* probability space $(Y, \mathscr{A}, l)$, where Y is the unit interval $[0, 1]$, $\mathscr{A}$ the class of Lebesgue measurable sets and l the Lebesgue measure. The random variable $U: Y \to \mathbb{R}$ defined by $U(y) = y$ then has uniform distribution on [0,1], i.e. $l[U \leqslant t] = (t \wedge 1) \vee 0$ for $t \in \mathbb{R}$. We refer to this distribution as $U[0, 1]$. More generally the *Hilbert cube* $\Omega = \prod_{i=1}^{\infty} Y_i$, where each Y_i is a copy of Y, with the product σ-field $\mathfrak{F}$ and product measure P, provides the canonical space for a countable sequence of independent random variables $U_1, U_2, \ldots$, each having uniform $[0, 1]$ distribution, defined by $U_i(\omega) = \omega_i$ for $\omega = (\omega_1, \omega_2, \ldots)$. Almost all statistical simulation is based – in theory – on this probability space. Random number generators in computers produce sequences of numbers which are, as closely as possible, statistically indistinguishable from samples $U_1, U_2, \ldots$, of i.i.d. $U[0, 1]$ random variables. Random variables with other distributions are then produced by various well-known transformations (see Press *et al.*, 1986, for example). For distributions on the real line the general transformation is as follows.

(23.1) *Proposition.* Let F be a distribution function on the real line, i.e, $F: \mathbb{R} \to [0, 1]$ is monotonically increasing and right-continuous with $F(-\infty)=0, F(+\infty)=1$. Let β denote the inverse function

(23.2) $$\beta(u)=\inf\{t:F(t)\geqslant u\}.$$

Let U be a random variable with $U[0,1]$ distribution, and define $X=\beta(U)$. Then $P[X\leqslant t]=F(t)$, i.e. F is the distribution function of X.

Proof. One only has to observe that $X\leqslant t$ if and only if $U\leqslant F(t)$ and $P[U\leqslant F(t)]=F(t)$, since U is $U[0,1]$. □

If $(X,\mathfrak{S},\mu)$ is a probability space, $(Y,\mathfrak{J})$ another measurable space and $\psi:X\to Y$ a measurable function, then we denote by $\mu\psi^{-1}$ the *image measure* on Y defined by $\mu\psi^{-1}(A)=\mu\{x:\psi(x)\in A\}$ for $A\in\mathfrak{J}$. With Proposition (23.1) in mind, one can then ask the following question: What is the most general class of measurable spaces $(Y,\mathfrak{J})$ such that if v is a probability measure on $(Y,\mathfrak{J})$ there exists a measurable function $\psi:[0,1]\to Y$ such that $l\psi^{-1}=v$? The answer to this question is not simple (see Chapter 7 of Bertsekas and Shreve (1978) for a clear account) but as a special case we have the following result, adequate for our purposes here.

(23.3) *Proposition.* Let E be a Borel subset of $\mathbb{R}^d$ and $\mathfrak{J}=\mathfrak{B}(E)$ the Borel σ-field in E (i.e. $\mathfrak{B}(E)=E\cap\mathfrak{B}(\mathbb{R}^d)$). Then there exists a Borel function $\phi:E\to[0,1]$ such that ϕ is one-to-one, $\phi(E)$ is a Borel subset of $[0,1]$ and $\phi^{-1}:\phi(E)\to E$ is Borel measurable.

(23.4) *Corollary.* If v is a measure on a Borel subset E of $\mathbb{R}^d$ then there exists a measurable function $\psi:[0,1]\to E$ such that $l\psi^{-1}=v$.

Proof of Corollary. Let ϕ be the function of Proposition (23.3), let $F(t)=v(\phi^{-1}([0,t]))$, and define $\beta:[0,1]\to[0,1]$ by (2.17). Now define $\chi:[0,1]\to E$ by $\chi(t)=\phi^{-1}(t)$ for $t\in\phi(E)$ and $\chi(t)=y_0$, where y_0 is an arbitrary point, for $t\notin\phi(E)$. Then $l\psi^{-1}=v$, where $\psi=\chi\circ\beta$. □

(23.5) *Remark*

This is a 'theoretical' rather than a practical result. The normal way of obtaining random vectors in $\mathbb{R}^d$ with specified distributions is by iterated application of the one-dimensional result, Proposition (23.1). For example, for a random vector $X\in\mathbb{R}^2$ one would generate a random variable having the marginal distribution of X_1 and then a further random variable having the conditional distribution of X_2 given X_1. This procedure uses a number d of $U[0,1]$ random variables to generate a random vector on $\mathbb{R}^d$, whereas Corollary (23.4) shows that it can in principle be done using only one. Either way, a random number generator is all we need to produce finite-dimensional random variables with arbitrary distributions.

24 Definition of the PDP

Let K be a countable set and d be a function mapping K into the natural numbers $\mathbb{N}$. For each $v \in K$, let E_v^0 be an open subset of $\mathbb{R}^{d(v)}$ and define E^0 as the disjoint union of these subsets:

$$E^0 = \bigcup_{v \in K} E_v^0 = \{(v, \zeta): v \in K, \zeta \in E_v^0\}.$$

For each $v \in K$, $\mathfrak{X}_v$ is a given locally Lipschitz continuous vector field in E_v^0, determining a flow $\phi_v(t, \zeta)$. On a point of notation, any function $h: E^0 \to \mathbb{R}$ is identified with its component functions $h_v: E_v^0 \to \mathbb{R}$ and we write $\mathfrak{X}h(x)$ in place of the more cumbersome $\mathfrak{X}_v h_v(\zeta)$ for $x = (v, \zeta) \in E^0$. For $x = (v, \zeta) \in E^0$ define

$$t_*(x) = \begin{cases} \inf\{t > 0: \phi_v(t, \zeta) \in \partial E_v^0\} \\ \infty \quad \text{if no such time exists.} \end{cases}$$

Here ∂E_v^0 is the boundary: $\partial E_v^0 = \bar{E}_v \backslash E_v^0$, where $\bar{E}_v$ is the closure of E_v^0. It will be verified below in section 27 that t_* is a Borel-measurable function. If $t_\infty(x)$ denotes the explosion time of the flow $\phi_v(\cdot, \zeta)$, then we assume that $t_\infty(x) = \infty$ when $t_*(x) = \infty$, thus effectively ruling out explosions. Now define

$$\partial^{\pm} E_v^0 = \{z \in \partial E_v^0 : z = \phi_v(\pm t, \zeta) \text{ for some } \zeta \in E_v^0, t > 0\}$$

and $\partial_1 E_v^0 = \partial^- E_v^0 \backslash \partial^+ E_v^0$. Let $E_v = E_v^0 \cup \partial_1 E_v^0$, $E = \bigcup_v E_v$ and

$$\Gamma^* = \bigcup_v \partial^+ E_v^0$$

The set E will be the state space for the PDP; it is a countable union of Borel subsets of Euclidean space. Let $\mathscr{E}$ denote the σ-field of subsets A of E taking the form $A = \bigcup_v A_v$, where A_v is a Borel set of E_v. As mentioned in section 22, E_v^0 could equally well be a subset of some non-flat manifold, but we will not pursue this minor generalization here. It is possible to define a metric ρ on E in such a way that $\rho(x_n, x) \to 0$ as $n \to \infty$ with $x_n = (v_n, \zeta_n)$, $x = (v, \zeta)$ if and only if there exists m such that $v_n = v$ for all $n \geqslant m$ and $\zeta_{m+k} \to \zeta$ as $k \to \infty$ in the usual topology in $\mathbb{R}^{d(v)}$. One's first thought is to define the distance between $x = (v, \zeta)$ and $x' = (v', \zeta')$ as 1 if $v \neq v'$ and $|\zeta - \zeta'|$ (the usual Euclidean distance) if $v = v'$; however, this fails to satisfy the triangle inequality, and it is necessary to rescale distances in $\mathbb{R}^{d(v)}$ so that they are bounded by 1. For instance, we can define $\chi(\zeta) = (2/\pi) \max_{1 \leqslant j \leqslant k} \tan^{-1}(|\zeta_i|)$ for $\zeta \in \mathbb{R}^k$ and then define $\rho(x, x') = 1$ if $v \neq v'$ and $\rho(x, x') = \chi(\zeta - \zeta')$ if $v = v'$. Then ρ satisfies the triangle inequality and corresponds to the notion of convergence described above. The metric ρ restricted to any component E_v is equivalent to the usual Euclidean metric, and hence (E, ρ) is separable and the Borel sets $\mathscr{E}^\rho$ corresponding to the topology generated by ρ coincide with $\mathscr{E}$ as defined above. This shows that $(E, \mathscr{E})$ is a Borel space (Bertsekas and Shreve, 1978, Definition 7.7), i.e. a Borel subset of a complete separable metric space.

The vector field $\mathfrak{X}$ determines the motion of the PDP between jumps. The jump mechanism is determined by two further functions, the *jump rate* λ and the *transition measure* Q. $(\mathfrak{X}, \lambda, Q)$ are called the *local characteristics* of the PDP. The jump rate $\lambda: E \to \mathbb{R}_+$ is a measurable function such that for each $x = (v, \zeta) \in E$ there exists $\varepsilon(x) > 0$ such that the function $s \to \lambda(v, \phi_v(s, \zeta))$ is integrable on $[0, \varepsilon(x)[$. The transition measure Q maps $E \cup \Gamma^*$ into the set $\mathscr{P}(E)$ of probability measures on $(E, \mathscr{E})$, with the properties that:

1. for each fixed $A \in \mathscr{E}$ the map $x \to Q(A; x)$ is measurable; and
2. $Q(\{x\}; x) = 0$.

(24.1) *Notation*

Let $f: E \cup \Gamma^* \to \mathbb{R}$ be a measurable function such that $f \in L_1(Q(\cdot; x))$ for each $x \in E \cup \Gamma^*$. Then we denote by Qf the function defined by

$$Qf(x) = \int_E f(y) Q(\mathrm{d}y; x).$$

It follows from assumption (1) above that Qf is a measurable function. In particular, if $f \in \mathscr{M}_b$, the set of bounded measurable functions on $E \cup \Gamma^*$, then $Qf \in \mathscr{M}_b$, i.e. Q is an operator mapping $\mathscr{M}_b$ into itself.

Let $(\Omega, \mathscr{A}, P)$ be the Hilbert cube (section 23), i.e. the canonical space for a sequence $U_1(\omega), U_2(\omega), \ldots$, of independent $U[0,1]$ distributed random variables. The sample path of a stochastic process $\{x_t(\omega), t > 0\}$ with values in E, starting from a fixed initial point $x = (v, \zeta) \in E$, is defined as follows. Let

$$(24.2) \qquad F(t,x) = I_{(t < t_*(x))} \exp\left(- \int_0^t \lambda(v, \phi_v(s,\zeta)) \mathrm{d}s \right).$$

This will be the survivor function of the first jump time T_1 of the process. Thus let

$$\psi_1(u,x) = \begin{cases} \inf\{t : F(t,x) \leqslant u\} \\ +\infty \quad \text{if the above set is empty} \end{cases}$$

and define $S_1(\omega) = T_1(\omega) = \psi_1(U_1(\omega), x)$. Then $P(T_1 > t) = F(t,x)$. Now let $\psi_2 : [0,1] \times (E \cup \Gamma^*) \to E$ be a measurable function such that $l\psi_2^{-1}(A)\ (= l\{u : \psi_2(u,x) \in A\}) = Q(A; x)$ for $A \in \mathscr{B}(E)$; the existence of such a function is guaranteed by Corollary (2.1) and the assumed measurability properties of the transition measure Q. The sample path $x_t(\omega)$ up to the first jump time is now defined as follows:

$$\text{if } T_1(\omega) = \infty: \qquad x_t(\omega) = (v, \phi_v(t,\zeta)), \quad t \geqslant 0$$

$$\text{if } T_1(\omega) < \infty: \qquad x_t(\omega) = (v, \phi_v(t,\zeta)), \quad 0 \leqslant t < T_1(\omega)$$

$$x_{T_1}(\omega) = \psi_2(U_2(\omega), (v, \phi_v(T_1(\omega)), \zeta))).$$

The process now restarts from $x_{T_1}(\omega)$ according to the same recipe. Thus if $T_1(\omega) < \infty$ we define

$$S_2(\omega) = \psi_1(U_3(\omega), x_{T_1}(\omega))$$
$$T_2(\omega) = T_1(\omega) + S_2(\omega).$$

Let $(v', \zeta') = x_{T_1}(\omega)$. If $T_2(\omega) = \infty$ then

$$x_t(\omega) = (v', \phi_{v'}(t - T_1(\omega), \zeta')), \quad t \geqslant T_1(\omega),$$

whereas if $T_2(\omega) < \infty$, then

$$x_t(\omega) = (v', \phi_{v'}(t - T_1(\omega), \zeta')), \quad T_1(\omega) \leqslant t < T_2(\omega)$$
$$x_{T_2}(\omega) = \psi_2(U_4(\omega), (v', \phi_{v'}(S_2(\omega), \zeta')));$$

and so on. We have thus defined a map from Ω into the set of sequences $(S_1(\omega), Z_1(\omega), \ldots, S_{k_0}(\omega))$, where $k_0 = \min\{k : S_k(\omega) = \infty\}$ ($= \infty$ if $S_k(\omega) < \infty$ for all k) and $Z_k(\omega) = x_{T_k}(\omega)$ as defined above.

We denote

(24.3)
$$N_t(\omega) = \sum_k I_{(t \geq T_k)}.$$

(24.4) *Assumption.* For every starting point $x \in E$, $\mathbb{E}N_t < \infty$ for all $t \in \mathbb{R}_+$.

This assumption implies in particular that $T_k(\omega) \to \infty$ almost surely. Assumption 2.1 is usually quite easily checked in applications, but it is hard to formulate general condition under which it holds, because of the complicated interaction between $\mathfrak{X}, \lambda, Q$ and the geometry of the boundary. The following example illustrates this point.

(24.5) *Example*

Let $E = [0, 1[\times [0, 2]$, $\mathfrak{X} = \partial/\partial x^1$ so that $\phi(t, x) = (x^1 + t, x^2), \lambda \equiv 0$, $Q(\mathrm{d}y; x) = \delta_{(1 - x^2/2, x^2/2)}(\mathrm{d}y)$ for $x = (1, x^2) \in \Gamma^* = \{1\} \times [0, 2]$. Then if $x = (0, 1)$ we have $T_1 = 1, T_2 = 1\frac{1}{2}, T_3 = 1\frac{3}{4}\ldots$, so that $T_\infty = 2$. Thus accumulation of jump times is possible even with $\lambda = 0$, if the sequence $t_*(x_{T_k})$ is not bounded away from zero.

We next give two useful sufficient conditions which imply satisfaction of Assumption (24.4) when the jump rate λ is bounded. In particular, condition (2), stating that the process always jumps to points 'uniformly inside E' as measured by the boundary hitting time, is satisfied in practically every application.

(24.6) *Proposition.* Suppose that λ is bounded, i.e. $\lambda(x) \leq c < \infty$ for all $x \in E$. Then Assumption (24.4) holds if either (1) or (2) below is satisfied.

1. There are no 'active' boundaries: $t_*(x) = \infty$ for all $x \in E$.
2. For some $\varepsilon > 0$, $Q(A_\varepsilon; x) = 1$ for all $x \in \Gamma^*$, where $A_\varepsilon = \{x \in E$: $t_*(x) \geq \varepsilon\}$.

Proof

1. From (24.2) we see that $F(t) > \mathrm{e}^{-ct}$. For $k = 1, 2, \ldots$, define $S'_k(\omega) = \ln(1/U_{2k-1}(\omega))/c$ (i.e. S'_k is constructed from the same $U[0, 1]$ random variable as $S_k(\omega)$, but has exponential distribution with parameter c). Then we see that $S'_k(\omega) < S_k(\omega)$, so that $T'_k = S'_1 + \cdots +$

$S'_k \leqslant T_k$ and $N'_t = \sum_k I_{t \geqslant T_k} \geqslant N_t$. But (S'_t) is just a Poisson process with rate c, so $\mathbb{E}N_t < \mathbb{E}N'_t = ct$.

2. This is most readily shown using the martingale theory of piecewise-deterministic processes developed in section 26 below. It follows from Proposition (26.7) that for any $n = 1, 2, \ldots$ and $t > 0$,

$$\mathbb{E}N_{t \wedge T_n} = \mathbb{E}\left\{\int_0^{t \wedge T_n} \lambda(x_s)\mathrm{d}s + \sum_{i=1}^{n} I_{(t \geqslant T_i)} I_{(x_{T_i-} \in \Gamma^*)}\right\}$$

Under the stated conditions, $\lambda \leqslant c$ and the time interval between successive boundary jumps is at least $\in$, so that the number of boundary hitting times in $[0, t]$ is majorized by $1 + t/\in$. Thus $\mathbb{E}N_{t \wedge T_n} \leqslant ct + 1 + t/\in$, and since $T_n \uparrow \infty$ a.s. it follows by monotone convergence that $\mathbb{E}N_t \leqslant ct + 1 + t/\in$. □

As a final stage in the construction of the PDP, we note that it may not be necessary to specify $Q(\cdot; x)$ on all of Γ^*. Indeed, let

$$\Gamma = \{x = (v, \zeta) \in \Gamma^* : P_y[T_1 = t] > 0 \quad \text{for some } t > 0, \quad \text{with } y = (v, \phi_v(-t, \zeta))\}.$$

It is easy to see that, with the same notation as above, $x \in \Gamma$ if and only if $P_y[T_1 = t] \to 1$ as $t \downarrow 0$. Points in $\Gamma^* \backslash \Gamma$ are, with probability 1, not hit by the process from any starting point, and it is therefore unnecessary to specify Q on $\Gamma^* \backslash \Gamma$. To illustrate this, consider the following example.

(24.7) *Example*
$E = [0, 1[$, $\mathfrak{X} = \mathrm{d}/\mathrm{d}x$, $\lambda(x) = 1/(1 - x)$. Thus $\Gamma^* = \{1\}$ and $\phi(t, x) = t + x$ for $t \leqslant t_*(x) = 1 - x$. However, from (2.18)

$$P_x(T_1 > t) = \frac{1 - x - t}{1 - x} I_{(t \leqslant (1 - x))}, \quad P_x(T_1 = t_*(x)) = 0.$$

Here T_1 is uniformly distributed on $[0, 1 - x]$ and $P_x[T_1 < 1 - x] = 1$. We do not need $Q(\mathrm{d}y; \{1\})$.

This completes the definition of the PDP. For later reference, we now collect together the various conditions and assumptions introduced above. Describing a process as a PDP will be taken to imply that these standard conditions are in force, unless explicit mention to the contrary is made.

(24.8) *Standard conditions*

1. For each $\nu \in K$, $\mathfrak{X}_\nu$ is a locally Lipschitz continuous vector field, with flow $\phi_\nu(t,\zeta)$. The explosion time $t_\infty(\zeta)$ is equal to $+\infty$ whenever $t_*(x) = \infty$ $(x = (\nu, \zeta))$.
2. $\lambda: E \to \mathbb{R}_+$ is a measurable function such that $t \to \lambda(\nu, \phi_\nu(t,\zeta))$ is integrable on $[0, \varepsilon(x)[$ for some $\varepsilon(x) > 0$, for each $x \in E$.
3. $Q: E \cup \Gamma \to \mathscr{P}(E)$ is a measurable function such that $Q(\{x\}; x) = 0$ for each $x \in E$.
4. Assumption (24.4) holds, i.e. $\mathbb{E}_x N_t < \infty$ for each $(t, x) \in \mathbb{R}_+ \times E$.

(24.9) *Remark.* Finally, it should be noted that no generality is lost if the state space E is taken simply as a subset of some Euclidean space $\mathbb{R}^d$ rather than a union of subsets E_ν as described above, or, equivalently, if the index set K reduces to a single point. There is no requirement that any component E_ν be connected, and if there are several components they can all be imbedded in some (possibly higher-dimensional) Euclidean space in an obvious way. The reason for specifying E as we have is that this is almost invariably the structure that arises in applications, and our formulation allows us to state the results in this and the following chapter in the most readily usable form. In Chapters 4 and 5, however, which are concerned with a general development of optimization theory, it is more convenient to use a compressed notation, eliminating explicit reference to the indicator variable ν and leaving it to the reader to transcribe the results into the form appropriate for specific applications.

25 The strong Markov property

The purpose of this section is to prove Theorem (25.5) below, showing that the process $(D_E, \mathfrak{F}, \mathfrak{F}_t, x_t, P_x)$ constructed in section 24 is a homogeneous strong Markov process. In doing this we exploit the strong connection between the PDP (x_t) and a related jump process. A *jump process* is a right-continuous process with piecewise-constant sample paths. Jump processes are analysed in detail in the Appendix and we will use results from that section here.

Let $D_E = D_E[0, \infty[$ denote the set of right-continuous functions z on $\mathbb{R}_+$ with values in E such that the left limit $\lim_{s \uparrow t} z(s)$ exists for all $t \in]0, \infty[$. Denote by $\check{x}_t$ the coordinate function $\check{x}_t(z) = z(t)$ for $z \in D_E$. Let $\mathfrak{F}_t^0$ denote the natural filtration, $\mathfrak{F}_t^0 = \sigma\{\check{x}_s, s \leqslant t\}$ and $\mathfrak{F}^0 = \vee_t \mathfrak{F}_t^0$.

With Assumption (24.4) in force, the construction in section 24 defines for each starting point $x \in E$ a measurable mapping $\psi_x : \Omega \to D_E$ such that $\check{x}_t(\psi_x(\omega)) = x_t(\omega)$. Let P_x denote the image measure, $P_x = P\psi_x^{-1}$. This defines a family of measures $\{P_x, x \in E\}$. Thus a PDP can be thought of as a process defined on Ω or as a Markov family defined on D_E. We shall, by convenient abuse of notation, generally denote it x_t whatever probability space it is defined on.

For any probability measure μ on E we define a measure P^μ on $(D_E, \mathfrak{F}^0)$ by

$$P^\mu A = \int_E P_x A \, \mu(\mathrm{d}x).$$

Now let $\mathfrak{F}_t^\mu$ be the completion of $\mathfrak{F}_t^0$ with all P^μ – null sets of $\mathfrak{F}^0$, and define

$$\mathfrak{F}_t = \bigcap_{\mu \in \mathscr{P}(E)} \mathfrak{F}_t^\mu,$$

where $\mathscr{P}(E)$ is the set of probability measures on E.

(25.1) *The associated jump process*

Fix $x \in E$ and consider the PDP x_t starting at x as constructed above. The *associated jump process* (η_t) takes values in $E \times \mathbb{Z}_+$ and is defined as

$$\text{(25.2)} \quad \eta_t = \begin{bmatrix} x \\ 0 \end{bmatrix}, t < T_1, \quad \eta_t = \begin{bmatrix} \eta_t^1 \\ \eta_t^2 \end{bmatrix} := \begin{bmatrix} x_{T_n} \\ n \end{bmatrix}, \quad T_n \leqslant t < T_{n+1}.$$

There is a one-to-one correspondence between the sample paths of (x_t) and (η_t): given the sample path $\{x_s, s \leqslant t\}$, the jump times are the finite set $\{T_n, n = 1, \ldots, k\} = \{s \in]0, t] : x_s \neq x_{s-}\}$, and the sample path $\{\eta_s, s \leqslant t\}$ is then defined by (25.2) while, conversely, given $\{\eta_s, 0 \leqslant s \leqslant t\}$ we know that $x_0 = \eta_0^1$ and $x_t = (v, \phi_v(t, \zeta))$ for $t \in [T_n, T_{n+1}[$, where $T_n = \inf\{t : \eta_t^2 = n\}$ and $\eta_{T_n} = ((v, \zeta), n)$. Note that (η_t^2) must be included here since it is possible that $P_x[x_{T_{n+1}} = x_{T_n}] > 0$ for some n, in which case T_{n+1} may not be 'detected' by examining the states $(x_{T_k}, k = 1, \ldots, n)$ alone. We are ready to prove the next result.

(25.3) *Theorem.* $\mathfrak{F}_t$ is right continuous: for any $t > 0$,

$$\mathfrak{F}_t = \mathfrak{F}_{t+} := \bigcap_{\varepsilon > 0} \mathfrak{F}_{t+\varepsilon}.$$

Proof. Fix $\mu \in \mathscr{P}(E)$, and let (x_t) be the PDP with initial measure μ and (η_t) the associated jump process. Let $\mathscr{G}_t^\mu$ be the P^μ-completion of the natural filtration of (η_t) in $D_{E \times \mathbb{Z}_+}$. Theorem A2.1 shows that the equality $\mathscr{G}_{t+}^\mu = \mathscr{G}_t^\mu$ holds for natural filtrations of right-constant processes and hence holds here since (η_t) is right constant. It follows easily that $\mathfrak{F}_t^\mu = \mathfrak{F}_{t+}^\mu$ and hence we have

$$\mathfrak{F}_{t+} = \bigcap_{\varepsilon > 0} \mathfrak{F}_{t+\varepsilon} = \bigcap_{\varepsilon > 0} \bigcap_{\mu \in \mathscr{P}(E)} \mathfrak{F}_{t+\varepsilon}^\mu$$

$$= \bigcap_{\mu \in \mathscr{P}(E)} \bigcap_{\varepsilon > 0} \mathfrak{F}_{t+\varepsilon}^\mu = \bigcap_{\mu \in \mathscr{P}(E)} \mathfrak{F}_t^\mu = \mathfrak{F}_t. \qquad \square$$

We now come to the main result of this section. For bounded measurable functions $f: E \to \mathbb{R}$, denote

(25.4) $$P_t f(x) = \mathbb{E}_x f(x_t),$$

where $\mathbb{E}_x$ denotes integration with respect to measure P_x.

(25.5) *Theorem.* The process $(x_t, t \geqslant 0)$ is a homogeneous strong Markov process, i.e. for any $x \in E$, $\mathfrak{F}_t$-stopping time T and bounded measurable function f,

(25.6) $$\mathbb{E}_x[f(x_{T+s}) I_{(T<\infty)} | \mathfrak{F}_T] = P_s f(x_T) I_{(T<\infty)}.$$

Proof. We will first show the simple Markov property, i.e. that (25.6) holds when $T = t$, a fixed non-negative time. By the construction of the process, we know that $P[S_{k+1} > s | \mathfrak{F}_{T_k}] = F(s, x_{T_k})$ when F is given by (24.2). Thus

$$P_x[T_{k+1} > t + s | \mathfrak{F}_t] I_{(T_k \leqslant t < T_{k+1})}$$

$$= \frac{F(t + s - T_k, x_{T_k})}{F(t - T_k, x_{T_k})} I_{(T_k \leqslant t < T_{k+1})}$$

$$= \exp\left(- \int_0^s \lambda(v, \phi_v(u + t - T_k, \zeta)) \mathrm{d}u \right) I_{(t+s-T_k < t_*(x_{T_k}))},$$

where $x_{T_k} = (v, \zeta)$. Denote also $x_t = (v(t), \zeta(t))$. On the set $(T_k \leqslant t < T_{k+1})$ we have $v(t) = v(T_k)$ and hence, using the semigroup property,

$$\phi_v(u + t - T_k, \zeta) = \phi_v(u, \phi_v(t - T_k, \zeta)) = \phi_{v(t)}(u, \zeta(t)),$$

and

$$t_*(x_t) = t_*(x_{T_k}) - (t - T_k),$$

so that $(t + s - T_k) < t_*(x_{T_k})$ is equivalent to $s < t_*(x_t)$. We thus see that

$$P_x[T_{k+1} > t + s | \mathfrak{F}_t] I_{(T_k \leqslant t < T_{k+1})} = F(s, x_t) I_{(T_k \leqslant t < T_{k+1})},$$

and hence, denoting $\tilde{T} = \inf\{s > t : x_s \neq x_{s-}\}$, $P_x[\tilde{T} > t + s | \mathfrak{F}_t] = F(s, x_t)$. Thus, conditional on $\mathfrak{F}_t$, the next jump time of the process has the same distribution as the first jump time of a PDP started at x_t. Since the remaining interarrival times and postjump states are determined by random variables $\{U_n, U_{n+1}, \ldots\}$ which are independent of the past, we have shown that if h is bounded measurable function on D_E then

$$\mathbb{E}_x[h(x_{t+s}, s \geqslant 0) | \mathfrak{F}_t] = \mathbb{E}_y[h(x_s, s \geqslant 0)]|_{y = x_t}$$

and (24.6) with $T = t$ is a special case of this.

To get the strong Markov property we use the characterization of jump process stopping times given in Theorem (A2.3). This carries over to stopping times of $\mathfrak{F}_t$ in view of the 1–1 correspondence between the sample paths of (x_t) and (η_t) mentioned earlier, and shows that if we define $Z_k = x_{T_k}$ and $\xi_k = (x, S_1, Z_1, S_2, Z_2, \ldots, S_k, Z_k)$ then associated with each stopping time T are functions $s_1, s_2 \ldots$, such that

$$T I_{(T \leqslant T_1)} = (s_1(x) \wedge T_1) I_{(T \leqslant T_1)}$$

and for $k = 2, 3, \ldots$,

$$T I_{(T_{k-1} < T \leqslant T_k)} = ((T_{k-1} + s_k(\xi_{k-1})) \wedge T_k) I_{(T_{k-1} < T \leqslant T_k)}.$$

This characterization shows that on the set $T < \infty$ there are three possible cases: either $T = 0$, or $T = T_k$ for some k, or $T = T_{k-1} + s_k(\xi_{k-1})$ for some k, and each of these sets is $\mathfrak{F}_T$-measurable. If we define $\tilde{T} = \inf(t > T : x_t \neq x_{t-})$ then the same calculations as above, applied to these three cases, show that $P_x[(\tilde{T} > T + s) \cap (T < \infty) | \mathfrak{F}_T] = F(s, x_T) I_{(T < \infty)}$. The same argument used for the simple Markov property now shows that

$$\mathbb{E}_x[h(x_{T+s}, s \geqslant 0) I_{(T < \infty)} | \mathfrak{F}_T] = E_y[h(x_s, s \geqslant 0)]|_{y = x_T} I_{(T < \infty)},$$

from which the strong Markov property (25.6) follows. □

(25.7) *Example*
This is another look at Example (14.9). We let $E = \mathbb{R}_+$ and T be an exponential time: $P(T > t) = e^{-t}$. For $x > 0$ let $x_t = x + t$; for $x = 0$ let $x_t = 0$ for $t \leqslant T$ and $x_t = t - T$ for $t \geqslant T$. Then it is easily seen that the strong Markov property fails at the stopping time T. However, (x_t) is not a PDP according to our definition. To 'realize' it as a PDP, take $K = \{1, 2\}, E_1 = \{0\}, E_2 = \mathbb{R}_+$, $\phi_2(t, \zeta) = t + \zeta$, $\lambda_1(0) = 1, \lambda_2(\zeta) = 0, Q(dy; (1, 0)) = \delta_{(2,0)}(dy)$. The process moves uniformly to the right when $v = 2$ and waits for an exponential time when $v = 1$. This effectively splits the point 0 into two points (1, 0) and (2, 0), and the resulting process *is* strong Markov.

26 The extended generator of the PDP

The main result of this section – and indeed one of the main results of PDP theory as a whole – is Theorem (26.14) which gives the extended generator of the PDP and characterizes precisely its domain. Recall from Definition (14.15) that if $\mathfrak{A}$ is the extended generator, then for $f \in \mathscr{D}(\mathfrak{A})$ the process C_t^f defined by

$$C_t^f = f(x_t) - f(x) - \int_0^t \mathfrak{A} f(x_s) \mathrm{d}s \tag{26.1}$$

is a local martingale. Our strategy here is to exploit the connection between the PDP (x_t) and the associated jump process (η_t) introduced in (25.1) and to use the characterization of the local martingales of a jump process given as Theorem A5.5 in the Appendix. This result implies that the local martingale C_t^f of (26.1) must take a very specific form and enables us to determine for what functions f it is possible to represent the process $f(x_t)$ as the sum of an absolutely continuous process and a local martingale, as required by (26.1)

For $A \in \mathscr{E}$ define processes p, p^* and $\tilde{p}$ as follows:

$$p(t, A) = \sum_{i=1}^{\infty} I_{(t \geqslant T_i)} I_{(x_{T_i} \in A)}$$

$$p^*(t) = \sum_{i=1}^{\infty} I_{(t \geqslant T_i)} I_{(x_{T_i -} \in \Gamma)}$$

$$\tilde{p}(t, A) = \int_0^t Q(A, x_s) \lambda(x_s) \mathrm{d}s + \int_0^t Q(A, x_{s-}) \mathrm{d}p^*(s). \tag{26.2}$$

Note that p, p^* are counting processes, $p^*(t)$ counting the number of jumps from the boundary of the process (x_t). Thus the second integral on the right of (26.2) is just a finite sum, almost surely.

(26.3) *Definition.* For the PDP (x_t), denote $\xi_k = (S_k, Z_k)$. A stochastic process g defined on Ω is said to be *predictable* if there exists a measurable function $g_1 = \mathbb{R}_+ \to \mathbb{R}$ and for $k = 2, 3, \ldots$, measurable functions $g_k: \mathbb{R}_+ \times (\mathbb{R}_+ \times E)^{k-1} \to \mathbb{R}$ such that

$$(26.4) \quad g(s, \omega) = g_1(s) I_{(s \leqslant T_1)} + \sum_{k=2}^{\infty} g_k(s, \xi_1, \ldots, \xi_{k-1}) I_{(T_{k-1} < s \leqslant T_k)}.$$

(26.5) *Remark.* This is the same definition as that given for general jump processes in the Appendix, Definition (A3.1), and further discussion will be found there.

(26.6) *Proposition.* $p^*(t)$ is a predictable process, and $\tilde{p}(t, A)$ is a predictable process for each $A \in \mathscr{E}$.

Proof. $p^*(t)$ is equal to $g(t)$ given by (26.4), with

$$g_1(t) = I_{(t = t_*(x))}, \quad g_k(t) = p^*(T_{k-1}) + I_{(t = t_*(x_{T_{k-1}}))}.$$

A similar representation holds for $\tilde{p}(t, A)$ with

$$g_1(t) = \chi(t, x), \quad g_k(t) = \tilde{p}(T_{k-1}, A) + \chi(t, x_{T_{k-1}}),$$

where

$$\chi(t, x) = \int_0^t Q(A; \phi(s, x)) \lambda(\phi(s, x)) \mathrm{d}s + Q(A; \phi(t_*(x), x)) I_{(t = t_*(x))}.$$

(26.7) *Proposition.* $\tilde{p}(t, A)$ is the compensator of $p(t, A)$, i.e. $q(t \wedge T_k, A)$ is an $\mathfrak{F}_t$-martingale for each k and $A \in \mathscr{E}$, where

$$q(t, A) = p(t, A) - \tilde{p}(t, A).$$

(In particular, $q(t, A)$ is a local martingale.)

Proof. This is a special case of Theorem (A4.4) and Proposition (A5.1). It is shown in Theorem (A4.4) that the compensator of $p(t \wedge T_1, A)$ is

$$(26.8) \qquad \tilde{p}(t \wedge T_1, A) = -\int_{]0, t \wedge T_1]} Q(A; x_{s-}) \frac{\mathrm{d}F(s)}{F(s-)},$$

where F is the survivor function of T_1. Here this is equal to $F(t,x)$ given by (24.2), and we see that (26.2) and (26.8) coincide for $t \leqslant T_1$ since $F(\cdot,x)$ is absolutely continuous on $[0,t_*(x)[$ and $\Delta F(t_*,x)/F(t_*-,x) = -1$. The general case follows by concatenation as in the proof of Proposition (A5.1).

We can now form *stochastic integrals* with respect to the local martingale q. The integrands are measurable functions $g: E \times \mathbb{R}_+ \times \Omega \to \mathbb{R}$ such that $g(x,\cdot,\cdot)$ is a predictable process (in the sense of Definition (26.3)) for each $x \in E$. Note that the counting process $p(t,A)$ can be identified with a random measure ν on $E \times \mathbb{R}_+$ defined by $\nu(A \times]0,t]) = p(t,A)$ and it is clear from the definition of p that

$$\nu = \sum_{i=1}^{\infty} \delta_{(x_{T_i}, T_i)},$$

where δ_γ denotes the Dirac measure (unit mass) at $\gamma \in E \times \mathbb{R}_+$. We therefore define

$$(26.9) \qquad \int_{E\times\mathbb{R}_+} g(x,s,\omega)p(\mathrm{d}s,\mathrm{d}x) = \int_{E\times\mathbb{R}_+} g\mathrm{d}\nu = \sum_i g(x_{T_i}, T_i, \omega).$$

We say that $g \in L_1(p)$ if $\mathbb{E}(\int |g|\mathrm{d}\nu) < \infty$ and $g \in L_1^{\mathrm{loc}}(p)$ if $gI_{(t<\sigma_n)} \in L_1(p)$ for some sequence of stopping times σ_n such that $\sigma_n \uparrow \infty$ a.s. Integrals with respect to $\tilde{p}$ are defined in a way that follows naturally from its definition (26.2), namely

$$(26.10) \quad \int_{E\times\mathbb{R}_+} g(x,s,\omega)\tilde{p}(\mathrm{d}t,\mathrm{d}x) = \iint_{\mathbb{R}_+ E} g(x,s,\omega)Q(\mathrm{d}x;x_s)\lambda(x_s)\mathrm{d}s + \iint_{\mathbb{R}_+ E} g(x,s,\omega)Q(\mathrm{d}x;x_{s-})\mathrm{d}p^*(s).$$

Note that the first integral on the right would be the same if x_s were replaced by x_{s-} since $x_s \neq x_{s-}$ only on a set of Lebesgue measure zero. Again, $g \in L_1(\tilde{p})$ in case $\mathbb{E}(\int |g|\mathrm{d}\tilde{p}) < \infty$ and $g \in L_1^{\mathrm{loc}}(\tilde{p})$ when $gI_{(s \leqslant \sigma_n)} \in L_1(\tilde{p})$ for σ_n as above. Stochastic integrals with respect to $q = p - \tilde{p}$ are defined simply as differences:

$$\int_{E\times\mathbb{R}_+} g\mathrm{d}q := \int_{E\times\mathbb{R}_+} g(x,s,\omega)q(\mathrm{d}s,\mathrm{d}x)$$

$$:= \int_{E\times\mathbb{R}_+} g(x,s,\omega)p(\mathrm{d}s,\mathrm{d}x) - \int_{E\times\mathbb{R}_+} g(x,s,\omega)\tilde{p}(\mathrm{d}s,\mathrm{d}x).$$

Finally, for a predictable integrand g we define a process $(M_t^g, t \geqslant 0)$ by

(26.11) $$M_t^g = \int_{E\times\mathbb{R}_+} gI_{(s\leqslant t)}\mathrm{d}q.$$

The following facts are proved in the Appendix.

(26.12) *Theorem*

1. $L_1(p) = L_1(\tilde{p})$ and $L_1^{\mathrm{loc}}(p) = L_1^{\mathrm{loc}}(p)$.
2. The process $(M_t^g, t \geqslant 0)$ is a local martingale if $g \in L_1^{\mathrm{loc}}(p)$ and is a martingale if $gI_{(s\leqslant t)} \in L_1(p)$ for all $t \geqslant 0$.
3. If (M_t) is an $\mathfrak{F}_t$-local martingale, then $M_t = M_t^g$ for all t, a.s., for some $g \in L_1^{\mathrm{loc}}(p)$.

(26.13) *Remark.* Part 1 is the local martingale representation theorem, Theorem (A5.5), applied to the jump process (η_t) associated with (x_t) defined by (25.2), and we are using the fact that the natural filtrations of (η_t) and (x_t) are the same.

We are now in a position to state the main result, identifying the extended generator $\mathfrak{A}$ of the PDP (x_t). We denote by $\mathscr{M}(E)$ the set of measurable real-valued functions on E, and define

$$\mathscr{M}_\Gamma(E) = \{f \in \mathscr{M}(E) : f(x) := \lim_{t\downarrow 0} f(\phi_\nu(-t, \zeta)) \text{ exists for all } x = (\nu, \zeta) \in \Gamma\}.$$

We also define for $x \in E$

$$s_*(x) = \inf\{t : F(t, x) = 0\} = \inf\{t : P_x(T_1 > t) = 0.$$

(26.14) *Theorem.* Let (x_t) be a PDP satisfying the standard conditions of section 24.2. Then the domain $\mathscr{D}(\mathfrak{A})$ of the extended generator $\mathfrak{A}$ of (x_t) consists of those functions $f \in \mathscr{M}_\Gamma(E)$ satisfying:

1. The function $t \to f(\phi_\nu(t, \zeta))$ is absolutely continuous on $[0, s_*(x)[$ for all $x = (\nu, \zeta) \in E$.
2. (Boundary condition)
$$f(x) = \int_E f(y)Q(\mathrm{d}y; x), \quad x \in \Gamma.$$
3. $\mathfrak{B}f \in L_1^{\mathrm{loc}}(p)$, where
$$\mathfrak{B}f(x, s, \omega) := f(x) - f(x_{s-}(\omega)).$$

For $f \in \mathscr{D}(\mathfrak{A})$, $\mathfrak{A}f$ is given by

$$\mathfrak{A}f(x) = \mathfrak{X}f(x) + \lambda(x)\int_E (f(y) - f(x))Q(\mathrm{d}y; x). \tag{26.15}$$

(26.16) *Remark.* It is clear that $\mathfrak{B}f$ is a predictable integrand in the sense described above. Indeed, on the interval $]T_{k-1}, T_k]$ this function is given by

$$\mathfrak{B}f(x, s, \omega) = f(x) - f(\phi_\nu(s, \zeta)) \quad (\nu, \zeta) = x_{T_{k-1}}.$$

From (26.8) its integral with respect to p is

$$\int_{E \times \mathbb{R}_+} \mathfrak{B}f(x, s, \omega)p(\mathrm{d}s, \mathrm{d}x) = \sum_k (f(x_{T_k}) - f(x_{T_k -})).$$

Then, in particular, $\mathfrak{B}f \in L_1^{\mathrm{loc}}(p)$ if for $n = 1, 2, \ldots,$

$$\mathbb{E}_x\left(\sum_{k=1}^{n} \left| f(x_{T_k}) - f(x_{T_k -})\right|\right) < \infty$$

or if

$$\mathbb{E}_x\left(\sum_{T_i \leqslant n} \left| f(x_{T_k}) - f(x_{T_k -})\right|\right) < \infty. \tag{26.17}$$

(Take $\sigma_n = T_n, n$, respectively.) Both of these conditions are satisfied if f is bounded (the second in view of Assumption (24.4)). Condition (26.17) ensures that $(M_t^{\mathscr{R}f})$ is a martingale (rather than a local martingale) in view of part 2 of Theorem (26.12).

(26.18) *Remark.* The formula (26.15) for $\mathfrak{A}$ is not surprising. Indeed, when $\lambda \equiv 0$, $\mathfrak{A} = \mathfrak{X}$ is just the 'generator' of the deterministic motion as discussed in section 22, whereas when $\mathfrak{X} = 0$, $\mathfrak{A}$ is the generator of a Markov jump process, the formula for which is easily obtained by heuristic arguments. The content of Theorem (26.14) lies in the fact that it is possible to characterize $\mathscr{D}(\mathfrak{A})$ so precisely, and in particular in showing that the boundary condition (part 2 of Theorem (26.14)) must be satisfied. There are few, if any, other significant classes of Markov processes for which such precise information about the generator is available.

Proof of Theorem (26.14). Let $(\tilde{\mathfrak{A}}, \mathscr{D}(\tilde{\mathfrak{A}}))$ denote the extended generator of (x_t). We want to show that $(\tilde{\mathfrak{A}}, \mathscr{D}(\tilde{\mathfrak{A}})) = (\mathfrak{A}, \mathscr{D}(\mathfrak{A}))$. Suppose first that f satisfies 1–3. Then $\mathfrak{B}f \in L_1^{\text{loc}}(\tilde{p})$ and from (26.10)

$$\int_{[o,t]\times E} \mathfrak{B}f\,\mathrm{d}\tilde{p} = \int_0^t \int_E (f(y) - f(x_s))Q(\mathrm{d}y; x_s)\lambda(x_s)\mathrm{d}s$$
$$+ \int_{[o,t]} \int_E (f(y) - f(x_{s-}))Q(\mathrm{d}y; x_{s-})\mathrm{d}p^*(s).$$

Now the support of p^* is contained in the countable set $\{s : x_{s-} \in \Gamma\}$ and hence the second integral vanishes in view of the boundary condition Theorem (26.14), part 2. Thus

$$\text{(26.19)} \qquad \int_{[o,t]\times E} \mathfrak{B}f\,\mathrm{d}q = \sum_{T_i \leqslant t} (f(x_{T_i}) - f(x_{T_i-}))$$
$$- \int_0^t \int_E (f(y) - f(x_s))Q(\mathrm{d}y; x_s)\lambda(x_s)\mathrm{d}s$$

and this is a local martingale, because of Theorem (26.14), part 3. Let T_k denote the last jump time prior or equal to t. Then

$$\sum_{T_i \leqslant t} (f(x_{T_i}) - f(x_{T_i-}))$$
$$= \left\{ f(x_t) - f(x_{T_k}) + \sum_{i=1}^{k} (f(x_{T_i})) - f(x_{T_{i-1}}) \right\}$$
$$- \left\{ f(x_t) - f(x_{T_k}) + \sum_{i=1}^{k} (f(x_{T_i-})) - f(x_{T_{i-1}}) \right\}.$$

The first bracketed term on the right is equal to $f(x_t) - f(x)$. To evaluate the second term, note that $x_{T_i-} = \phi_\nu(T_i - T_{i-1}, \zeta)$, where $(\nu, \zeta) = x_{T_{i-1}}$. Thus from Theorem (26.14), part 1,

$$f(x_{T_i-}) - f(x_{T_{i-1}}) = \int_{T_{i-1}}^{T_i} \frac{\mathrm{d}}{\mathrm{d}s} f(\phi_\nu(s, \zeta))\mathrm{d}s$$
$$= \int_{T_{i-1}}^{T_i} \mathfrak{X}f(x_s)\mathrm{d}s.$$

The second term is therefore equal to

$$\int_0^t \mathfrak{X} f(x_s)\mathrm{d}s$$

and we see from (26.19) that

$$C_t^f := f(x_t) - f(x) - \int_0^t \mathfrak{A} f(x_s)\mathrm{d}s = M_t^{\mathfrak{B} f}$$

is a local martingale, where $\mathfrak{A}$ is given by (26.15). Thus $f \in \mathscr{D}(\tilde{\mathfrak{A}})$ and $\tilde{\mathfrak{A}} f = \mathfrak{A} f$.

Conversely, suppose that $f \in D(\tilde{\mathfrak{A}})$. Then the process

$$M_t := f(x_t) - f(x) - \int_0^t h(x_s)\mathrm{d}s$$

is a local martingale, where $h = \tilde{\mathfrak{A}} f$. From Theorem (26.12), part 3, it must be the case that $M_t = M_t^g$ for some predictable integrand $g \in L_1^{\mathrm{loc}}(p)$, and M_t^g is given by

$$M_t^g = \sum_{T_i \leqslant t} g(x_{T_i}, T_i, \omega) - \int_0^t \int_E g(y, s, \omega) Q(\mathrm{d}y; x_s)\lambda(x_s)\mathrm{d}s \tag{26.20}$$

$$- \iint_{[0,t]]E} g(y, s, \omega) Q(\mathrm{d}y; x_{s-})\mathrm{d}p^*(s).$$

Since M_t and M_t^g agree, their jumps ΔM_t and ΔM_t^g must agree; these only occur when $t = T_i$ for some i and are given by

$$\Delta M_t = f(x_t) - f(x_{t-})$$

$$\Delta M_t^g = g(x_t, t, \omega) - \int_E g(y, t, \omega) Q(\mathrm{d}y; x_{t-}) I_{(x_{t-} \in \Gamma)}.$$

Thus $g(x_t, t, \omega) = f(x_t) - f(x_{t-})$ on the set $(x_{t-} \notin \Gamma)$, which implies that $g(x, t, \omega) = f(x) - f(x_{t-})$ for all (x, t) except perhaps a set to which the process 'never jumps', i.e. a set $G \subset \mathbb{R}_+ \times E$, such that

$$\mathbb{E}_z \int_G p(\mathrm{d}t, \mathrm{d}x) = 0 \quad \text{for all } z \in E.$$

Now suppose $z = x_{t-} \in \Gamma$.. Then equating ΔM_t and ΔM_t^g gives

$$f(x_t) - f(z) = g(x_t, t, \omega) - \int_E g(y, t, \omega) Q(\mathrm{d}y; z)$$

and hence

$$(26.21) \qquad f(x) - f(z) = g(x, t, \omega) - \int_E g(y, t, \omega) Q(\mathrm{d}y; z)$$

except on a set $A \in \mathscr{E}$ such that $Q(A; z) = 0$. Integrating both sides of (26.21) with respect to $Q(\mathrm{d}x; z)$, we obtain (using the fact that $Q(\cdot\,; z)$ is a probability measure)

$$\int_E f(x) Q(\mathrm{d}x; z) - f(z) = \int_E g \mathrm{d}Q - \int_E g \mathrm{d}Q = 0.$$

Thus f satisfies the boundary condition Theorem (26.14), part 2. Now define (for fixed z)

$$\tilde{g}(x, t, \omega) = g(x, t, \omega) - (f(x) - f(z)).$$

Then from (26.21), and using Theorem (26.14), part 2,

$$\tilde{g}(x, t, \omega) = \int_E (f(y) - f(z) + \tilde{g}(y, t, \omega)) Q(\mathrm{d}y; z) = \int_E \tilde{g}(y, t, \omega) Q(\mathrm{d}y; z).$$

However, the right-hand side does not depend on x, and hence $\tilde{g}(x, t, \omega) = \alpha(t, \omega)$ for some predictable process α. The general expression for g is thus

$$g(x, t, \omega) = f(x) - f(x_{t-}) + \alpha(t, \omega) I_{(x_{t-} \in \Gamma)}.$$

Inserting this expression in (26.20) we find that M_t^g does not depend on α, and we can therefore take $\alpha \equiv 0$, thus obtaining $g = \mathfrak{B} f$; hence Theorem (26.14), part 3, is satisfied.

Finally, consider the sample paths of M_t, $M_t^{\mathfrak{B} f}$ for $t < T_1(\omega)$, starting at $x = (v, \zeta) \in E$. We have

$$M_t = f(v, \phi_v(t, \zeta)) - f(x) + \int_0^t h(v, \phi_v(s, \zeta)) \mathrm{d}s,$$

while, because $p = p^* = 0$ on $[0, T_1[$,

$$M_t^{\mathfrak{B} f} = - \int_0^t \int_E (f(y) - f(v, \phi_v(s, \zeta))) Q(\mathrm{d}y; v, \phi_v(s, \zeta)) \lambda(v, \phi_v(s, \zeta)) \mathrm{d}s.$$

By the definition of $s_*(x)$, $P_x(T_1 > c) > 0$ for any $c < s_*(x)$, so since $M_t = M_t^{\mathscr{B} f}$ for all t a.s. it must be the case that the function $t \to f(v, \phi_v(, \zeta))$ is absolutely continuous on $[0, s_*(x)[$, i.e. Theorem (26.14), part 1 is satisfied. Calculations as before now show that for all $t \geqslant 0$,

$$M_t^{\mathfrak{B}f} = f(x_t) - f(x) - \int_0^t \mathfrak{A}f(x_s)\mathrm{d}s$$

with $\mathfrak{A}$ given by (26.15). Hence $f \in \mathscr{D}(\mathfrak{A})$ and $\mathfrak{A}f = \tilde{\mathfrak{A}}f$. This completes the proof. □

27 Further Markov properties of the PDP

This section is included for completeness, and may be skipped at a first reading: the results in it are not used elsewhere in the book.

In the development of the general theory of Markov processes, various classes of processes such as Feller processes, Ray processes and 'right' processes have been introduced which turn out to have useful properties. Since many general results appear in the literature pertaining to these classes, it is germane to ask whether such results apply to PDPs. In this section we show that, roughly speaking, a PDP is not a Feller process unless there are no active boundaries, but is always a right process. The latter property has been used, for example, by Kaspi and Mandelbaum (1992) in an analysis of stationary distributions of queueing networks.

The Feller and right properties refer in part to conditions satisfied by the semigroup $P_t f(x) = \mathbb{E}_x f(x_t)$ of the process (x_t), and we start by obtaining a useful representation for the semigroup associated with a PDP (x_t).

For notational convenience we will in this section take K, the index set of the indicator variable v, to be a singleton and suppress reference to this variable. As pointed out in Remark (24.9), no loss of generality is entailed. First, we establish a technical point already alluded to in section 24.

(27.1) *Lemma.* Suppose that the standard conditions (24.8) are satisfied. Then $t_*: E \rightarrow \mathbb{R}_+ \cup \{\infty\}$ is Borel measurable.

Proof. Let $F = (E^O)^c$. Then F is closed and t_* can equivalently be defined as

$$t_*(x) = \inf\{t : d(\phi(t,x), F) = 0\},$$

where d is the distance function $d(y, F) = \inf_{z \in F} |y - z|$. It is easily shown that $y \rightarrow d(y, F)$ is continuous (Clarke, 1983, Proposition 2.4.1) and hence $(t,x)\to d(\phi(t,x),F)$ is continuous. Now for $t\in\mathbb{R}_+$,

$$\{x:t_*(x)\leqslant t\}=\bigcap_{n=1}^{\infty}\bigcup_{r\in\mathbb{Q}\cap[0,t]}\left\{x:d(\phi(r,x),F)<\frac{1}{n}\right\}$$

and this is a Borel set since each of the sets on the right is open. □

Fix $f\in B(E)$ and for $\psi\in B(\mathbb{R}_+\times E)$ define

$$G\psi(t,x)=\mathbb{E}_x[f(x_t)I_{(t<T_1)}+\psi(t-T_1,x_{T_1})I_{(t\geqslant T_1)}].$$

$G\psi$ can be expressed in explicit terms as follows:

(27.2) $$G\psi(t,x)=I_{(t<t_*(x))}f(\phi(t,x))\mathrm{e}^{-\Lambda(t,x)}$$
$$+I_{(t\geqslant t_*(x))}\int_E\psi(t-t_*(x),y)Q(\mathrm{d}y;\phi(t_*(x),x)\mathrm{e}^{-\Lambda(t_*(x),x)}$$
$$+\int_0^{t\wedge t_*(x)}\int_E\psi(t-s,y)Q(\mathrm{d}y;\phi(s,x))\lambda(\phi(s,x)\mathrm{e}^{-\Lambda(s,x)}\mathrm{d}s,$$

where

$$\Lambda(t,x)=\int_0^t\lambda(\phi(s,x))\mathrm{d}s,\quad t\leqslant t_*(x).$$

It follows from this representation together with Lemma (27.1) and the Fubini theorem that $G\psi(t,x)$ is a Borel measurable function of (t,x). Thus G is an operator mapping $B(\mathbb{R}^+\times E)$ into itself. The next result is actually a special case of a more general result (32.35) established in Chapter 3.

(27.3) *Lemma.* Under the standard conditions of section (24.8)

1. $G^n\psi(t,x)=\mathbb{E}_x[f(x_t)I_{(t<T_n)}+\psi(t-T_n,x_{T_n})I_{(t\geqslant T_n)}\}\quad n=1,2,\dots$
2. For each $(t,x)\in\mathbb{R}_+\times E$ and $\psi\in B(\mathbb{R}_+\times E)$,

$$\lim_{n\to\infty}G^n\psi(t,x)=P_tf(x).$$

Proof. Statement 1 is true by definition when $n=1$. Suppose it holds for $n=k$; then

$$G^{k+1}\psi(t,x)=G^k(G\psi)(t,x)$$
$$=\mathbb{E}_x[f(x_t)I_{(t<T_k)}+G\psi(t-T_k,x_{T_k})I_{(t\geqslant T_k)}],$$

by the induction hypothesis. Now by the strong Markov property and the construction of the process,

$$\begin{aligned}
&\mathbb{E}_x[f(x_t)I_{(t<T_{k+1})} + \psi(t-T_{k+1}, x_{T_{k+1}})I_{(t\geqslant T_{k+1})}|\mathfrak{F}_{T_k}]I_{(t\geqslant T_k)}\\
&= \mathbb{E}_x[f(\phi(t-T_k, x_{T_k}))I_{(t-T_k<S_{k+1})}\\
&\quad + \psi((t-T_k)-S_{k+1}, x_{T_k+S_{k+1}})I_{(t-T_k\geqslant S_{k+1})}|\mathfrak{F}_{T_k}]I_{(t\geqslant T_k)}\\
&= G\psi(t-T_k, x_{T_k})I_{(t\geqslant T_k)}.
\end{aligned}$$

Therefore

$$\begin{aligned}
G^{k+1}\psi(t,x) &= \mathbb{E}_x\{f(x_t)I_{(t<T_k)} + \mathbb{E}_x[f(x_t)I_{(t<T_{k+1})}\\
&\quad + \psi(t-T_{k+1}, x_{T_{k+1}})I_{(t\geqslant T_{k+1})}|\mathfrak{F}_{T_k}]I_{(t\geqslant T_k)}\}\\
&= \mathbb{E}_x[f(x_t)I_{(t<T_{k+1})} + \psi(t-T_{k+1}, x_{T_{k+1}})I_{(t\geqslant T_{k+1})}]
\end{aligned}$$

which establishes 1 for $n=k+1$. The result follows.

For statement 2

$$\begin{aligned}
&|G^n\psi(t,x) - P_tf(x)|\\
&\quad = \mathbb{E}_x[f(x_t)I_{(t<T_n)} + \psi(t-T_n, x_{T_n})I_{(t\geqslant T_n)}] - \mathbb{E}_x[f(x_t)]|\\
&\quad \leqslant \mathbb{E}_x[(|f(x_t)| + |\psi(t-T_n, x_{T_n})|)I_{(t\geqslant T_n)}]\\
&\quad \leqslant 2cP_x(t\geqslant T_n),
\end{aligned}$$

where c is a bound for f and for ψ. By Assumption (24.4), $T_n\to\infty$ P_x-a.s. for all $x\in E$, and this implies that $P_x(t\geqslant T_n)\to 0$ for all $t\in\mathbb{R}_+$. □

(27.4) *Corollary.* For each $t\in\mathbb{R}_+$, P_t maps $B(E)$ into $B(E)$.

Proof. This is immediate from part 2 of Lemma (27.3). □

Let us now examine the Feller property. (x_t) is a *Feller process* if the semigroup P_t maps $C_b(E)$ into $C_b(E)$, i.e. the function $x\to P_tf(x)$ is continuous for each $t\in\mathbb{R}_+$ if f is bounded and continuous (of course, P_tf is automatically bounded for $f\in C_b(E)$). This is not always true for PDPs, as the following example shows.

(27.5) *Example*
Let $E=[0,1[$, $\mathfrak{X}=\mathrm{d}/\mathrm{d}x$, $\lambda=0$, $Q(\mathrm{d}y;1)=\delta_{\{0\}}(\mathrm{d}y)$. Then $x_t=(t+x)$ mod 1. Thus if we take $f(x)=x$ and $t=\frac{1}{2}$, then $P_tf(x)=(x+\frac{1}{2})I_{(x<1/2)}+(x-\frac{1}{2})I_{(x\geqslant 1/2)}$, which is not continuous at $x=\frac{1}{2}$.

It is clear from this example that (x_t) is generally not a Feller process if the boundary is hit with positive probability, i.e., if the active boundary Γ is non-empty: on the other hand, if the boundary is never hit then (x_t) *is* a Feller process under reasonable continuity assumptions.

(27.6) *Theorem.* If $t_*(x) = \infty$ for all $x \in E$, $\lambda \in C_b(E)$ and the function $x \to Qf(x)$ is continuous for $f \in C_b(E)$, then (x_t) is a Feller process.

Proof. Take $\psi \in C_b(\mathbb{R}_+ \times E)$ and $f \in C_b(E)$. It is clear from the representation (27.2) that $G\psi$ is then continuous under the stated conditions, and hence $G^n\psi \in C_b(\mathbb{R}_+ \times E)$ for $n = 1, 2, \ldots$. From the proof of Lemma (27.3) we have $|G^n(t,x) - P_t f(x)| \leqslant 2cP_x(t \geqslant T_n)$. Let $\bar{\lambda}$ be an upper bound for $\lambda(x)$. Now refer to the construction of the sample path of (x_t) from a sequence U_1, U_2 of i.i.d. $U[0,1]$ random variables described in section 2.4, and let $T'_n = S'_1 + \cdots + S'_n$, where $S'_k = \bar{\lambda}^{-1} \ln((U_{2k-1})^{-1})$. ($U_{2k-1}$ is the random variable involved in constructing $S_k = T_k - T_{k-1}$). Then clearly $S'_k \leqslant S_k$ for each k and hence $T'_n \leqslant T_n$. Thus $T_n \leqslant t$ implies $T'_n \leqslant t$ and therefore $P_x(T_n \leqslant t) \leqslant P_x(T'_n \leqslant t)$. But, for any x, the sequence T'_n is a Poisson process with rate $\bar{\lambda}$. It follows that $P_x(t \geqslant T_n) \to 0$, and hence $G^n(t,x) \to P_t f(x)$, as $n \to \infty$ uniformly in $x \in E$. This shows that $x \to P_t f(x)$ is continuous for any $t \geqslant 0$, which is the Feller property. □

A *right process* was originally defined by P.A. Meyer (1966) as a process satisfying two *hypothèses droites* denoted HD1 and HD2. The definition has subsequently undergone various subtle modifications at the hands of Getoor (1975) and Sharpe (1988), and the reader can find a complete exposition in Sharpe's book. A subclass of right processes is the class of *Borel right processes* defined in section 20 of Sharpe (1988). To qualify as a Borel right process, a process (x_t) must have the following properties.

(27.7)
(a) The state space E is Lusinian.

(b) The semigroup P_t maps $B(E)$ into $B(E)$

(c) The sample paths $t \to x_t(\omega)$ are a.s. right continuous.

(d) If f is an α-excessive function for P_t then the sample path $t \to f(x_t(\omega))$ is a.s. right continuous.

A topological space is *Lusinian* if it is homeomorphic to a Borel subset of a compact metric space. If P_t is a Markov semigroup then a function f is *α-super-mean-valued* ($\alpha \geqslant 0$) if it is non-negative and $e^{-\alpha t}P_t f \leqslant f$ for all $t \geqslant 0$. f is *α-excessive* if it is α-super-mean-valued and $e^{-\alpha t}P_t f \uparrow f$ as $t \downarrow 0$.

(27.8) *Theorem.* A PDP satisfying the standard conditions of section (24.8) is a Borel right process.

Proof. As pointed out by Sharpe (1988, p. 370) any locally compact Hausdorff space with countable base (LCCB) is Lusinian, and certainly the PDP state space E as defined in section 24 is LCCB. This establishes Property (27.7)(a). Property (b) was shown above in Corollary (27.4), and Property (c) is true by construction of the process. Now let f be an α-excessive function. Then the process $M_t := e^{-\alpha t} f(x_t)$ is a non-negative supermartingale. Indeed, for $t > s$, by the Markov and α-super-mean-value properties,

$$\mathbb{E}_x[M_t | \mathfrak{F}_s] = e^{-\alpha s} e^{-\alpha(t-s)} P_{t-s} f(x_s) \leqslant e^{-\alpha s} f(x_s) = M_s.$$

We conclude from the optional sampling theorem (13.7) that, for any $t \geqslant 0$, $f(x) \geqslant \mathbb{E}_x e^{-\alpha(t \wedge T_1)} f(x_{t \wedge T_1})$. Fix $x \in E$ and choose t sufficiently small that the functions $s \to \lambda(\phi(s,x))$ is integrable on $[0,t]$ ((24.8), part 2). Then

$$\begin{aligned} &\mathbb{E}_x[e^{-\alpha(t \wedge T_1)} f(x_{t \wedge T_1})] \\ &\quad = \mathbb{E}_x[e^{-\alpha t} f(\phi(t,x)) I_{(t < T_1)} + e^{-\alpha T_1} f(x_{T_1}) I_{(t \geqslant T_1)}] \\ &\quad = f(\phi(t,x)) e^{-(\alpha t + \Lambda(t,x))} \\ &\qquad + \int_0^t \int_E f(y) Q(dy; \phi(s,x)) \lambda(\phi(s,x)) e^{-(\alpha s + \Lambda(s,x))} ds. \end{aligned}$$

By dominated convergence, the second integral on the right converges to 0 as $t \downarrow 0$, and $e^{-(\alpha t + \Lambda(t,x))} \to 1$ as $t \downarrow 0$. Since f is α-excessive, we therefore have

$$f(x) = \lim_{t \downarrow 0} \mathbb{E}_x e^{-\alpha(t \wedge T_1)} f(x_{t \wedge T_1}) = \lim_{t \downarrow 0} f(\phi(t,x)).$$

Using the semigroup property of the flow ϕ we conclude that for any $x \in E$ the function $t \to f(\phi(t,x))$ is right-continuous. In view of the construction of the sample paths of (x_t) this implies that the function $t \to f(x_t(\omega))$ is right-continuous, a.s. This is Property (27.7)(d). □

28 Notes and references

21 Poisson and renewal processes are covered in most textbooks on applied probability, for example Cox and Miller (1964), Çinlar (1975) or Heyman and Sobel (1990). The method of supplementary variables dates back at least to Cox (1955); it is closely related to the idea of semi-Markov processes (Çinlar, 1975). The literature on queues is voluminous: a survey and extensive list of references are given in the chapter by R.B. Cooper in Heyman and Sobel (1990), while an excellent recent general reference is Asmussen (1987). The insurance models (21.11) and (21.12) are due to Dassios and Embrechts (1989) and Waters *et al.* (1991), respectively. Standard texts on insurance and risk theory are Beard *et al.* (1984) and Bühlmann (1970); treatments slanted towards modern stochastic process theory are Gerber (1979) and Grandell (1991). The capacity expansion model (21.13) is due to Davis *et al.* (1987b), based on a study of water resource provision by Erlenkotter, Sethi and Okada (1981). The repair model (21.15) is from Schornagel (1987), and is studied by Costa and Davis (1989) and Costa (1991).
22 Ordinary differential equations are discussed in many texts on analysis and control theory, for example Coddington and Levinson (1955), Hestenes (1966), Hirsch and Smale (1974). For geometric notions of vector fields and first-order differential operators see Boothby (1984).
23 Generation of random numbers and random variables with specified distributions is covered in Press *et al.* (1986) and by B. Schmeiser in Chapter 7 of Heyman and Sobel (1990). A recent text on simulation is Ripley (1987).
24 The definition of the PDP is essentially that of Davis (1984a, b). Other authors have introduced very similar constructs, for example the *Markov Decision Drift Processes* of van der Duyn Schouten (1983) or the similarly named processes described by Yushkevich (1983, 1987). Gnedenko and Kovalenko (1966) defined the piecewise-linear Markov process, which is similar to the PDP except that the deterministic motion is less general, specifically in order to study problems in queueing systems. The piecewise-linear process has been further studied by V.V. Kalashnikov (see, for example, Kalashnikov, 1975). Vermes (1980) developed a theory of control for piecewise-linear processes, predating the theory presented in Chapter 4 of this book.
25 The martingale approach to stochastic jump processes was

developed in series of papers in the mid-1970s: Boel, Varaiya and Wong (1975), Chou and Meyer (1975), Jacod (1975), Davis (1976). Some of this material (but not all of it) can be found in the books of Brémaud (1981), Elliott (1982) and Davis (1984a). The Appendix of this book is based on Davis (1976).

26 This section is based on Davis (1976). The constructive definition (26.3) of predictability is new but implicit in earlier work: results in Boel *et al.* (1975) imply that it is equivalent to the usual definition of predictability in the *théorie générale des processus*. This term refers to stochastic process theory as developed by the 'Strasbourg school' of P.-A. Meyer and co-workers; see Dellacherie and Meyer (1975/80) or Elliott (1982).

27 The main reference for this section is Sharpe (1988), where the right process is discussed in detail. A further very informative text covering general Markov process theory is Williams (1979). As mentioned earlier, an application of the right process property of PDPs to the characterization of stationary distributions is given by Kaspi and Mandelbaum (1992).

CHAPTER 3

Distributions and expectations

In Chapter 2 we isolated the piecewise-deterministic process (x_t) as a general stochastic model and identified its generator $\mathfrak{A}$. This chapter is concerned with calculating distributions of the process and the expectations of functionals of it. Distributions and expectations are of course closely related, in that if we are able to calculate $\mathbb{E}_x f(x_t)$ for bounded measurable functions f then taking $f = I_F$ for measurable sets F in the state space gives the distribution $P_x[x_t \in F] = \mathbb{E}_x[I_F(x_t)]$. For this reason we concentrate mainly on computing expectations in the first three sections of the chapter. The key to these computations is the differential formula (Theorem (31.3)) below. This formula is in fact an immediate corollary of Theorem (26.14) giving the generator $\mathfrak{A}$ of the process, but its importance can hardly be overstated. It enables us to state conditions on a function $v(t, x)$ under which the process $M_t^v := v(t, x_t)$ is a martingale. If these conditions are satisfied then in particular M_t^v has constant expectation, so the equality $\mathbb{E}_x M_0^v = v(0, x) = \mathbb{E}_x v(\tau, x_\tau)$ is valid for fixed times τ and, by the optional sampling theorem, for certain classes of stopping times. By choosing these times judiciously we get the expectations we want.

Section 32 is mainly concerned with evaluating

$$V(x) = \mathbb{E}_x \left\{ \int_0^\infty l(x_t)\mathrm{d}t + \int_0^\infty c(x_{t-})\mathrm{d}p_t^* \right\},$$

where l, c are given functions and p_t^* is the process counting boundary jumps of (x_t). This functional appears as a 'cost function' in many of the optimization problems discussed later in the book, and is sufficiently general to cover a wide variety of apparently different functionals as special cases. For example, the functional $\mathbb{E}_x f(x_t)$ is a special case, though this is not perhaps obvious at first sight. Most of section 31 is devoted to introducing various transformations of

PDPs with the help of which such connections can be recognized. In section 32 we discover that (under suitable conditions) V is the unique solution of a set of integro-differential equations. In some simple cases – for example in renewal theory – these equations can be solved directly. To handle more complex cases we introduce in section 32.2 a general recursive method by which numerical solutions may be obtained. In either case the details are inevitably problem-specific, and three applications are considered in some detail in section 33. The remaining section, section 34, concerns the important topic of stationary distributions. The main result here is an elucidation of the relationship between stationary distributions for the process (x_t) and for the *imbedded Markov chain* (z_k), where $z_k = x_{T_k}$ is the sequence of 'post-jump' states of (x_t). Thanks to this result, most questions about stationary distributions for (x_t) can be rephrased as questions about the simpler discrete-time process (z_k).

31 The differential formula and transformations of PDPs

Throughout this section, (x_t) is a PDP on a state space E satisfying the standing assumptions (24.8). Recall that one of the conditions for a function $f: E \to \mathbb{R}$ to belong to the domain of the generator is that the map $t \to f(\phi_\nu(t, \zeta))$ be absolutely continuous and possess a limit as the boundary Γ is approached. These are the most general conditions, but in applications we frequently encounter functions exhibiting greater degrees of smoothness. It will be helpful to collect together the definitions of various classes of functions at this point. In view of the existence of limits at the boundary, it is convenient to work with functions defined on $E \cup \Gamma$ rather than just on E.

(31.1) *Definition.* Let $\mathscr{M}$ denote the class of measurable functions $f: E \cup \Gamma \to \mathbb{R}$, and $\mathscr{M}(P) = \{f \in \mathscr{M}$: for each $x = (\nu, \zeta) \in E$, the map $t \to f(\phi_\nu(t, \zeta))$ is 'P' on $[0, s_*(x)]\}$. Here P denotes some property of measurable functions $h: \mathbb{R} \to \mathbb{R}$ and $s_*(x) = \inf\{t: P_x[T_1 > t] = 0\}$. Now define:

1. $\mathscr{C} = \mathscr{M}$(continuous);
2. $\mathscr{AC} = \mathscr{M}$(absolutely continuous);
3. $\mathscr{L} = \mathscr{M}$(locally Lipschitz continuous); and
4. $\mathscr{C}^1 = \mathscr{M}$(continuously differentiable).

Clearly, $\mathscr{C}^1 \subset \mathscr{L} \subset \mathscr{AC} \subset \mathscr{C} \subset \mathscr{M}$. The subscript b denotes 'bounded', so for example $\mathscr{M}_b = B(E \cup \Gamma)$ is the set of bounded measurable functions on $E \cup \Gamma$.

Recall that $\mathfrak{A}$ denotes the extended generator of the process, given by (26.15) and that $\mathfrak{B}$ is the operator mapping functions into integrands defined by

$$\mathfrak{B}f(t, x, \omega) = f(x) - f(x_{t-}(\omega)).$$

A function f is in $L_1^{\text{loc}}(p)$ if for some sequence of stopping times $\sigma_n \uparrow \infty$

$$\mathbb{E}_x \sum_i |f(x_{T_i \wedge \sigma_n}) - (f(x_{T_i \wedge \sigma_n -})| < \infty.$$

In particular, this always holds if f is bounded or if for each $t \geqslant 0$

$$\mathbb{E}_x \sum_{T_i \leqslant t} |f(x_{T_i}) - f(x_{T_i -})| < \infty. \tag{31.2}$$

The stochastic integral $\int \mathfrak{B} f \mathrm{d}q$ defined as in (26.11) is a local martingale for $f \in L_1^{\text{loc}}(p)$ and is a martingale under condition (31.2)

We shall need a further operator $\mathfrak{C}$, defined as follows

$$\mathfrak{C}f(z) = \int_E f(y) Q(\mathrm{d}y; z) - f(z), \quad z \in \Gamma.$$

Thus a function f satisfies the boundary conditions 2 of Theorem (26.14) if and only if $\mathfrak{C}f(z) = 0$ for all $z \in \Gamma$. Finally, recall that $p^*(t)$ is the predictable process counting jumps from the boundary

$$p^*(t) = \sum_{T_i \leqslant t} I_{(x_{T_i -} \in \Gamma)}.$$

We can now state the following result, which is a simple corollary of Theorem (26.14) but which plays a fundamental role in PDP theory.

(31.3) *Theorem.* (The PDP Differential Formula). Suppose $f \in \mathscr{AC}$ and $\mathfrak{B} f \in L_1^{\text{loc}}(p)$. Then for each $t \geqslant 0$

$$f(x_t) - f(x_0) = \int_0^t \mathfrak{A} f(x_s) \mathrm{d}s + \int_0^t \int_E \mathfrak{B} f(s, x, \omega) q(\mathrm{d}s, \mathrm{d}x) + \int_0^t \mathfrak{C} f(x_{s-}) \mathrm{d}p^*(s). \tag{31.4}$$

The stochastic integral term is a martingale under condition (31.2)

Proof. f satisfies all the conditions of Theorem (26.14) except the boundary condition. The result follows from equations (26.19)–(26.21) in the proof of this theorem. □

An immediate consequence of (31.4) is that if $\mathfrak{A}f(x)=0$ for all $x\in E$ and $\mathfrak{C}f(z)=0$ for all $z\in\Gamma$ then the process $M_t := f(x_t)$ is a local martingale; in particular, it is martingale, and therefore has constant expectation equal to $f(x)$ under measure P_x, if (31.2) is satisfied. This is useful in computing various expectations and probabilities, as we shall see below, but first we need to consider various transformations of PDPs in order to unify various apparently diverse cases.

(31.5) *Time variation*

It may happen that the local characteristics of a PDP contain explicit time variation, i.e. take the form $g(t,x)$, $\lambda(t,x)$, $Q(\mathrm{d}y;t,x)$. A simple expedient removes this: we define an augmented state $\tilde{x}=(x^0,x)$ on the state space $\tilde{E}=\mathbb{R}_+\times E$ and construct a PDP as before with local characteristics

$$\tilde{g}(\tilde{x})=\begin{bmatrix}1\\ g(x^0,x)\end{bmatrix},\quad \tilde{\lambda}(\tilde{x})=\lambda(x^0,x),\quad \tilde{Q}(\{x^0\}\times A;\tilde{x})=Q(A;x^0,x).$$

Then $x_t^0=x^0+t$ so that in particular $x_t^0=t$ when the starting point is $(0,x)$. The generator of $\tilde{x}_t$ is just

$$\tilde{\mathfrak{A}}f(\tilde{x})=\frac{\partial}{\partial x^0}f(x^0,x)+\mathfrak{A}f(x^0,x),$$

where $\mathfrak{A}$ is given by the usual formula (26.15) in terms of the original g,λ,Q. Often we do not explicitly mention the state augmentation when dealing with time-varying systems, and simply write the generator as

$$\tilde{\mathfrak{A}}f(t,x)=\frac{\partial}{\partial t}f(t,x)+\mathfrak{A}f(t,x);$$

but it should be remembered that, paradoxically, time-varying coefficients are a special case of time-invariant coefficients. This is why we only deal with time-invariant transition functions in PDP theory.

There is another reason for introducing the above augmentation, and that is to provide a finite time horizon. Start with a PDP with (time-invariant) local characteristics (g, λ, Q) on a state space E, fix a time $t_f > 0$, and form a new state space $\hat{E} = ([0, t_f] \times E) \cup \{\Delta\}$, where $\{\Delta\}$ is an isolated single-point set. (Δ is often referred to as the 'cemetery state' for reasons that will be clear below.) The active boundary of $\tilde{E}$ is $\tilde{\Gamma} := ([0, t_f] \times \Gamma) \cup (\{t_f\} \times E)$ and we define new local characteristics $\tilde{g} := \tilde{\lambda}, \tilde{Q}$ for $\tilde{x} = (x^0, x) \in \tilde{E}$ as follows: for $0 \leqslant x^0 < t_f$

$$\tilde{g}(\tilde{x}) = \begin{bmatrix} 1 \\ g(x) \end{bmatrix}, \quad \tilde{\lambda}(\tilde{x}) = \lambda(x), \quad \tilde{Q}(\{x^0\} \times A; \tilde{x}) = Q(A; x)$$

while on the boundary at $x^0 = t_f$

$$Q(\{\Delta\}; (t_f, x)) = 1.$$

In the cemetery state, $\tilde{g}(\Delta) = \tilde{\lambda}(\Delta) = 0$. The careful reader may object at this point that a singleton set is not a valid state space component in accordance with the definitions of section 24; this is true, but the same effect could be achieved by taking Δ as an interior point of some isolated open set throughout which $\tilde{g} = \tilde{\lambda} = 0$. Either way, the process $\tilde{x}_t$ corresponding to $(\tilde{g}, \tilde{\lambda}, \tilde{Q})$ is indistinguishable from the 'ordinary' process x_t for $t < t_f$ and $\tilde{x}_{t_f-} = x_{t_f}$; at time t_f, $\tilde{x}_t$ jumps to Δ and stays there since the jump rate out of Δ is zero. Thus $P_{(x^0, x)}(\tilde{x}_t = \Delta \text{ for all } t \geqslant t_f) = 1$. We say that $\tilde{x}_t$ is 'x_t killed at time t_f'.

(31.6) *Killing at finite rate*

Often in applications one wants to compute expected discounted rewards or costs which take the form

$$v(x) = \mathbb{E}_x \int_0^\infty e^{-\delta t} l(x_t) dt$$

for a PDP x_t, where $l: E \to \mathbb{R}_+$ is a given function and δ, the *discount rate*, is strictly positive. Note that the integrand is bounded by k/δ if l is bounded by k, so the expectation is always well defined in this case. If $l(x_t)$ represents a monetary payment per unit time then the integral is the cash value of this payment stream at time zero (i.e. the cash amount needed at time 0 to finance the payment stream out of interest at rate δ) and in many applications this is indeed the interpretation of $v(x)$; in other applications $v(x)$ is simply a convenient *performance index* which is finite while depending on the whole (infinite-horizon) trajectory of x_t. By introducing 'killing at rate δ'

one can avoid explicit mention of the discount rate in such expressions. The general formulation is as follows.

Let x_t be a PDP on state space E and define $E_\Delta = E \cup \{\Delta\}$, where Δ is an isolated point. Let U be a $U[0,1]$ random variable, independent of x_0 (for all measures P_x). Let $\delta: E \to \mathbb{R}_+$ be a measurable function satisfying the same condition as the jump rate λ, i.e.

(31.7) for each $x = (v, \zeta) \in E$ there exists $\varepsilon > 0$ such that the function $t \to \delta(v, \phi_v(t, \zeta))$ is integrable on $[0, \varepsilon[$.

Define a random time τ by

$$\tau = \inf\left\{t > 0 : \exp\left(-\int_0^t \delta(x_s)\mathrm{d}s\right) \leqslant U\right\}$$

and define the killed process $\tilde{x}_t$ as

$$\tilde{x}_t = \begin{cases} x_t, & t < \tau \\ \Delta, & t \geqslant \tau. \end{cases}$$

We claim that $\tilde{x}_t$ is a PDP on E_Δ whose local characteristics for $x \in E$ are $\tilde{g}(x) = g(x)$, $\tilde{\lambda}(x) = \lambda(x) + \delta(x)$ and

$$\left.\begin{aligned} \tilde{Q}(A; x) &= \frac{\lambda(x)}{\lambda(x) + \delta(x)} Q(A; x) \quad A \in \mathscr{E} \\ \tilde{Q}(\{\Delta\}; x) &= \frac{\delta(x)}{\lambda(x) + \delta(x)}. \end{aligned}\right\} \tag{31.8}$$

At the cemetery state we have $\tilde{g}(\Delta) = \tilde{\lambda}(\Delta) = 0$. Indeed, starting at $x = (v, \zeta) \in E$, the first jump time $\tilde{T}_1$ of the process is $T_1 \wedge \tau_1$, where T_1 is the first jump time of x_t and τ_1 is independent of T_1 with survivor function

$$P_x(\tau_1 > t) = \exp\left(-\int_0^t \delta(v, \phi_v(s, \zeta))\mathrm{d}s\right).$$

Thus

$$\begin{aligned} P_x(\tilde{T}_1 > t) &= P_x(T_1 \wedge \tau_1 > t) \\ &= P_x[(T_1 > t) \cap (\tau_1 > t)] \\ &= \exp\left(-\int_0^t \lambda(\phi_v(s, \zeta))\mathrm{d}s\right)\exp\left(-\int_0^t \delta(\phi_v(s, \zeta))\mathrm{d}s\right) I_{(t < t_*(x))}, \end{aligned}$$

which shows that the jump rate of the new process is $\lambda + \delta$. A simple argument using Bayes' rule shows that $P_x[T_1 \wedge \tau_1 = T_1 | T_1 \wedge \tau_1 = s] = \lambda(s)/(\lambda(s) + \delta(s))$, where $\lambda(s) := \lambda(v, \phi_v(s, \zeta))$, etc.) for $s < t_*$, from which (31.8) follows. It is also easy to see that the process restarts from x_{T_1} according to the same recipe if $T_1 < \tau_1$.

This construction is called 'killing at (state-dependent) rate $\delta(x)$'. The most important special case is killing at constant rate $\delta(x) \equiv \delta$; then the killing time τ is simply an exponentially distributed random variable with parameter δ, independent of (x_t).

The generator $\tilde{\mathfrak{A}}$ of the killed process is given by standard formula (26.15) in terms of the new local characteristics $(\tilde{g}, \tilde{\lambda}, \tilde{Q})$ and acting on functions $f: E_\Delta \to \mathbb{R}$, i.e.

$$\tilde{\mathfrak{A}} f(x) = \mathfrak{X} f(x) + (\lambda(x) + \delta(x)) \left\{ \int_E \frac{\lambda(x)}{\lambda(x) + \delta(x)} f(y) Q(\mathrm{d}y; x) + \frac{\delta(x)}{\lambda(x) + \delta(x)} f(\Delta) - f(x) \right\}, \quad x \in E$$

$$\tilde{\mathfrak{A}} f(\Delta) = 0.$$

By convention, functions $f: E \to \mathbb{R}_+$ are extended to E_Δ by setting $f(\Delta) = 0$; with this convention $\mathscr{D}(\tilde{\mathfrak{A}}) = \mathscr{D}(\mathfrak{A})$ and from the above

$$\begin{aligned} \tilde{\mathfrak{A}} f(x) &= \mathfrak{X} f(x) + \lambda(x) \int_E f(y) Q(\mathrm{d}y; x) - (\lambda(x) + \delta(x)) f(x), \quad x \in E \\ &= \mathfrak{A} f(x) - \delta(x) f(x). \end{aligned}$$

Thus killing at rate δ just adds the zeroth-order term $-\delta f$ to the generator. The main result relating killing to expectations of functionals is the following.

(31.9) *Theorem.* Let x_t be a standard PDP and $\tilde{x}_t$ be the killed process with killing rate $\delta: E \to \mathbb{R}_+$ satisfying (31.7), constructed as described above. Let $l: E \to \mathbb{R}_+$ and $c: \Gamma \to \mathbb{R}_+$ be measurable functions. Then for $x \in E$

$$\begin{aligned} \text{(31.10)} \qquad & \mathbb{E}_x \left\{ \int_0^\infty \exp\left(- \int_0^t \delta(x_s) \mathrm{d}s \right) l(x_t) \mathrm{d}t + \int_0^\infty \exp\left(- \int_0^t \delta(x_s) \mathrm{d}s \right) c(x_{t-}) \mathrm{d}p_t^* \right. \\ & \left. = \mathbb{E}_x \left\{ \int_0^\infty l(\tilde{x}_t) \mathrm{d}t + \int_0^\infty c(\tilde{x}_{t-}) \mathrm{d}\tilde{p}_t^* \right\}. \right. \end{aligned}$$

Here l, c are extended to $\boldsymbol{E}_\Delta$ by taking $l(\Delta) = c(\Delta) = 0$, and $\tilde{p}_t^* = p_{t\wedge\tau}^*$ is the boundary jump counting process for $\tilde{x}_t$. Both sides of (31.10) may be equal to $+\infty$.

Proof. It is clear from the construction that (x_t), τ are not independent (unless δ is constant) but that the conditional distribution of τ given $\mathfrak{F}_\infty = \sigma\{x_t, t \geqslant 0\}$ is $P_x[\tau > t | \mathfrak{F}_\infty] = \beta(t)$, where

$$\beta(t) = \exp\left(-\int_0^t \delta(x_s)\mathrm{d}s\right)$$

with conditional density function $\delta(x_t)\beta(t)$ on $[0, \infty[$ with $P_x[\tau = \infty | \mathfrak{F}_\infty] = \beta(\infty)$. Thus

$$\begin{aligned}
\mathbb{E}_x\left\{\int_0^\infty l(\tilde{x}_t)\mathrm{d}t\right\} &= \mathbb{E}_x\left\{\int_0^\tau l(x_t)\mathrm{d}t\right\} \\
&= \mathbb{E}_x\left\{\mathbb{E}_x\left[\int_0^\tau l(x_t)\,\mathrm{d}t \,\middle|\, \mathfrak{F}_\infty\right]\right\} \\
&= \mathbb{E}_x\left\{\beta(\infty)\int_0^\infty l(x_t)\mathrm{d}t + \int_0^\infty \delta(x_s)\beta(s)\int_0^s i(x_t)\mathrm{d}t\,\mathrm{d}s\right\} \\
&= \mathbb{E}_x\left\{\int_0^\infty \beta(t)l(x_t)\mathrm{d}t\right\}.
\end{aligned}$$

Similarly,

$$\begin{aligned}
\mathbb{E}_x\left\{\int_0^\infty c(\tilde{x}_{t-})\mathrm{d}\tilde{p}_t^*\right\} &= \mathbb{E}_x\left\{\sum_{i=1}^\infty c(x_{T_i-})I_{(x_{T_i-}\in\Gamma)}I_{(T_i<\tau)}\right\} \\
&= \mathbb{E}_x\left\{\sum_{i=1}^\infty c(x_{T_i-})I_{(x_{T_i-}\in\Gamma)}P_x[\tau > T_i | \mathfrak{F}_\infty]\right\} \\
&= \mathbb{E}_x\left\{\sum_{i=1}^\infty c(x_{T_i-})I_{(x_{T_i-}\in\Gamma)}\beta(T_i)\right\} \\
&= \mathbb{E}_x\left\{\int_0^\infty \beta(t)c(x_{t-})\mathrm{d}p_t^*\right\}.
\end{aligned}$$

This completes the proof. □

The commonest case of (31.10) is that in which $\delta(x) \equiv \delta$, a constant, when the formula becomes

(31.11)
$$\mathbb{E}_x\left\{\int_0^\infty \mathrm{e}^{-\delta t} l(x_t)\mathrm{d}t + \int_0^\infty \mathrm{e}^{-\delta t} c(x_{t-})\mathrm{d}p_t^*\right\}$$
$$= \mathbb{E}_x\left\{\int_0^\infty l(\tilde{x}_t)\mathrm{d}t + \int_0^\infty c(\tilde{x}_{t-})\mathrm{d}\tilde{p}_t^*\right\}.$$

The point of Theorem (31.9) is simply that such expressions written with the discount factor $\mathrm{e}^{-\delta t}$ are actually no more general than similar expressions without the discount factor, since the killed process – which is still a PDP – incorporates discounting automatically. The expectations in (31.10) (or (31.11)) may not be finite even if $\delta(x) \geqslant \delta_0 > 0$ and l, c are bounded, because p_t^* might increase at too fast a rate. There is however a simple sufficient condition, satisfied in the vast majority of applications, which prevents this happening.

(31.12) *Assumption.* $Q(A_\varepsilon; z) = 1$ for every $z \in \Gamma$, for some $\varepsilon > 0$, where $A_\varepsilon = \{x \in E : t_*(x) \geqslant \varepsilon\}$.

This assumption ensures that boundary jumps take the process uniformly into the interior of E as measured by boundary hitting time. (It is violated by Example (24.5).)

(31.13) *Proposition.* Suppose that c, l and the jump rate λ are bounded, that $\delta(x) \geqslant \delta_0 > 0$, and that Assumption (31.12) holds. Then the expectations in (31.10) are bounded functions of $x \in E$.

Proof. Let K denote a bound for c, l. Then clearly

$$\int_0^\infty \exp\left(-\int_0^t \delta(x_s)\mathrm{d}s\right) l(x_t)\mathrm{d}t \leqslant \frac{K}{\delta_0},$$

so the Lebesgue integral term is no problem. For the boundary jump term,

(31.14)
$$\int_0^\infty \exp\left(-\int_0^t \delta(x_s)\mathrm{d}s\right) c(x_{t-})\mathrm{d}p_t^*$$
$$= \sum_{i=1}^\infty \exp\left(-\int_0^{T_i} \delta(x_s)\mathrm{d}s\right) c(x_{T_i-}) I_{(x_{T_i-} \in \Gamma)}$$
$$\leqslant K \sum_{i=1}^\infty \mathrm{e}^{-\delta_0 T_i} I_{(x_{T_i-} \in \Gamma)}$$
$$= K \sum_{i=1}^\infty \mathrm{e}^{-\delta_0 T_i^*},$$

where T_i^* are the boundary jump times. Taking $T_0^* = 0$ and $S_i^* = T_i^* - T_{i-1}^*$ for $i \geqslant 1$ we have

(31.15) $$\mathbb{E}_x[e^{-\delta_0 T_i^*}] = \mathbb{E}_x(e^{-\delta_0 T_{i-1}^*}\mathbb{E}_x[e^{-\delta_0 S_i^*} | \mathfrak{F}_{T_{i-1}^*}]).$$

Now $S_i^* \geqslant S$, where S denotes the interarrival time following T_{i-1}^* and we also know that with probability one $x_{T_{i-1}^*} \in A_\varepsilon$, in view of Assumption (31.12). Thus denoting $X_{T_{i-1}^*} = (v, \zeta)$ we have by construction of the process

$$p := P_x[S > \varepsilon | \mathfrak{F}_{T_{i-1}^*}] = \exp\left(- \int_0^\varepsilon \lambda(\phi_v(s, \zeta)\mathrm{d}s \right) \geqslant e^{-K\varepsilon}$$

and hence

$$\begin{aligned}\mathbb{E}_x[e^{-\delta_0 S} | \mathfrak{F}_{T_{i-1}^*}] &= \mathbb{E}_x[e^{-\delta_0 S} I_{(S>\varepsilon)} + e^{-\delta_0 S} I_{(S \leqslant \varepsilon)} | \mathfrak{F}_{T_{i-1}^*}] \\ &\leqslant e^{-\delta_0 \varepsilon} p + 1(1-p) = 1 - p(1 - e^{-\delta_0 \varepsilon}) \\ &\leqslant 1 - e^{-K\varepsilon}(1 - e^{-\delta_0 \varepsilon}) =: \rho < 1.\end{aligned}$$

From this and (31.15) we obtain

$$\mathbb{E}_x[e^{-\delta_0 T_i^*}] \leqslant \rho \mathbb{E}_x[e^{-\delta_0 T_{i-1}^*}]$$

and hence

$$\mathbb{E}_x[e^{-\delta_0 T_i^*}] \leqslant \rho^i, \quad i = 1, 2 \ldots.$$

We now see from (31.14) that

$$\mathbb{E}_x\left\{ \int_0^\infty \exp\left(- \int_0^t \delta(x_s)\mathrm{d}s \right) c(x_{t-})\mathrm{d}p_t^* \right\} \leqslant K \sum_{i=1}^\infty \rho^i = \frac{K\rho}{1-\rho}.$$

Since this constant is independent of $x \in E$, the proof is complete. □

(31.16) *Functionals depending on the jumps of the process*

In some applications one encounters functionals of the form

(31.17) $$J(x) = \mathbb{E}_x \sum_{i=1}^\infty e^{-\delta T_i} b(x_{T_i}, x_{T_i -}).$$

These are 'cost functions' in which one pays an amount $b(y, x)$ if the process jumps from x to y. Equation (31.17) then represents the mean total discounted cost. If the state space E has just one component then b could, for example, take the form $b(y,x)=\tilde{b}(|x-y|)$, i.e. the cost paid depends on the magnitude of the jump. A slight extension of Proposition (31.13) shows that $J(x)$ is finite if b is bounded, $\delta>0$ and Assumption (31.12) holds. Note however that we can write (31.17) as

$$J(x)=\mathbb{E}_x\int_0^\infty\int_E \mathrm{e}^{-\delta t}b(y,x_{t-})p(\mathrm{d}t,\mathrm{d}y)$$

and the above conditions imply that $f\in L_1(p)$, where $f(t,y,\omega)=\mathrm{e}^{-\delta t}b(y,x_{t-}(\omega))$. From the integration theory developed in the Appendix we know that then $f\in L_1(\tilde{p})$ and that

$$J(x)=\mathbb{E}_x\int_0^\infty\int_E \mathrm{e}^{-\delta t}b(y,x_{t-})\tilde{p}(\mathrm{d}t,\mathrm{d}y),$$

where $\tilde{p}$ is the compensator of the process p, given by (26.2) as

$$\tilde{p}(t,A)=\int_0^t Q(A,x_s)\lambda(x_s)\mathrm{d}s+\int_0^t Q(A,x_{s-})\mathrm{d}p_s^*.$$

Hence

$$J(x)=\mathbb{E}_x\left\{\int_0^\infty \mathrm{e}^{-\delta t}\int_E b(y,x_t)Q(\mathrm{d}y,x_t)\lambda(x_t)\mathrm{d}t\right.$$
$$\left.+\int_0^\infty \mathrm{e}^{-\delta t}\int_E b(y,x_{t-})Q(\mathrm{d}y,x_{t-})\mathrm{d}p_t^*\right\}$$

(31.18) $$=\mathbb{E}_x\left\{\int_0^\infty \mathrm{e}^{-\delta t}l(x_t)\mathrm{d}t+\int_0^\infty \mathrm{e}^{-\delta t}c(x_{t-})\mathrm{d}p_t^*\right\},$$

where

$$l(x)=\lambda(x)\int_E b(y,x)Q(\mathrm{d}y,x),\quad x\in E,$$

$$c(x)=\int_E b(y,x)Q(\mathrm{d}y,x),\qquad x\in\Gamma.$$

Thus costs of the form (31.18) are sufficiently general to cover $J(x)$ defined by (31.17) as a special case. This is why we consider expressions similar to (31.18) in Theorem (31.9), omitting any explicit consideration of interior jumps.

32 Expectations

32.1 *Expectations and integro-differential equations*

In this section we explore the connection between the expectations of certain functionals of a PDP (x_t) and systems of integro-differential equations associated with its generator. Solving the latter then provides a systematic way of computing these expectations. Our primary result will be for the functional

$$V(x) = \mathbb{E}_x\left\{\int_0^\infty l(x_t)\mathrm{d}t + \int_0^\infty c(x_{t-})\mathrm{d}p_t^*\right\} \tag{32.1}$$

already discussed in the last section. As we saw there, several functionals taking apparently different forms are actually special cases of (32.1). Throughout, (x_t) will be PDP satisfying the standard assumptions (24.8). Our main result is the following.

(32.2) *Theorem.* Let $l:E\to\mathbb{R}_+$ and $c:\Gamma\to\mathbb{R}_+$ be measurable functions. Suppose that for each $x=(v,\zeta)\in E$ the function $t\to l(v,\phi_v(t,\zeta))$ is integrable on some interval $[0,\varepsilon]$, $\varepsilon>0$ and that the function V defined by (32.1) is bounded. Then $V\in\mathscr{A}\mathscr{C}$ and satisfies

$$\mathfrak{A}V(x) + l(x) = 0, \quad x\in E \tag{32.3}$$

with the boundary condition

$$\mathfrak{C}V(x) + c(x) = 0, \quad x\in\Gamma. \tag{32.4}$$

(32.5) *Remark.* Let us specify precisely what (32.3) means. Recall that $\mathfrak{A}V = \mathfrak{X}V + \lambda(QV - V)$. For $V\in\mathscr{A}\mathscr{C}$ the function $t\to V(v,\phi_v(t,\zeta))$ is differentiable almost everywhere and $\mathfrak{X}V(x')$ denotes the value of this derivative at points $x'=(v,\phi_v(t,\zeta))$, where it exists. Equation (32.3) says that this derivative is equal to $-l(x')-\lambda(x')(QV(x')-V(x'))$ at all such points.

(32.6) *Corollary.* If $\lambda, l\in\mathscr{C}$ and $Qf\in\mathscr{C}$ for all $f\in\mathscr{C}$, then $V\in\mathscr{C}^1$. If λ, l and Qf are locally bounded for $f\in\mathscr{C}$ (i.e. for each $x=(v,\zeta)\in E$ there exists $\varepsilon>0$ such that $t\to h(v,\phi_v(t,\zeta))$ is bounded on $[0,\varepsilon]$, $h=\lambda, l, Qf$), then $V\in\mathscr{L}$.

Proof of Theorem (32.2). Fix $x=(v,\zeta)\in E$. We will use the notation $f(t):=f(v,\phi_v(t,\zeta))$ for $f\in\mathscr{M}, t\geqslant 0$ (so $f(0)=f(x)$, but this should cause no confusion). Take $t_1 \in]0, s_*(x)[$ such that the functions $s \to \lambda(s), s \to l(s)$ are integrable on $[0, t_1]$. By the strong Markov property, the function V defined by (32.1) satisfies

$$V(x) = \mathbb{E}_x\left\{\int_0^{T_1 \wedge t} l(x_s)\mathrm{d}s + V(x_{T_1 \wedge t})\right\}$$
$$= \mathbb{E}_x\left\{\int_0^{T_1 \wedge t} l(s)\mathrm{d}s + V(t)I_{(t<T_1)} + QV(T_1)I_{(t \geqslant T_1)}\right\}$$

for any $t \leqslant t_1$, where T_1 is the first jump time of the process. The density function of T_1 on the interval $[0, t]$ is $\lambda(s)\mathrm{e}^{-\Lambda(s)}$, where $\Lambda(s) = \int_0^s \lambda(u)\mathrm{d}u$ and hence

$$\text{(32.7)} \qquad V(x) = \mathrm{e}^{-\Lambda(t)}\int_0^t l(s)\mathrm{d}s + \int_0^t \lambda(s)\mathrm{e}^{-\Lambda(s)}\int_0^s l(u)\mathrm{d}u\,\mathrm{d}s$$
$$+ \mathrm{e}^{-\Lambda(t)}V(t) + \int_0^t \lambda(s)\mathrm{e}^{-\Lambda(s)}QV(s)\mathrm{d}s.$$

After changing the order of integration in the second term, (32.7) becomes

$$V(0)(=: V(x)) = \int_0^t \mathrm{e}^{-\Lambda(s)}(l(s) + \lambda(s)QV(s))\mathrm{d}s + \mathrm{e}^{-\Lambda(t)}V(t),$$

which is equivalent to

$$\text{(32.8)} \qquad V(t) = \exp\left(\int_0^t \lambda(s)\mathrm{d}s\right)V(0) + \int_0^t \exp\left(\int_s^t \lambda(u)\mathrm{d}u\right)g(s)\mathrm{d}s,$$

where $g(s) = -(l(s) + \lambda(s)QV(s))$. If V is bounded, g is integrable on $[0, t_1]$ and an elementary calculation using (32.8) shows that V satisfies

$$\text{(32.9)} \qquad V(t) - V(0) = \int_0^t \lambda(s)V(s)\mathrm{d}s + \int_0^t g(s)\mathrm{d}s.$$

Thus V is absolutely continuous on $[0, t_1]$ and its derivative is $(\lambda(t)V(t) + g(t))$. In view of Remark (32.5), this is equivalent to (32.3). The statements of Corollary (32.6) also follow directly from the representation (32.9). Let us now show that V satisfies the boundary condition (32.4). Take $z = (\nu, \zeta') \in \Gamma$, fix $t > 0$ and let $x = (\nu, \zeta) =$

$(v, \phi_v(-t, \zeta'))$. By the strong Markov property

$$V(x) = \mathbb{E}_x\left\{\int_0^{T_1} l(s)\mathrm{d}s + c(z)I_{(T_1 = t_*(x))} + QV(x_{T_1-})\right\}.$$

Using the distribution function of T_1 as before it is easily shown that the expectations of the three terms on the right converge to $0, c(z)$ and $QV(z)$ respectively as $t \downarrow 0$. Thus $V(z) := \lim_{t \downarrow 0} V(v, \phi_v(-t, \zeta'))$ exists and (3.13) is satisfied since $\mathfrak{C}V(z) = QV(z) - V(z)$. This completes the proof. □

Written out in full (32.3) and (32.4) constitute a system of integro-differential equations (which may be quite complicated if the state space E has several components) satisfied by V. We would like to know that these equations have a unique solution, so that any solution f that we compute must coincide with V. The following is a general result in this direction.

(32.10) *Theorem.* Suppose $f \in \mathscr{A}\mathscr{C}$ satisfies

$$\mathfrak{A}f(x) + l(x) = 0, \quad x \in E \tag{32.11}$$

$$\mathfrak{C}f(z) + c(z) = 0, \quad z \in \Gamma. \tag{32.12}$$

Suppose further that $\mathbb{E}_x f(x_t) \to 0$ as $t \to \infty$ and that $\mathfrak{B}f \in L_1(p)$. Then

$$f(x) = \mathbb{E}_x\left\{\int_0^\infty l(x_s)\mathrm{d}s + \int_0^\infty c(x_{s-})\mathrm{d}p_s^*\right\}. \tag{32.13}$$

Proof. Since $f \in \mathscr{A}\mathscr{C}$ and $\mathfrak{B}f \in L_1(p)$ the differential formula is applicable, and for any $t > 0$,

$$f(x_t) - f(x_0) = \int_0^t \mathfrak{A}f(x_s)\mathrm{d}s + \int_0^t\int_E \mathfrak{B}f\,\mathrm{d}q + \int_0^t \mathfrak{C}f(x_{s-})\mathrm{d}p_s^*. \tag{32.14}$$

Now $\mathfrak{A}f(x_s) = -l(x_s)$ in view of (3.17), and since $dp_s^* = 0$ except when $x_{s-} \in \Gamma$ we have from (3.18)

$$\begin{aligned}\int_0^t \mathfrak{C}f(x_{s-})\mathrm{d}p_s^* &= \int_E^t (Qf(x_{s-}) - f(x_{s-}))\mathrm{d}p_s^* \\ &= -\int_0^t c(x_{s-})\mathrm{d}p_s^*.\end{aligned}$$

Since $\mathfrak{B}f \in L_1(p)$, the stochastic integral in (32.14) is a martingale, and taking expectations we get

$$f(x) = \mathbb{E}_x f(x_t) + \mathbb{E}_x\left\{\int_0^t l(x_s)\mathrm{d}s + \int_0^t c(x_{s-})\mathrm{d}p_s^*\right\}.$$

The term in braces on the right is monotone increasing in t since $c, l \geqslant 0$; now (32.13) follows by monotone convergence and the assumption that $\lim_{t\to\infty} \mathbb{E}_x f(x_t) = 0$. □

(32.15) *Corollary.* Suppose the PDP x_t satisfies the conditions of Theorem (32.3) and that V defined by (32.1) is bounded. Then if $\mathbb{E}_x V(x_t) \to 0$ as $t \to \infty$, V is the unique solution in $\mathscr{AC}$ of (32.11) and (32.12).

Let us now turn to some special cases, the first of which is the finite time horizon problem of evaluating the expectation

$$\text{(32.16)} \quad V_1(t,x) = \mathbb{E}_{(t,x)}\left\{\int_t^{t_f} l(s,x_s)\,\mathrm{d}s + \int_t^{t_f} c(s,x_{s-})\mathrm{d}p_s^* + \Phi(x_{t_f})\right\},$$

where $\mathbb{E}_{(t,x)}$ is the expectation over the time interval $[t, t_f]$ with starting point $x_t = x$. Note that using the augmented process $\tilde{x} = (x^0, x)$ described in section (31.5) above this is given by $V_1(t. x) = \tilde{V}(t,x)$, where

$$\tilde{V}(\tilde{x}) = \mathbb{E}_{\tilde{x}}\left\{\int_0^\infty \tilde{l}(\tilde{x}_s)\mathrm{d}s + \int_0^\infty \tilde{c}(\tilde{x}_{s-})\mathrm{d}\tilde{p}_s^*\right\},$$

$\tilde{l}(\tilde{x}) = l(x^0, x), \tilde{c}(x^0, z) = c(x^0, z)$ for $x^0 < t_f, z \in \Gamma, \tilde{c}(t_f, z) = \Phi(z)$ and $\tilde{c}(\Delta) = \tilde{l}(\Delta) = 0$, where Δ is the cemetery state. Equations (32.11) and (32.12) become

$$\text{(32.17)} \qquad \frac{\partial}{\partial t} f(t,x) + \mathfrak{A} f(t,x) + l(t,x) = 0, \quad x \in E, \quad t \in [0, t_f[,$$

$$\text{(32.18)} \qquad \mathfrak{C} f(t,z) + c(t,z) = 0, \qquad z \in \Gamma, \quad t \in [0, t_f[,$$

$$\text{(32.19)} \qquad f(t_f, x) = \Phi(x), \qquad x \in E,$$

where, as is conventional, we have written $x^0 = t$. In this case the condition $\mathbb{E}_x \tilde{V}(\tilde{x}_t) \to 0$ as $t \to \infty$ is automatically satisfied since $\tilde{x}_t = \Delta$ for all $t \geqslant \Delta$, and V_1 is bounded if l, c, Φ are bounded and

$$\text{(32.20)} \qquad \sup_{x \in E} \mathbb{E}_x N_{t_f} < \infty.$$

(This condition is not required if, as in many applications, $c \equiv 0$.) The following result is then an immediate consequence of Corollary (32.15).

(32.21) *Proposition.* Suppose the local characteristics of the PDP (x_t) satisfy the conditions of Theorem (32.2), that l, c, Φ are bounded functions as described above and that (32.20) is satisfied or $c \equiv 0$. Then the function V_1 defined by (32.16) is in $\mathscr{A}\mathscr{C}$ and is the unique solution in $\mathscr{A}\mathscr{C}$ of equations (32.17)–(32.19).

A similar argument applies to the discounted case. For notational simplicity we will discuss this for a constant discount rate $\delta > 0$, but the arguments for a state-dependent discount rate $\delta(x)$ with $\delta(x) \geqslant \delta_0 > 0$ are exactly the same. We consider the functional

$$V_2(x) = \mathbb{E}_x\left\{\int_0^\infty e^{-\delta t} l(x_t)dt + \int_0^\infty e^{-\delta t} c(x_{t-})dp_t^*\right\} \tag{32.22}$$

with l, c satisfying $0 \leqslant l, c \leqslant K$. From Theorem (31.9) we know that this is equal to

$$\tilde{V}(x) = \mathbb{E}_x\left\{\int_0^\infty l(\tilde{x}_t)dt + \int_0^\infty c(\tilde{x}_{t-})dp_t^*\right\}$$

for the killed process $\tilde{x}_t$, and that $\tilde{V}$ is bounded if $c \equiv 0$ or if Assumption (31.12) holds. Now for $t > 0$

$$\mathbb{E}_x \tilde{V}(\tilde{x}_t) = e^{-\delta t}\mathbb{E}_x V_2(x_t)$$

so that $\mathbb{E}_x \tilde{V}(\tilde{x}_t) \to 0$ as $t \to \infty$ under either of the conditions just mentioned. The generator of the killed process is $\tilde{\mathfrak{A}} f = \mathfrak{A} f - \delta f$ so (32.11) and (32.12) become

$$\mathfrak{A} f(x) - \delta f(x) + l(x) = 0, \quad x \in E, \tag{32.23}$$

$$\mathfrak{C} f(z) + c(z) = 0, \quad z \in \Gamma. \tag{32.24}$$

From Corollary (32.15) and the above discussion we obtain Proposition (32.25)

(32.25) *Proposition.* Suppose that the local characteristics of the PDP (x_t) satisfy the conditions of Theorem (32.3), that l, c are bounded and that either $c \equiv 0$ or Assumption (31.12) holds. Then the function $V_2(x)$ given by (32.22) is in $\mathscr{A}\mathscr{C}$, satisfies (32.23) and (32.24) and is the unique solution in $\mathscr{A}\mathscr{C}$ of these equations.

32.2 *Recursive computations*

We have now characterized the expectations of various functionals of a PDP in terms of systems of equations (32.3) and (32.4) involving the generator $\mathfrak{A}$ of the process. Some examples of solution of these equations are given in the next section. Because $\mathfrak{A}$ includes an integral term (i.e. is not a local operator) the equations are not differential or partial differential equations but rather systems of integro-differential equations. The way they are solved depends very much on the boundary conditions arising in specific problems, and because these can take many different forms, depending on the jump behaviour of the process between different components of the state space, direct methods of solution are inevitably highly problem-specific. In this section, however, we examine a recursive scheme which provides the basis for a widely applicable numerical technique for computing discounted or finite-horizon costs.

Let $c:\Gamma \to \mathbb{R}$ and $l, \psi: E \to \mathbb{R}$ be bounded measurable functions and consider the functional

(32.26)

$$v(x) = \mathbb{E}_x\left\{ \int_0^{T_1} e^{-\delta s} l(x_s) ds + e^{-\delta T_1} c(x_{T_1 -}) I_{(T_1 = t_*(x))} + e^{-\delta T_1} \psi(x_{T_1}) \right\},$$

where T_1 is the first jump time of the PDP (x_t) and $\delta > 0$. Clearly, v is bounded. By the strong Markov property we have, for any $t \in [0, t_*(x)[$,

$$v(x) = \mathbb{E}_x\left\{ \int_0^{T_1 \wedge t} e^{-\delta s} l(x_s) ds + e^{-\delta(T_1 \wedge t)} v(x_{T_1 \wedge t}) \right\}.$$

By the same calculations as in the proof of Theorem (32.2) we find that this is equivalent to

$$(32.27) \quad v_x(0) = e^{-\Lambda(t,x)} v_x(t) + \int_0^t e^{-\Lambda(s,x)}(l_x(s) + \lambda_x(s) Q\psi_x(s)) ds,$$

where we now use the notation $l_x(s) := (v, \psi_v(s, \zeta))$ for $x = (v, \zeta)$ and similarly for other functions. Now write $z = (v, \psi_v(t, \zeta))$ and $r = -t$ so that $x = (v, \psi_v(r, z))$. In terms of these variables (32.27) becomes

$$(32.28) \quad v_z(r) = \exp\left(\int_0^r (\delta + \lambda_z(u)) du \right) v_z(0) - \int_0^r \exp\left(\int_0^u (\delta + \lambda_z(s)) ds \right) (l_z(u) + Q\psi_z(u)) du.$$

The difference between this and the situation studied in Theorem (32.2) is that here the non-local operator Q acts only on the exogenously-given function ψ; as in Theorem (32.2), expression (32.28) is just the standard variation-of-constants formula for the unique solution of the scalar ODE

$$\text{(32.29)} \qquad \frac{\mathrm{d}}{\mathrm{d}r} v_z(r) = (\delta + \lambda_z(r)) v_z(r) - (l_z(r) + \lambda_z(r) Q\psi_z(r)).$$

This gives us the following results.

(32.30) *Proposition.* Suppose l, c, ψ are bounded and measurable, and define v by (32.26). Then $v \in \mathscr{AC}$ and is the unique bounded solution of the equations

$$\text{(32.31)} \qquad \mathfrak{X} f(x) + \lambda(x)(Q\psi(x) - f(x)) - \delta f(x) + l(x) = 0, \quad x \in E,$$

$$\text{(32.32)} \qquad f(z) = c(z) + Q\psi(z), \qquad z \in \Gamma.$$

Proof. Equation (32.31) is just (32.29) written in different notation and with v replaced by f, so we have already shown that v satisfies (32.31) and a direct calculation shows that v satisfies the boundary condition (32.32). Conversely, if f satisfies (32.31) and (32.32) then certainly $f(x) = v(x)$ at points $x \in E$ such that $t_*(x) < \infty$, since (32.29) has a unique solution, and f and v agree on Γ. Now take a point $x \in E$ such that $t_*(x) = \infty$, and for convenience of notation write $\alpha(t) = (\delta + \lambda_x(t))$, $\beta(t) = (l_x(t) + \lambda_x(t) Q\psi_x(t))$. Then all solutions of (32.31) along the integral curve $\{v, \psi_v(t, x)) : t \in \mathbb{R}_+\}$ are given by the variation-of-constants formula for (32.29) which in our notation becomes

$$f_x(t) = \exp\left(\int_0^t \alpha(s)\mathrm{d}s \right) \left(\gamma - \int_0^t \exp\left(- \int_0^s \alpha(u)\mathrm{d}u \right) \beta(s)\mathrm{d}s \right).$$

The family of solutions is parametrized by the 'initial condition' $\gamma = f_x(0) = f(x)$.

Since β is bounded and $\alpha(s) \geqslant \delta > 0$ it is easy to see that the only choice of γ for which $f_x(t)$ is bounded is

$$\gamma_0 = \int_0^{\infty} \exp\left(- \int_0^s \alpha(u)\mathrm{d}u \right) \beta(s)\mathrm{d}s,$$

the corresponding solution $f_x(t)$ being given by

$$f_x(t) = \int_0^\infty \exp\left(-\int_t^s \alpha(u)\mathrm{d}u\right)\beta(s)\mathrm{d}s.$$

Since v is a bounded solution of (32.31) it follows that $f(x) = \gamma_0 = v(x)$. This completes the proof.

Let us write $v = G\psi$, where v is given by (32.26). Then G is an operator mapping $\mathscr{M}_b$ into $\mathscr{AC}$. Using the distribution function of T_1 it can be expressed explicitly as

$$G\psi(x) = \int_0^{t_*(x)} \mathrm{e}^{-(\delta+\Lambda(s))}(l(s) + \lambda(s)Q\psi(s))\mathrm{d}s +$$
$$\mathrm{e}^{-(\delta+\Lambda(t_*))}(c(t_*(x)) + Q\psi(t_*(x))).$$

As we have seen above, G maps $\mathscr{C}$ into $\mathscr{C}^1$ if $l, \lambda \in \mathscr{C}$ and Q maps $\mathscr{C}$ into $\mathscr{C}$. Iterates $G^2, G^3, \ldots,$ of G are defined by $G^k\psi = G(G^{k-1}\psi)$, $k = 2, 3, \ldots$. By the strong Markov property we can show that

$$G^2\psi(x) = \mathbb{E}_x\left\{\int_0^{T_2} \mathrm{e}^{\delta t} l(x_t)\mathrm{d}t + \int_0^{T_2} c(x_{t-})\mathrm{d}p_t^* + \mathrm{e}^{-\delta T_2}\psi(x_{T_2})\right\}$$

and an induction argument then shows that for any $k \geqslant 1$

(32.33)

$$G^k\psi(x) = \mathbb{E}_x\left\{\int_0^{T_k} \mathrm{e}^{-\delta t} l(x_t)\mathrm{d}t + \int_0^{T_k} c(x_{t-})\mathrm{d}p_t^* + \mathrm{e}^{-\delta T_k}\psi(x_{T_k})\right\}.$$

(32.34) *Proposition.* For each $x \in E$ and $\psi \in \mathscr{M}_b$,

$$\lim_{k\to\infty} G^k\psi(x) = \mathbb{E}_x\left\{\int_0^\infty \mathrm{e}^{-\delta t} l(x_t)\mathrm{d}s + \int_0^\infty \mathrm{e}^{-\delta t} c(x_{t-})\mathrm{d}p_t^*\right\}.$$

Proof. By assumption $T_k \to \infty$ a.s. P_x for any $x \in E$ as $k \to \infty$. Thus if ψ is any bounded function, $\mathrm{e}^{-\delta T_k}\psi(x_{T_k}) \to 0$ a.s. and hence $\mathbb{E}_x[\mathrm{e}^{-\delta T_k}\psi(x_{T_k})] \to 0$ by bounded convergence. Since $c, l \geqslant 0$ the two integrals on the right of (32.33) are increasing functions of k. The result follows by monotone convergence. □

Combining this result with Proposition (32.25), we see that it provides, in principle, a way of computing the unique solution V_2 of the system of integro-differential equations (32.23) and (32.24). Computing $G\psi$ does not involve solving integro-differential equations, but only one-dimensional ODEs solved separately along each integral curve of the vector field. Whether or not this is a viable computational method depends on how large k has to be before $G^k\psi(x)$ is close to $V_2(x)$. This in turn depends on how large the discount rate δ is (compared to λ and the rate of accumulation of boundary jumps.) Think of $V_2(x)$ as a cost expression for a killed process with killing time τ satisfying $P_x(\tau > t) = e^{-\delta t}$. If j is such that, for example, $P_x(T_j < \tau) < 0.01$ then clearly the evolution of the process beyond T_j has little effect on the average cost, and $G^j\psi$ will be close to V_2 for any bounded ψ. In applications it is usually possible in this way to make rough *a priori* estimates of the required value of j.

Similar arguments apply to finite horizon calculations, i.e. to the solution of (32.17)–(32.19) for the cost functional $V_1(t, x)$ of (32.16). Here we fix $t \in [0, t_f]$ and define

$$G\psi(t,x) = \mathbb{E}_{(t,x)}\left\{ I_{(t_f < T_1)}\left(\int_t^{t_f} l(s, x_s)\mathrm{d}s + \Phi(x_{t_f}) \right) \right.$$

$$\left. + I_{(t_f \geq T_1)}\left(\int_t^{T_1} l(s, x_s)\mathrm{d}s + c(x_{T_1 -})I_{(T_1 = t + t_*(x))} + \psi(T_1, x_{T_1}) \right) \right\}.$$

This is then the unique solution of the equations

$$\frac{\partial}{\partial t} f(t,x) + \mathfrak{X} f(t,x) + \lambda(t,x)Q\psi(x) - \lambda(x)f(t,x) + l(t,x) = 0,$$

$$(t,x) \in [0, t_f] \times E$$

$$f(t,z) = c(z) + Q\psi(t,z), \quad z \in \Gamma, \quad t \in [0, t_f[$$

$$f(t_f, z) = \Phi(x), \qquad x \in E,$$

which, again, is a system of ODEs solved separately along each integral curve. As in Proposition (32.34) we can show that $G^k\psi(t,x) \to V_1(t,x)$ as $k \to \infty$. We already saw a special case of this in Lemma (27.3).

As a final remark, note that the computational method suggested here is tailor-made for parallel implementation: computing $G\psi$ involves solving hundreds, if not thousands, of completely independent ODEs along a suitably chosen collection of integral curves of the vector field $\mathfrak{X}$; the results of these calculations must then be shared before the next iteration $G(G\psi)$ is computed. The application to permanent health insurance discussed in section (33.18) shows exactly how these calculations work. Something very close to linear speed-up should be achievable if it is possible to do them on parallel machines.

33 Applications

This section contains applications of the preceding results to three of the examples of section 21: renewal theory (21.7), the insurance risk model (21.11) and the permanent health insurance model (21.12). The first of these is in a way a bad advertisement for PDP theory in that the results are readily obtained by more direct means. However, the example demonstrates the equivalence between a special-purpose methods of applied probability and general-purpose methods of PDP theory and has illustrative value in showing how some well-known results emerge from our general approach. For the two insurance examples, on the other hand, PDP theory seems to provide by far the best – if not the only – approach. This is even more true of problems involving optimization, but these will be discussed in later chapters.

(33.1) *Renewal theory* (continued from (21.7))

Recall that a renewal process is a point process $N_t = \sum_i I_{(t \geqslant T_i)}$, where $T_1 = S_1$, $T_i = S_1 + S_2 + \cdots + S_i$ for $i \geqslant 2$ and the S_i are independent non-negative random variables with common density function ψ. If $\mu = \mathbb{E}S_1 < \infty$ then the strong law of large numbers, Theorem (12.7), implies that $T_n/n \to \mu$ a.s. as $n \to \infty$. The same conclusion holds if $\mu = \infty$; in this case $T_n \geqslant S_1^k + \cdots + S_n^k$, where $S_i^k = S_i \wedge k$, so that $\mu_k := \mathbb{E}S_1^k < \infty$ and

$$\liminf_{n \to \infty} \frac{1}{n} T_n \geqslant \mu_k.$$

As $k \to \infty$, $\mu_k \to \infty$ and we conclude that $\lim_{n \to \infty} T_n/n = \infty$. This argument shows in particular that, whether or not μ is finite, $T_n \to \infty$ with probability one.

The renewal process is a PDP if we take $x_t = (\nu_t, \zeta_t)$, where $\nu_t = N_t$ is the number of renewals and $\zeta_t = \tau_t$ is the time since the last renewal.

The generator is

$$\mathfrak{A}f_\nu(\zeta)=\frac{\mathrm{d}}{\mathrm{d}\zeta}f_\nu(\zeta)+\lambda(\zeta)(f_{\nu+1}(0)-f_\nu(\zeta)),$$

where we write $f_\nu(\zeta)=f(\nu,\zeta)$ and the jump rate is $\lambda(\zeta)=\psi(\zeta)/\Psi(\zeta)$ with $\Psi(\zeta)=\int_\zeta^\infty\psi(t)\mathrm{d}t$. Fix $t_1>0$ and suppose that a family of functions $v_\nu(t,\zeta)$ satisfies

$$\frac{\partial}{\partial t}v_\nu(t,\zeta)+\mathfrak{A}v_\nu(t,\zeta)=0\quad(t,\zeta)\in[0,t_1]\times\mathbb{R}_+ \tag{33.2}$$

$$v_\nu(t_1,\zeta)=\nu. \tag{33.3}$$

If the terminal function $\Phi(x)=\nu$ in (33.3) were bounded, we could conclude immediately from Proposition (32.21) that

$$v_\nu(t,\zeta)=\mathbb{E}_{(t,(\nu,\zeta))}\nu_{t_1}, \tag{33.4}$$

where the subscript indicates that the process starts in state (ν,ζ) at time t. We will show below that (33.4) is true even though Φ is not bounded. Thus the original problem of computing $m(t)=\mathbb{E}N_t$ is now embedded in the more general problem of computing this expectation when, at times zero, ν renewals have already occurred and (if one thinks in terms of replacing light bulbs) the light bulb operating at time zero has age ζ at that time. It is clear that $v_{\nu+1}(t,\zeta)=1+v_\nu(t,\zeta)$ so that $v_\nu(t,\zeta)=\nu+h(t,\zeta)$, where $h(t,\zeta)=v_0(t,\zeta)$. Equation (33.2) becomes

$$\frac{\partial h}{\partial t}(t,\zeta)+\frac{\partial h}{\partial\zeta}(t,\zeta)+\lambda(\zeta)[1+h(t,0)-h(t,\zeta)]=0 \tag{33.5}$$

with the boundary condition $h(t_1,\zeta)=0$. Define $z(t)=h(t,t)$; then $\mathrm{d}z(t)/\mathrm{d}t=(\partial h/\partial t+\partial h/\partial\zeta)(t,t)$ and hence z satifies

$$\frac{\mathrm{d}}{\mathrm{d}t}z(t)=\lambda(t)z(t)-\lambda(t)(1+h(t,0)),\quad z(t_1)=0.$$

This is a linear ODE, whose solution is

$$z(t)=\int_t^{t_1}\exp\left(-\int_t^s\lambda(u)\mathrm{d}u\right)\lambda(s)(1+h(s,0))\mathrm{d}s.$$

Now

$$\exp\left(-\int_t^s\lambda(u)\mathrm{d}u\right)=\exp\left(\int_t^s\frac{\dot\Psi(u)}{\Psi(u)}\mathrm{d}u\right)=\exp\left(\log\left(\frac{\Psi(s)}{\Psi(t)}\right)\right)=\frac{\Psi(s)}{\Psi(t)},$$

so that, using the facts that $\Psi(0) = 1$ and $\lambda(s) = \psi(s)/\Psi(s)$ we see that

$$z(0) = \int_0^{t_1} \psi(s)(1 + h(s,0))\mathrm{d}s.$$

This is equivalent to the renewal equation (21.8) since $z(0) = m(t)$ and $h(s,0) = m(t_1 - s)$.

The complete solution of (33.5) can now be obtained in the following way: fix $s \in \mathbb{R}_+$ and define $z_s(t) = h(t_1 - s + t, t)$. Then from (33.5) we obtain

$$(33.6) \qquad \frac{\mathrm{d}}{\mathrm{d}t} z_s(t) = \lambda(t) z_s(t) - \lambda(t)[1 + h(t_1 - s + t, 0)]$$
$$= \lambda(t) z_s(t) - \lambda(t)[1 + m(s - t)]$$

and

$$x_s(s) = 0.$$

Thus, once $\mu(\cdot)$ has been computed, the value of $h(t,\zeta)$ can be computed for any (t,ζ) by solving (33.6) for the appropriate value of s.

It remains to verify that (33.4) holds with $v_v(t,\zeta) := v + h(t,\zeta)$. Denoting $v(t,x) = v_v(t,\zeta)$ with $x = (v,\zeta)$, the argument of Proposition (32.21) shows that

$$v(0,x) = v(t,x_t) - \int_0^t \int_E \mathfrak{B} v \,\mathrm{d}q.$$

Now certainly $\mathfrak{B} v I_{(s \leqslant t_1 \wedge T_n)} \in L_1(p)$, so that for any n,

$$(33.7) \qquad v(0,x) = \mathbb{E} v(t_1 \wedge T_n, x_{t_1 \wedge T_n})$$
$$= \mathbb{E} v_{t_1 \wedge T_n} + \mathbb{E} h(t_1 \wedge T_n, \zeta_{t_1 \wedge T_n}),$$

since $T_n \to \infty$ a.s., $t_1 \wedge T_n = t_1$ for sufficiently large n, a.s. Thus $v_{t_1 \wedge T_n} \to v_{t_1}$ and $h(t_1 \wedge T_n, \zeta_{t_1 \wedge T_n}) \to h(t_1, \zeta_{t_1}) = 0$ a.s. Applying the monotone and bounded convergence theorems respectively to the two terms in (33.7) we conclude that $v(0,x) = \mathbb{E} v_{t_1}$, as claimed.

(33.8) *Insurance models* (continued from (21.11))

As described in section (21.11), the claims on an insurance company are amounts of money Y_i which are positive i.i.d. random variables with density function g, arriving at Poisson (rate λ) times T_i. The premium income is c per unit time. Denoting $\kappa = \mathbb{E} Y_1$ we assume that the *net profit condition* holds, namely

$$\frac{\lambda\kappa}{c} < 1. \tag{33.9}$$

In a time interval t, the premium income is ct while the expected monetary value of claims is $\kappa\lambda t$, so (33.9) is clearly a necessary condition for a viable insurance business. Let

$$z_t = ct - \sum_i Y_i I_{(t \geq T_i)};$$

then

$$\frac{1}{n} z_{T_n} = \frac{c}{n} T_n - \frac{1}{n}(Y_1 + \cdots + Y_n)$$

so that by the strong law of large numbers, Theorem (12.7),

$$\lim_{n\to\infty} \frac{1}{n} z_{T_n} = \frac{c}{\lambda} - \kappa = \frac{c}{\lambda}\left(1 - \frac{\lambda\kappa}{c}\right) > 0.$$

Since $T_n \to \infty$ a.s. this shows that under condition (33.9) $z_n \to \infty$ a.s. as $t \to \infty$.

Let us first consider the 'classical' case in which the company is ruined at the first time τ_0 at which its reserves are negative. If the initial reserves are u then $\tau_0 = \inf\{t : u + z_t < 0\}$ and we define

$$x_t = \begin{cases} u + z_t, & t < \tau_0 \\ \Delta, & t \geq \tau_0, \end{cases}$$

where Δ is a cemetary state. This is a PDP on the state space $\mathbb{R}_+ \cup \{\Delta\}$ with generator

$$\mathfrak{A} f(x) = c'(x) + \lambda\left(\int_0^x f(x-y)g(y)\mathrm{d}y - f(x)\right), \quad x \in \mathbb{R}_+,$$

where $f'(x) = \mathrm{d}f(x)/\mathrm{d}x$ and we recall that by convention $f(\Delta) = 0$. Consider the equation $\mathfrak{A} f(x) = 0$. Since the integral is a convolution integral this equation is readily solved by the Laplace transform method and the solution is unique up to a multiplicative constant. Taking $f(0) = 1$ the solution has Laplace transform

$$\hat{f}(s) = \int_0^\infty \mathrm{e}^{-sx} f(x)\mathrm{d}x = \frac{1}{s + \dfrac{\lambda}{c}(\hat{g}(s) - 1)}.$$

It follows from standard theorems on Laplace transforms that

$$\kappa = \mathbb{E} Y_1 = -\frac{\mathrm{d}}{\mathrm{d}s}\hat{g}(s)|_{s=0} = \lim_{s\downarrow 0}\frac{1}{s}(\hat{g}(s) - 1)$$

and hence that

$$\lim_{x\to\infty} f(x) = \lim_{s\downarrow 0} s\hat{f}(s) = \frac{1}{1-\dfrac{\lambda\kappa}{c}} =: l.$$

Thus in particular f is bounded. The function f is related to the probability of eventual ruin as follows.

(33.10) *Proposition*

$$P_u[\tau_0 < \infty] = 1 - \left(1 - \frac{\lambda\kappa}{c}\right) f(u).$$

In particular, $P_0[\tau_0 < \infty] = \lambda\kappa/c$.

Proof. There are no boundary conditions associated with this PDP and therefore since $\mathfrak{A}f = 0$ the process $f(x_t)$ is a bounded martingale. Thus for any $t \geqslant 0$,

(33.11)
$$\begin{aligned} f(u) &= \mathbb{E}_u f(x_0) = \mathbb{E}_u f(x_t) \\ &= \mathbb{E}_u[f(x_t)|\tau_0 > t]P_u[\tau_0 > t] \\ &\quad + \mathbb{E}_u[f(x_{\tau_0})|\tau_0 \leqslant t]P_u[\tau_0 \leqslant t] \\ &= \mathbb{E}_u[f(x_t)|\tau_0 > t]P_u[\tau_0 > t], \end{aligned}$$

since $x_{\tau_0} = \Delta$ and $f(\Delta) = 0$. Since $x_t = u + z_t$ on $(\tau_0 > t)$ and $\lim_{t\to\infty} z_t = \infty$ a.s. it follows easily that

$$\lim_{t\to\infty} \mathbb{E}_u[f(x_t)|\tau_0 > t] = f(\infty) = l.$$

The result now follows from (33.11), since

$$\lim_{t\to\infty} P_u[\tau_0 > t] = P_u[\tau_0 = \infty].$$

In the exponential claims case, $g(x) = \mathrm{e}^{-\alpha/\kappa}/\kappa$. The Laplace transform $\hat{f}(s)$ is easily inverted and we obtain

$$P_u[\tau_0 < \infty] = \frac{\kappa\lambda}{c}\mathrm{e}^{-u/\kappa l}.$$

We would now like to study the situation described in section (21.11), where the company can borrow money at interest rate β to cover its debts; for simplicity we will do this only for exponential claims. The *absolute ruin time* is $\tau_a = \inf\{t : x_t = -c/\beta\}$, where x_t is the process described in section (21.11). As a prelude, we need to generalize the above results to cover reserve-dependent premium income, $c = c(x) > 0$, where c is bounded, continuously differentiable and satisfies

(33.12) $$C(x) = \int_0^x \frac{1}{c(y)} \mathrm{d}y < \infty \quad \text{for all } x > 0.$$

The equation $\mathfrak{A}f = 0$ now becomes

(33.13) $$c(x)f'(x) + \lambda\left(\int_0^x f(x-y)\frac{1}{\kappa}\mathrm{e}^{-y/\kappa}\mathrm{d}y - f(x)\right) = 0.$$

Differentiating this with respect to x gives

$$c(x)f''(x) + (c' + \alpha c(x) - \lambda)f'(x) = 0.$$

This is easily solved for $f'(x)$ given a value for $f'(0)$; choosing $f'(0) = 1/c(0)$ the solution is

(33.14) $$f(x) = \int_0^x \frac{1}{c(t)} \exp\left(-\frac{1}{\kappa}t + \int_0^t \lambda c^{-1}(t)\mathrm{d}y\right)\mathrm{d}t + \frac{1}{\lambda},$$

where the constant $1/\lambda$ is fixed by (33.13).

(33.15) *Proposition.* Suppose condition (33.12) is satisfied. If $P_u[\tau_0 = \infty] > 0$ then

(33.16) $$P_u[\tau_0 < \infty] = \frac{\displaystyle\int_u^\infty \frac{1}{c(x)} \exp\left(-\frac{1}{\kappa}x + \lambda C(x)\right)\mathrm{d}x}{\displaystyle\int_0^\infty \frac{1}{c(x)} \exp\left(-\frac{1}{\kappa}x + \lambda C(x)\right)\mathrm{d}x + \frac{1}{\lambda}}.$$

Proof. It follows by the same argument as in the proof of Proposition (33.10) that

$$P_u[\tau_0 < \infty] = 1 - \frac{f(u)}{\lim_{x\to\infty} f(x)},$$

where f solves (33.13). The result now follows from (33.14).

The obvious way to apply this result to the absolute ruin problem is to change coordinates to $\zeta = x + c/\beta$, so that the effective premium rate is $c(\zeta) = c$ for $\zeta \geqslant c/\beta$ and $c(\zeta) = \beta\zeta$ for $\zeta \in [0, c/\beta]$. However, this fails to satisfy condition (33.12), and we need a limiting argument. If we define $\tau_\varepsilon = \inf\{t : x_t \leqslant -c/\beta + \varepsilon\}$ for $\varepsilon > 0$ then a simple argument shows that $P_u[\tau_\varepsilon < \infty] \to P_u[\tau_a < \infty]$ as $\varepsilon \downarrow 0$. To calculate $P_u[\tau_\varepsilon < \infty]$ we can change coordinates to $\zeta = x + c/\beta - \varepsilon$ and apply Proposition (33.15): the effective premium rate in the new coordinates is $c(\zeta) = c$ for $\zeta \geqslant c/\beta - \varepsilon$ and $c(\zeta) = \beta(\zeta + \varepsilon)$ for $\zeta \in [0, c/\beta - \varepsilon]$, so that c^{-1} is bounded, and 'simple ruin' for the model in the new coordinates is 'ε-absolute ruin' for the original model. The function $C(\zeta)$ of (33.12) is given by

$$C(\zeta) = \begin{cases} \dfrac{1}{\beta}\log\left(1 + \dfrac{\zeta}{\varepsilon}\right), & \zeta \leqslant \dfrac{c}{\beta} - \varepsilon \\ \dfrac{1}{\beta}\log\left(\dfrac{c}{\varepsilon\beta}\right) + \dfrac{1}{c}\left(\zeta - \dfrac{c}{\beta} + \varepsilon\right), & \zeta \geqslant \dfrac{c}{\beta} - \varepsilon. \end{cases}$$

The numerator and denominator of (33.16) both converge to ∞ as $\varepsilon \downarrow 0$, but the ratio is bounded. Evaluating (33.16) and taking the limit one obtains (after some rather detailed computations) the following result.

(33.17) *Proposition.* The absolute ruin probability is given by

$$P_u[\tau_a < \infty] = \begin{cases} \dfrac{1}{A} \mathrm{e}^{-(1 - \lambda\kappa/c)u/\kappa}, u > 0 \\ \dfrac{1}{A}\left\{1 + \dfrac{1 - \lambda\kappa/c}{\kappa} \displaystyle\int_u^0 \mathrm{e}^{-x/\kappa}\left(1 + \dfrac{\beta x}{c}\right)^{\lambda/\beta - 1} \mathrm{d}x\right\}, -\dfrac{c}{\beta} < u \leqslant 0, \end{cases}$$

where

$$A = 1 + \frac{1 - \lambda\kappa/c}{\kappa} \int_{-c/\beta}^0 \mathrm{e}^{-x/\kappa}\left(1 + \frac{\beta x}{c}\right)^{\lambda/\beta - 1} \mathrm{d}x.$$

Of course, the integrals in these expressions have to be evaluated. This can be done by numerical integration or, if λ/β is large, by normal approximation (Dassios and Embrechts, 1989, section 32)

(33.18) *Permanent health insurance*

This continues the discussion of section (21.12). The state space E for the PHI process $x_t = (v_t, \zeta_t)$ is shown in Fig. 2.7 which for convenience

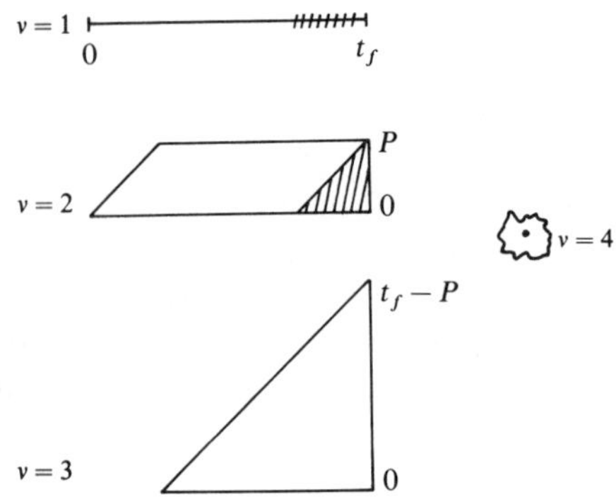

Fig. 3.1. *State space for the PHI process.*

we reproduce here (Fig. 3.1) with a little more detail. Note in particular that the policy is valueless in the shaded region since no claims could possibly be made from points in this region. Write $f_v(\zeta) = f(v, \zeta)$ for a real-valued function $E \to \mathbb{R}$. It will be clear from the description of the process in section (21.12), that the generator $\mathfrak{A}$ is given by

$$\mathfrak{A} f_1(t) = \frac{\mathrm{d}}{\mathrm{d}t} f_1(t) + \rho(t+t_0) f_2(t,0) - (\rho(t+t_0) + \mu(t+t_0)) f_1(t)$$

$$\mathfrak{A} f_2(t,y) = \frac{\partial}{\partial t} f_2(t,y) + \frac{\partial}{\partial y} f_2(t,y) + \sigma(t+t_0,y) f_1(t)$$
$$- (\sigma(t+t_0,y) + \pi(t+t_0,y)) f_2(t,y)$$
$$\mathfrak{A} f_3(t,y) = \frac{\partial}{\partial t} f_3(t,y) + \frac{\partial}{\partial y} f_3(t,y) + \sigma(t+t_0,y+p) f_1(t)$$
$$- (\sigma(t+t_0,y+p) + \pi(t+t_0,y+p)) f_3(t,y).$$

As usual, $f_4(\Delta) = 0$. The boundary conditions are

$$f_2(t,p) = f_3(t,0), \quad t \in [p, t_f]$$

and $f_1 = f_2 = f_3 = 0$ at $t = t_f$. Now define

$$V(x) = \mathbb{E}_x \int_0^{t_f} l(x_s)\mathrm{d}s$$

for $x = (v, (t, \zeta))$, where $l(x) = 1000 I_{(v=3)}$. Thus cost accumulates at the rate of £1000 per year whenever the current sickness period duration exceeds p, so that $V(x)$ is the mean expected total claim per £1000 of income insured. Calculating V is an application of Proposition (32.21): $V(x)$ is the unique solution of (32.17)–(32.19) which, specialized to our case, becomes

$$\mathfrak{A}V(x) + l(x) = 0 \quad x \in E$$

together with the boundary conditions

$$\left.\begin{aligned} V_2(t,p) &= V_3(t,0) && t \in [p, t_f] \\ V_1(t_f) &= 0 && \\ V_2(t_f, y) &= 0 && y \in [0, p] \\ V_3(t_f, y) &= 0 && y \in [0, t_f - p]. \end{aligned}\right\} \tag{33.19}$$

This is quite a complicated set of equations, particularly in view of the non-constant μ, π, ρ, σ (we assume these are bounded continuous functions, so that the solution V is in $\mathscr{C}^1$). They are, however, readily solved using the iterative method of section (32.2). For $\psi \in \mathscr{C}_b$ we define

$$G\psi(x) = \mathbb{E}_x \left\{ \int_t^{T_1 \wedge t_f} l(x_s)\mathrm{d}s + \psi(s_{T_1}) I_{(T_1 < t_f)} \right\}.$$

Then, as shown in Proposition (32.34), $G^k\psi(x) \to V(x)$ as $k \to \infty$, and $w = G\psi$ satisfies

$$\frac{\mathrm{d}}{\mathrm{d}t} w_1(t) + \rho\psi_2(t,0) - (\rho + \mu) w_1(t) = 0, \qquad w_1(t_f - p) = 0, \tag{33.20}$$

$$\frac{\partial}{\partial t} w_2(t,y) + \frac{\partial}{\partial y} w_2(t,y) + \sigma\psi_1(t) - (\sigma + \pi) w_2(t,y) = 0, \qquad w_2(t,p) = \psi_3(t,0), \tag{33.21}$$

$$\frac{\partial}{\partial t} w_3(t,y) + \frac{\partial}{\partial y} w_3(t,y) + \sigma\psi_1(t) - (\sigma + \pi) w_3(t,y) + 1000 = 0, \qquad w_3(t_f, y) = 0, \tag{33.22}$$

where we have omitted the arguments in ρ, μ, etc., for brevity. We can take the boundary condition as stated in (33.20) rather than the original $w_1(t_f) = 0$ because we know that $w_1(t) = 0$ for $t \in [t_f - p, t_f]$. Also $w_2(t, y) = 0$ for $t \in [t_f - p, t_f]$ and $y \in [0, t - t_f + p]$ since, as remarked earlier, no claims can be made starting from these points.

The three equations are now uncoupled. Equation (33.20) is a scalar linear ODE which can be solved backwards from the terminal condition at $t = t_f - p$. For equation (33.21), define $\tilde{w}_2(s, y) = w_2(s + y, y)$, where $s \in [0, t_f - p]$ and $y = [0, p]$ (This misses out the shaded area of the state space, but we know the solution is zero there.) Because the partial derivatives in (33.21) are the directional derivative along the 45° lines, (33.21) becomes (since $t = s + y$)

$$(33.21') \qquad \frac{\mathrm{d}}{\mathrm{d}y}\tilde{w}_2(s, y) + \sigma\psi_1(s + y) - (\sigma + \pi)\tilde{w}_2(s, y) = 0$$

$$\tilde{w}_2(s, p) = \psi_3(s + p, 0).$$

In this equation, s appears only as a parameter, and the equation can be solved *separately for each* s for y ranging from p to 0 with the given starting condition at $y = p$.

A similar construction applies so equation (33.22), where again we define $\tilde{w}_3(s, y) = w_3(s + y, y)$; this time $s \in [p, t_f]$ and $y \in [0, t_f - s]$. The equation for $\tilde{w}_3$ is

$$(33.22') \qquad \frac{\mathrm{d}}{\mathrm{d}y}\tilde{w}_3(s, y) + \sigma\psi_1(s + y) - (\sigma + \pi)\tilde{w}_3(s, y) + 1000 = 0$$

$$\tilde{w}_3(s, t_f - s) = 0.$$

This is solved for each s, separately, with y ranging from $t_f - s$ down to 0 with a zero boundary condition at $y = t_f - s$.

These equations are readily solved using numerical integration. Figure 3.2 shows $(G^k\psi)_1(t)$ for $k = 10, 20, 30$, with $\psi \equiv 0$ for an example in which $t_f = 25$ (years), $p = 0.5$, $\mu = \pi = 0.001$, $\rho = 0.3$, $\sigma = 2.8$. (It is not realistic to suppose that these rates are constant, but the numerical method is just the same if they depend on (t, y); this example is for illustrative purposes.) The time interval $[0, t_f]$ was divided into 100 steps and equations (33.21) and (33.22) solved at each grid points s. The values $w_1(s)$, $\tilde{w}_2(s, 0) = w_2(s, 0)$ and $\tilde{w}_3(s, 0) = w_3(s, 0)$ provide the data for calculating the second iteration $G^2\psi = Gw$. Calculation was continued until convergence took place in about 28 iterations. The numerical method as implemented was rather crude but executed in only a second or two on a workstation, using a short Pascal program. It could easily be made more accurate, but there is a limit to the degree of accuracy that is worthwhile, in that the calculation ultimately depends on the hazard rates, μ, π, ρ, σ, and these cannot be estimated with complete certainty. The results are rather sensitive to the value of the recovery rate σ, and getting good estimates for this may in fact be the major problem in pricing PHI.

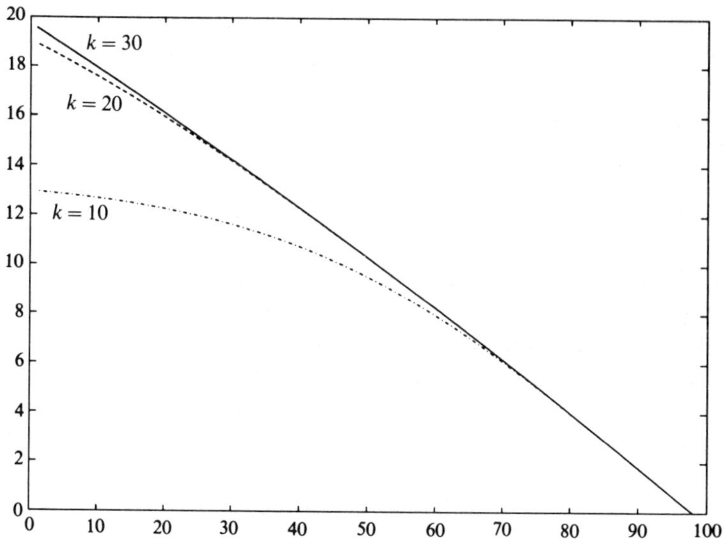

Fig. 3.2. *Iterative calculation of expected payout for the PHI model.*

34 Stationary distributions

(34.1) Definitions and examples

Let (x_t, P_x) be a Markov family with state space $(E, \mathscr{E})$ on a measurable space $(\Omega, \mathfrak{F})$, with transition function $p(t, x, A)$. A measure μ on E is an *invariant measure* for p if for all $t \in \mathbb{R}_+, A \in \mathscr{E}$,

$$\mu(A) = \int_E p(t, x, A)\mu(\mathrm{d}x). \tag{34.1}$$

μ is a *stationary distribution* of the process (x_t) if it is an invariant measure for p and is a probability measure, i.e. satisfies $\mu(E) = 1$. Of course, if (3.54) is satisfied by a finite measure μ then the normalized measure $\mu'(A) = \mu(A)/\mu(E)$ is stationary distribution, but it may be the case that (34.1) is only satisfied by measures μ with $\mu(E) = \infty$, as we shall see below.

If μ is any probability measure on E we can define a measure P^μ on $(\Omega, \mathfrak{F})$ by

$$P^\mu(A) = \int_E P_x(A)\mu(\mathrm{d}x).$$

Under P^μ, the process (x_t) is a Markov process with transition function p and initial distribution μ (i.e. $P^\mu(x_0 \in A) = \mu(A), A \in \mathscr{E}$). Denoting by $\mathbb{E}^\mu$ the expectation with respect to P^μ we see from (34.1) that μ is a stationary distribution if and only if for all bounded measurable $f: E \to \mathbb{R}$,

$$\mathbb{E}^\mu f(x_t) = \int_E f(y)\mu(\mathrm{d}y). \tag{34.2}$$

Thus the random variable x_t has distribution μ for any $t \geqslant 0$, and it follows easily that (x_t) is a stationary process under measure P^μ.

The same definitions apply to Markov families in discrete time. Here we denote the one-step transition function by $p(x, A) := p(1, x, A)$; a necessary and sufficient condition for μ to be an invariant measure is then

$$\mu(A) = \int_E p(x, A)\mu(\mathrm{d}x).$$

i.e. that (34.1) holds with $t = 1$. It then follows from the Chapman–Kolmogorov equation that (34.1) holds for $t = 2, 3, \ldots$. It is instructive to consider a couple of examples.

(34.3) *Birth and death process*
Here the state space is $\mathbb{Z}_+ = \{0, 1, \ldots\}$ and the transition function is $p(k, \{k+1\}) = p$ and $p(k, \{k-1\}) = q = (1-p), k \geqslant 1$, for some number $p \in]0, 1/2[$. Transitions from the state 0 are governed by $p(0, \{1\}) = p$, $p(0, \{0\}) = q$. Thus a 'birth' occurs with probability p and a 'death' with probability $q > p$; the process remains at zero until a birth occurs. This process has the unique stationary distribution

$$\mu(\{k\}) = p_k := \left(\frac{p}{q}\right)^k \left(1 - \frac{p}{q}\right). \tag{34.4}$$

Indeed, with this choice of μ the left hand side of (34.2) with $t = 1$ is given by

$$(qf(0) + pf(1))\left(1 - \frac{p}{q}\right) + \sum_{k \geqslant 1} (qf(k-1) + pf(k+1))p_k,$$

and a short computation shows that this is equal to $\sum_{k \geqslant 1} f(k)p_k$, which is the right-hand side of (34.2); thus μ is a stationary distribution. A direct argument, which we do not give here, shows that if p'_k is a stationary distribution then p'_k must coincide with p_k given by (34.4).

(34.5) *Symmetric random walk*
In this case the state space is $\mathbb{Z} = \{0, \pm 1, \dots\}$, the transition function is given by $p(k, \{k+1\}) = p(k, \{k-1\}) = \frac{1}{2}$, and is very easy to see that 'counting measure' $\mu(\{k\}) = 1$ is invariant for this transition function. This measure has infinite total mass and cannot therefore be normalized to give a stationary distribution. In fact, there is no stationary distribution for this process. If $\mu(\{k\}) = p_k$ were stationary then (34.2) would be satisfied, i.e. for all bounded f,

$$\sum_{k\in\mathbb{Z}} f(k)p_k = \frac{1}{2}\sum_{k\in\mathbb{Z}} f(k+1)p_k + \frac{1}{2}\sum_{k\in\mathbb{Z}} f(k-1)p_k,$$

which is equivalent to

$$\sum_{k\in\mathbb{Z}} f(k)(2p_k - p_{k+1} - p_{k-1}) = 0.$$

This is satisfied for all f only if $2p_k - p_{k+1} - p_{k-1} = 0$ for all $k\in\mathbb{Z}$, and this recursive relationship implies that p_k is given in terms of p_0 and p_1 by

$$p_k = p_0 + k(p_1 - p_0), \quad k \geqslant 2.$$

The only choice of p_0, p, for which p_k is bounded is $p_1 = p_0$ (μ then being a multiple of the counting measure), and in particular there is no choice of p_0, p_1 for which μ is a probability measure. Thus no stationary distribution can exist.

Let us now return to continuous-time processes. There is a simple characterization of stationary distributions in terms of the semigroup P_t and strong generator $\hat{\mathfrak{A}}$ of the process. Note that (34.2) can be written in terms of P_t as follows

(34.6) $$\int_E P_t f(x)\mu(\mathrm{d}x) = \int_E f(x)\mu(\mathrm{d}x).$$

A class of functions $\mathscr{D} \subset B(E)$ is said to be *separating* if, for probability measures μ^1 and μ^2 on E, $\mu^1 = \mu^2$ whenever $\int_E f\,\mathrm{d}\mu^1 = \int_E f\,\mathrm{d}\mu^2$ for all $f\in\mathscr{D}$.

(34.7) *Proposition.* Suppose $\mathscr{D}(\hat{\mathfrak{A}}) \cap B(E)$ is separating. Then μ is a stationary distribution if and only if

$$\int_E \hat{\mathfrak{A}} f(x)\mu(\mathrm{d}x) = 0 \quad \text{for all } f \in \mathscr{D}(\hat{\mathfrak{A}}). \tag{34.8}$$

Proof. It μ is stationary then from (34.6), for $t > 0$,

$$\int_E \frac{1}{t}(P_t f(x) - f(x))\mu(\mathrm{d}x) = 0.$$

Since for $f \in \mathscr{D}(\hat{\mathfrak{A}})$, $(P_t f - f)/t \to \hat{\mathfrak{A}} f$ in the uniform norm, equation (34.8) follows by dominated convergence. Conversely, suppose (34.8) holds. If $f \in \mathscr{D}(\hat{\mathfrak{A}})$ then $P_s f \in \mathscr{D}(\hat{\mathfrak{A}})$, for $s \geqslant 0$ and from the Dynkin formula

$$P_t f(x) - f(x) = \int_0^t \hat{\mathfrak{A}} P_s f(x)\mathrm{d}s.$$

Thus by the Fubini theorem and (34.8)

$$\int_E (P_t f(x) - f(x))\mu(\mathrm{d}x) = \int_0^t \int_E \hat{\mathfrak{A}} P_s f(x)\mu(\mathrm{d}x)\mathrm{d}s = 0,$$

so that (34.6) holds for all $f \in \mathscr{D}(\hat{\mathfrak{A}})$. If μ^t denotes the distribution of x_t then this implies

$$\int_E f(x)\mu^t(\mathrm{d}x) = \int_E f(x)\mu(\mathrm{d}x), \quad f \in \mathscr{D}(\hat{\mathfrak{A}}) \cap B(E).$$

Now $\mu^t = \mu$ follows from the assumption that $\mathscr{D}(\hat{\mathfrak{A}}) \cap B(E)$ is separating. □

In general it is not a simple matter to determine whether a given process has a stationary distribution, and if so whether that distribution is unique. One approach is to write the 'pairing' of a bounded measurable function f and probability measure μ as

$$\langle f, \mu \rangle := \int_E f(x)\mu(\mathrm{d}x)$$

and to introduce the *adjoint operator* $\hat{\mathfrak{A}}^*$ defined by

$$\langle \hat{\mathfrak{A}} f, \mu \rangle = \langle f, \hat{\mathfrak{A}}^* \mu \rangle, \quad f \in \mathscr{D}(\hat{\mathfrak{A}}),\ \mu \in \mathscr{P}(E).$$

Since $\mathscr{D}(\hat{\mathfrak{A}})$ is dense in $\mathscr{B}_0$, (34.8) is then equivalent to

$$\hat{\mathfrak{A}}^* \mu = 0, \tag{34.9}$$

and we can regard (34.9) as an 'equation' for μ. This is in fact the direct generalization of the normal procedure in the discrete time countable state space case, where solving the equivalent of (34.9) reduces to a problem in linear algebra, but getting the right interpretation for (34.9) and solving it in a general setting is a complicated task. Our strategy in the remainder of this section is to use the special structure of PDPs to reduce it, at least in some cases, to more manageable proportions.

(34.2) Stationary distributions for PDPs

For the remainder of this section, (x_t) will be a standard PDP satisfying the conditions of section (24.8). For simplicity of notation (see Remark (24.9)) we will not distinguish different components of the state space explicitly, writing for example $\phi(t, x)$ for the deterministic flow starting at $x \in E \subset \mathbb{R}^d$. The following additional standing assumptions will be made. Recall that $N_t = \sum_i I_{(t \geqslant T_i)}$ is the number of jumps of the process in the time interval $[0, t]$.

(34.10) *Assumptions.* (a) For all $x \in E$ such that $t_*(x) = \infty$,

$$\lim_{t \to \infty} \int_0^t \lambda(\phi(s, x))\mathrm{d}s = \infty.$$

(b) For any $t > 0$ and probability measure μ on E,

$$\int_E \mathbb{E}_x[N_t]\mu(\mathrm{d}x) < \infty.$$

Assumption (a) implies that $P_x[T_1 < \infty] = 1$, which is certainly necessary if the distribution of x_t, starting at $x_0 = x$, is to converge to some stationary distribution, while Assumption (b) is a slight strengthening of Assumption (24.3).

We first show that a condition similar to that of Proposition (34.7) is automatically satisfied by PDPs.

(34.11) *Proposition.* Under the standard conditions in section (24.8) the set $\mathscr{D}(\mathfrak{A}) \cap B(E)$ is separating.

Proof. Take $A \in \mathscr{E}$ and apply Proposition (32.21) with $c = l = 0$ and $\Phi = I_A$ to conclude that $f_t(x) := P_x[x_t \in A]$ belongs to $\mathscr{D}(\mathfrak{A}) \cap B(E)$ for each $t > 0$ and $\lim_{t \downarrow 0} f_t(x) = I_A(x)$. Hence for any probability measure μ

we have

$$\mu(A) = \lim_{t \downarrow 0} \int f_t(x)\mu(\mathrm{d}x) < \infty$$

and the result follows. □

Define a discrete-time process $(z_k, k = 0, 1, \ldots)$ by $z_0 = x_0$ and $z_k = x_{T_k}$ for $k \geqslant 1$. It is clear from the construction of the PDP that (z_k) is a Markov process with stationary transition probability

$$(34.12) \quad p(A, x) = P_x[x_{T_1} \in A] = \int_0^{t_*(x)} Q(A; \phi(t, x))\lambda(\phi(t, x))\mathrm{e}^{-\Lambda(t,x)}\mathrm{d}t$$
$$+ \mathrm{e}^{-\Lambda(t_*(x),x)} Q(A; \phi(t_*(x), x)),$$

where

$$\Lambda(t, x) = \int_0^t \lambda(\phi(s, x))\mathrm{d}s.$$

The main aim of this section is to show that there is a certain relation between stationary distributions for the imbedded Markov chain (z_k) and the (continuous time) PDP (x_t). Since conditions for the existence of stationary distributions are more easily stated in the discrete time setting, this will provide us with a way of getting some information about the PDP without relying on general continuous-time theory.

Let Γ be the *active boundary* of the state space, as defined in section 24, and for $x \in E \cup \Gamma$ define $t_*^-(x) = \inf\{t > 0 : \phi(-t, x) \notin E\}$ and

$$\Gamma_h = \{\phi(-s, x) : x \in \Gamma \quad \text{and} \quad 0 < s < h \wedge t_*^-(x)\}, \quad h > 0.$$

(34.13) *Proposition.* Let $f : \Gamma \to \mathbb{R}$ be a bounded measurable function and, for $t > 0$, define

$$I_t(h) = \frac{1}{h}\int_0^t f(\phi(t_*(x_s), x_s))I_{\Gamma_h}(x_s)\mathrm{d}s.$$

Then

$$(34.14) \qquad \lim_{h \downarrow 0} I_t(h) = \int_0^t f(x_{s-})\mathrm{d}p_s^* \quad \text{a.s.}$$

Proof. Note first that $\phi(t_*(x), x) \in \Gamma$, so $I_t(h)$ is well defined. With probability one, the right-hand side of (34.14) is a finite sum, and it is not hard to see that $I_t(h)$ is exactly equal to this sum for all sufficiently small h. The result follows. □

(34.15) *Theorem.* Suppose that μ is a stationary distribution for (x_t). Then there exists a finite measure σ on Γ such that for all $t > 0$

$$\mathbb{E}^{\mu}\frac{1}{t}\int_0^t f(x_{s-})\mathrm{d}p_s^* = \int_{\Gamma} f(x)\sigma(\mathrm{d}x). \tag{34.16}$$

Proof. Note that $I_t(h) \leqslant K(N_t + 1)$, where K is a bound for f, and $\mathbb{E}^{\mu}N_t < \infty$ (34.12). By Proposition (34.13) and dominated convergence,

$$\frac{1}{t}\mathbb{E}^{\mu}I_t(h) \to \frac{1}{t}\mathbb{E}^{\mu}\int_0^t f(x_{s-})\mathrm{d}p_s^* \quad \text{as} \quad h \to 0. \tag{34.17}$$

However, using the Fubini theorem and property (34.2) of the stationary distribution μ,

$$\begin{aligned}\frac{1}{t}\mathbb{E}^{\mu}I_t(h) &= \frac{1}{th}\int_0^t \mathbb{E}^{\mu}[f(\phi(t_*(x_s), x_s))I_{\Gamma_h}(x_s)]\mathrm{d}s.\\ &= \frac{1}{th}\int_0^t\int_E f(\phi(t_*(x), x))I_{\Gamma_h}(x)\mu(\mathrm{d}x)\mathrm{d}s\\ &= \frac{1}{h}\int_E f(\phi(t_*(x), x))I_{\Gamma_h}(x)\mu(\mathrm{d}x).\end{aligned}$$

This shows that the right-hand side of (34.17) is the same for any $t > 0$. For a Borel set A of Γ, define

$$\sigma(A) = \frac{1}{t}\mathbb{E}^{\mu}\int_0^t I_A(x_{s-})\mathrm{d}p_s^*. \tag{34.18}$$

By standard properties of the integral the set function $\sigma(\cdot)$ is countably additive, and

$$\sigma(\Gamma) = \frac{1}{t}\mathbb{E}^{\mu}p_t^* \leqslant \frac{1}{t}\mathbb{E}^{\mu}N_t < \infty.$$

Thus σ is a finite measure, and (34.16) follows by approximation of f by simple functions.

The interpretation of σ is that for any Borel set A of Γ the average number of times per unit time that the process (x_t) hits the boundary at a point in A is $\sigma(A)$. Because of stationarity, this can be evaluated as in (34.17) using any $t > 0$.

(34.19) *Theorem.* Suppose μ is a stationary distribution for (x_t) and let σ be the corresponding boundary measure defined by (34.18). Then

(34.20) $$\int_E \mathfrak{A}f(x)\mu(\mathrm{d}x) + \int_\Gamma \mathfrak{C}f(z)\sigma(\mathrm{d}z) = 0$$

for all $f \in \mathscr{A}\mathscr{C}$, such that $\mathfrak{B}f \in L_1(p)$, where $\mathfrak{A}$ is the extended generator and $\mathfrak{C}f = (Qf - f)$.

Proof. The differential formula (Theorem 31.3) shows that for f satisfying the stated conditions,

$$f(x_t) - f(x_0) = \int_0^t \mathfrak{A}f(x_s)\mathrm{d}s + \int_0^t \int_E \mathfrak{B}f(s, x)q(\mathrm{d}s,\mathrm{d}x) + \int_0^t \mathfrak{C}f(x_{s-})\mathrm{d}p_s^*$$

and the second term on the right is a martingale. Taking expectations with respect to $\mathbb{E}^\mu$, where μ is a stationary distribution, and using (34.2) and Theorem (34.15) gives

$$\begin{aligned} 0 &= \mathbb{E}^\mu \int_0^t \mathfrak{A}f(x_s)\mathrm{d}s + \mathbb{E}^\mu \int_0^t \mathfrak{C}f(x_{s-})\mathrm{d}p_s^* \\ &= t\int_E \mathfrak{A}f(x)\mu(\mathrm{d}s) + t\int_\Gamma \mathfrak{C}f(z)\sigma(\mathrm{d}z). \end{aligned}$$

This is (34.20). □

We are now in a position to study the relationship between the sets Π_{PDP}, Π_{MC} of stationary distributions for, respectively, the process (x_t) and the imbedded Markov chain (z_k). Define

$$\Pi^*_{\mathrm{PDP}} = \left\{\mu \in \Pi_{\mathrm{PDP}}: \int_E \lambda(x)\mu(\mathrm{d}x) < \infty\right\}.$$

Our first result is the following.

(34.21) *Theorem.* Suppose $\mu \in \Pi^*_{\mathrm{PDP}}$. Then

(34.22) $$\int_E \lambda(x)\mu(\mathrm{d}x) + \sigma(\Gamma) > 0$$

and $\pi \in \Pi_{MC}$, where

$$(34.23) \qquad \pi(A) = \frac{\int_E \lambda(x)Q(A;x)\mu(\mathrm{d}x) + \int_\Gamma Q(A;z)\sigma(\mathrm{d}z)}{\int_E \lambda(x)\mu(\mathrm{d}x) + \sigma(\Gamma)}.$$

Proof. If $\int_E \lambda(x)\mu(\mathrm{d}x) + \sigma(\Gamma) = 0$, then

$$\mathbb{E}^\mu\left(\int_0^\infty \lambda(x_s)\mathrm{d}s + \int_0^\infty \mathrm{d}p_s^*\right) = 0,$$

implying that $\lambda(x_t) = 0$ and $p_t^* = 0$ for almost every t, a.s. This however contradicts Assumption (34.10)(a), so (34.22) must hold. We denote $D = \int_E \lambda\, \mathrm{d}\mu + \sigma(\Gamma)$. Let $p(A,x)$ be the transition function for (z_k), defined by (34.12). It will be shown at the end of this proof that $p(A,\cdot)$ satisfies for each $A \in \mathscr{E}$,

$$(34.24) \qquad \mathfrak{A}p(A,x) = \lambda(x)\int_E (p(A,y) - Q(A;x))Q(\mathrm{d}y;x).$$

Using Theorem (34.19) with $f(x) = p(A,x)$, we have

$$0 = \int_E \lambda(x)\int_E p(A,y)Q(\mathrm{d}y;x)\mu(\mathrm{d}x) - \int_E \lambda(x)Q(A;x)\mu(\mathrm{d}x)$$
$$+ \int_\Gamma\left(\int_E p(A,y)Q(\mathrm{d}y;z) - p(A,z)\right)\sigma(\mathrm{d}z),$$

i.e.

$$(34.25) \qquad \frac{1}{D}\int_E p(A,y)\int_E \lambda(x)Q(\mathrm{d}y;x)\mu(\mathrm{d}x) + \frac{1}{D}\int_E p(A,y)\int_\Gamma Q(\mathrm{d}y;z)\sigma(\mathrm{d}z)$$
$$= \frac{1}{D}\left(\int_E \lambda(x)Q(A;x)\mu(\mathrm{d}x) + \int_\Gamma p(A,z)\sigma(\mathrm{d}z)\right).$$

Now for $z \in \Gamma$ we see, using (34.12), that

$$p(A,z) = \lim_{t \downarrow 0} p(A;\phi(-t,z)) = Q(A;z).$$

Therefore (34.25) is equivalent to

$$\int_E p(A,y)\pi(\mathrm{d}y) = \pi(A),$$

where π is defined by (34.23); but any π satisfying this is a stationary distribution for (z_k), i.e. $\pi \in \Pi_{MC}$. It remains to verify (34.24).

A short calculation using the fact that $t^*(\phi(t,x)) = t^*(x) - t$ for $t < t_*(x)$ shows that

(34.26)

$$p(A, \phi(t,x)) = e^{\Lambda(t,x)}(p(A,x) - \int_0^t Q(A; \phi(s,x))\lambda(\phi(s,x))e^{-\Lambda(s,x)}\,ds)$$

and hence if the integrand is a continuous function of s,

$$\begin{aligned}
\mathfrak{X}p(A,x) &= \lim_{t\downarrow 0} \frac{1}{t}(p(A,\phi(t,x)) - p(A,x)) \\
&= \lim_{t\downarrow 0} \frac{1}{t}(e^{\Lambda(t,x)} - 1)p(A,x) \\
&\quad - \lim_{t\downarrow 0} e^{\Lambda(t,x)} \frac{1}{t}\int_0^t Q(A; \phi(t,s))\lambda(\phi(s,x))e^{-\Lambda(s,x)}ds \\
&= \lambda(x)(p(A,x) - Q(A;x)).
\end{aligned}$$

Without the continuity assumption, (34.26) shows that $t \to p(A, \phi(t,x))$ is absolutely continuous with derivative given almost everywhere by $\mathfrak{X}p$ as above. Thus

$$\begin{aligned}
\mathfrak{A}p(A,x) &= \mathfrak{X}p(A,x) + \lambda(x)\int_E (p(A,y) - p(A,x))Q(dy;x) \\
&= \lambda(x)\int_E p(A,y)Q(dy;x) - \lambda(x)Q(A;x),
\end{aligned}$$

and this is (34.24). □

(34.27) *Example. The $M/M/1$ queue*
This is a special case of the $M/G/1$ queue discussed at (21.9) and we use the state space defined there. The arrivals are at Poisson times with rate λ and each arrival brings a service time requirement Y_i; the random variables Y_i are i.i.d with exponential distribution $P[Y_i > y] = e^{-\beta y}$. Thus λ/β is the mean processing time requirement arriving per unit time. Asmussen (1987) shows that if $\lambda < \beta$, the unique

stationary distribution is

$$\mu(0,0) = \frac{\lambda}{\beta}$$

$$\mu(\{1\} \times]0,a]) = \frac{\lambda}{\beta}(1 - e^{-(\beta-\lambda)a}) =: m(a).$$

($\mu(0,0)$ is the probability that the queue is empty.) The active boundary is $\Gamma = \{(1,0)\}$ and with $x = (v,\zeta)$,

$$\begin{aligned}\lim_{h\downarrow 0}\frac{1}{h}\int_{\Gamma_h} f(v, \phi_v(t_*(\zeta),\zeta))\mu(\mathrm{d}x) &= \lim_{h\downarrow 0} f(1,0)\frac{1}{h}\mu(\{1\} \times]0,h])\\ &= f(1,0)m'(0)\\ &= f(1,0)\frac{\lambda}{\beta}(\beta-\lambda) = \sigma(\{(1,0)\}).\end{aligned}$$

Thus, from (34.18), $\mathbb{E}^\mu p_t^* = \lambda(\beta-\lambda)t/\beta$ so that $\lambda(\beta-\lambda)/\beta$ is the mean number of times per unit time that the queue empties out. For fixed β, this has a maximum value of $\beta/4$ when $\lambda = \beta/2$: when λ/β is close to 1 the queue is hardly ever empty, while when λ/β is small it is empty most of the time. The maximum rate of transitions between empty and non-empty occurs under 'medium-traffic' conditions.

We can now apply Theorem (34.21). We have

$$\int_E \lambda(x)\mu(\mathrm{d}x) + \sigma(\Gamma) = \lambda + \frac{\lambda}{\beta}(\beta-\lambda) = \frac{\lambda}{\beta}(2\beta-\lambda)$$

and we find from (34.23) that the stationary distribution for the imbedded Markov chain (z_k) is

$$\pi(0,0) = \frac{\beta}{2\beta-\lambda}$$

$$\pi(\{1\} \times]0,a]) = \frac{\beta}{2\beta-\lambda}(1 - e^{-(\beta-\lambda)a}) = \frac{\beta^2}{\lambda(2\beta-\lambda)}m(a).$$

Since $\mu(0,0) = (\beta-\lambda)/\beta$ while $\pi(0,0) = (\beta-\lambda)/(\beta+(\beta-\lambda))$, clearly $\pi(0,0) < \mu(0,0)$, so that weight has been shifted from empty to non-empty; but the *conditional* stationary distribution given that the queue is non-empty is the same for μ and π. The difference between the distributions arises because the imbedded Markov chain $z_k = x_{T_k}$ takes no account of *how long* the process x_t stays in various regions of the state space.

(34.28) *Example*

This example is designed to show that, for a process (x_t) with stationary distribution $\mu \in \Pi_{\text{PDP}}$, the chain (z_k) may fail to have a stationary distribution if $\mu \notin \Pi^*_{\text{PDP}}$. Take a process with $E = \mathbb{Z}, \phi(t,x) = x$, $Q(\{x+1\};x) = Q(\{x-1\};x) = \frac{1}{2}$ and some jump rate function $\lambda(x)$. Thus the process stays at x for an exponentially-distributed time with mean $1/\lambda(x)$ and then jumps to $(x+1)$ or $(x-1)$. Clearly the imbedded Markov chain z_k is a symmetric random walk, and we showed in the section on symmetric random walk in (34.5) that this has no stationary distribution. However, we claim that (x_t) has a stationary distribution if λ satisfies (writing $\lambda_k = \lambda(k)$)

(34.29) $$\sum_{k=-\infty}^{\infty} \frac{1}{\lambda_k} < \infty.$$

This condition implies that $\lambda_k \to \infty$ as $|k| \to \infty$, so the process 'speeds up' when it is far from the origin. From Proposition (34.7), $\mu \in \Pi_{\text{PDP}}$ if and only if $\int_E \hat{\mathfrak{A}} f(x)\mu(dx) = 0$ for all $f \in \mathscr{D}(\hat{\mathfrak{A}})$. Certainly $\mathscr{D}(\hat{\mathfrak{A}})$ is separating in this case, since all functions f of finite support are in $\mathscr{D}(\hat{\mathfrak{A}})$, and $\hat{\mathfrak{A}} f(x) = \lambda(x)[\frac{1}{2}f(x+1) + \frac{1}{2}f(x-a) - f(x)]$. Therefore, denoting $\mu_k = \mu(\{k\})$, $\mu \in \Pi_{\text{PDP}}$ if and only if for all $f \in \mathscr{D}(\hat{\mathfrak{A}})$,

$$\begin{aligned} 0 &= \sum_{k=-\infty}^{\infty} \lambda_k \mu_k [\tfrac{1}{2} f(k+1) + \tfrac{1}{2} f(k-1) - f(k)] \\ &= \sum_{k=-\infty}^{\infty} f(k)[\tfrac{1}{2} c_{k+1} + \tfrac{1}{2} c_{k-1} - c_k], \end{aligned}$$

where we write $c_k = \lambda_k \mu_k$. This can only be satisfied if

$$\tfrac{1}{2} c_{k+1} + \tfrac{1}{2} c_{k-1} - c_k = 0 \qquad \text{for all} \quad k \in \mathbb{Z}.$$

By the same argument as in the symmetric random walk section of section (34.5), the only globally non-negative solution of this system of equations is $c_k = c$, constant. Thus $\mu_k = c/\lambda_k$ and μ_k is a probability distribution if

$$1 = \sum_k \mu_k = c \sum_k \frac{1}{\lambda_k}.$$

Thus if (34.29) holds, (x_t) has the unique stationary distribution

$$\mu(\{x\}) = \left(\lambda_x \sum_k \frac{1}{\lambda_k}\right)^{-1}.$$

However, this distribution is not in Π^*_{PDP}, because

$$\int_E \lambda(x)\mu(\mathrm{d}x) = \sum_k \lambda_k \mu_k = \sum_k c = \infty.$$

This example shows that existence of a stationary distribution for (z_k) may fail if $\mu \notin \Pi^*_{\text{PDP}}$, but we do not know whether $\mu \notin \Pi^*_{\text{PDP}}$ is a *necessary* condition for the existence of $\pi \in \Pi_{\text{MC}}$.

We now consider the inverse problem of recovering a stationary distribution μ for the process (x_t) given a stationary distribution π for the chain (z_k). Again, some restrictions are required, and we define

$$\Pi^*_{\text{MC}} = \left\{\pi \in \Pi_{\text{MC}}: \int_E \int_0^{t_*(x)} \mathrm{e}^{-\Lambda(t,x)}\,\mathrm{d}t\pi(\mathrm{d}x) < \infty\right\}.$$

Note that $\Pi^*_{\text{MC}} = \Pi_{\text{MC}}$ if $\lambda(x) \geqslant \lambda_0 > 0$. For $\pi \in \Pi^*_{\text{MC}}$ we now define a probability measure μ on E by

$$\mu(A) = \frac{\displaystyle\int_E \int_0^{t_*(x)} I_A(\phi(t,x))\mathrm{e}^{-\Lambda(t,x)}\mathrm{d}t\pi(\mathrm{d}x)}{\displaystyle\int_E \int_0^{t_*(x)} \mathrm{e}^{-\Lambda(t,x)}\mathrm{d}t\pi(\mathrm{d}x)}, \quad A \in \mathscr{E}. \tag{34.30}$$

(34.31) *Theorem.* If $\pi \in \Pi^*_{\text{MC}}$ then μ defined by (34.30) belongs to Π_{PDP}.

Proof. From Proposition (34.7), and in view of Proposition (34.11), it suffices to show that

$$\int_E \mathfrak{A} f(x)\mu(\mathrm{d}x) = 0 \qquad \text{for all} \quad f \in \mathscr{D}(\mathfrak{A}). \tag{34.32}$$

Proposition (34.7) is stated terms of the strong generator $\hat{\mathfrak{A}}$, but since $\mathfrak{A}$ is an extension of $\hat{\mathfrak{A}}$, (34.32) is also a sufficient condition. Now for $f \in \mathscr{D}(\mathfrak{A})$,

$$\int_0^{t_*(x)} \mathfrak{A} f(\phi(t,x))\mathrm{e}^{-\Lambda(t,x)}\,\mathrm{d}t$$

$$= \int_0^{t_*(x)} \left(\frac{\mathrm{d}}{\mathrm{d}t} f(\phi(t,x)) \right) - \lambda(\phi(t,x) f(\phi(t,x)) \mathrm{e}^{-\Lambda(t,x)} \, \mathrm{d}t$$

$$+ \int_0^{t_*(x)} Qf(\phi(t,x))\lambda(\phi(t,x))\mathrm{e}^{-\Lambda(t,x)} \, \mathrm{d}t$$

$$= \int_0^{t_*(x)} \frac{\mathrm{d}}{\mathrm{d}t}(f(\phi(t,x))\mathrm{e}^{-\Lambda(t,x)})\mathrm{d}t + \int_0^{t_*(x)} Qf(\phi(t,x))\lambda(\phi(t,x))\mathrm{e}^{-\Lambda(t,x)}\mathrm{d}t$$

$$= f(\phi(t_*(x),x))\mathrm{e}^{-\Lambda(t_*(x),x} - f(x) + \int_0^{t_*(x)} Qf(\phi(t,x))\lambda(\phi(t,x))\mathrm{e}^{-\Lambda(t,x)}\mathrm{d}t$$

$$= \mathbb{E}_x[f(x_{T_1})] - f(x),$$

where in the last step we use the boundary condition $Qf(z) = f(z)$ for $z \in \Gamma$. Now $\pi \in \Pi_{\mathrm{MC}}$ precisely when

$$\int_E (\mathbb{E}_x[f(x_{T_1})] - f(x))\pi(\mathrm{d}x) = 0.$$

Thus for $\pi \in \Pi_{\mathrm{MC}}$ and $f \in \mathscr{D}(\mathfrak{A})$

$$\int_E^{t_*(x)} \int_0 \mathfrak{A} f(\phi(t,x))\,\mathrm{e}^{-\Lambda(t,x)}\,\mathrm{d}t\pi(\mathrm{d}x) = 0.$$

It follows that (34.32) holds when μ is defined by (34.30). □

(34.33) *Example*

This example shows that if $\pi \in \Pi_{\mathrm{MC}}$ but $\pi \notin \Pi^*_{\mathrm{MC}}$ then there may be no stationary distribution $\mu \in \Pi_{\mathrm{PDP}}$. The example is based on the birth-and-death process discussed in (34.3). The state space for (x_t) is $\mathbb{E} = \mathbb{Z}_+$, with $\phi(t,x) = x$, $\lambda(x)$ some rate function to be chosen, and Q given by

$$Q(\{x+1\};x) = p, \quad Q(\{x-1\};x) = q, \qquad x \geqslant 1$$

$$Q(\{1\};0) = p, \quad Q(\{0\};0) = q,$$

with $p < \frac{1}{2}$, $q = 1 - p$. The Markov chain $z_k = x_{T_k}$ is thus the birth-and-death process of section (34.3), with unique stationary distribution $\pi_k = (p/q)^k(1 - p/q)$. The generator of (x_t) is $\mathfrak{A}f(x) = \lambda(x)(pf(x+1) + qf(x-1) - f(x))$ for $x \geqslant 1$, $\mathfrak{A}f(0) = \lambda(0)p(f(1) - f(0))$ and a measure μ is a stationary distribution if

$$0 = \int_E \mathfrak{A} f \,\mathrm{d}\mu = \mu_0 \lambda_0 p(f(1)) - f(0)) + \sum_{k=1}^{\infty} \mu_k \lambda_k (pf(k+1) + qf(k-1) - f(x)) \tag{34.34}$$

for all $f \in \mathscr{D}(\mathfrak{A})$, where $\mu_k = \mu(\{k\})$. Denoting $c_k = \mu_k \lambda_k$ this equality becomes, after some rearrangement,

$$f(0)[-pc_0 + qc_1] + \sum_{k=1}^{\infty} (pc_{k-1} + qc_{k+1} - c_k) f(k) = 0.$$

If this is to hold for all $f \in \mathscr{D}(\mathfrak{A})$ then the coefficient of $f(k)$ must vanish for all $k \in \mathbb{Z}_+$, and we find that this implies

$$c_k = c_0 \left(\frac{p}{q}\right)^k.$$

Now choose (for example) $\lambda_k = (p/q)^k$; then $\mu_k = c_k/\lambda_k = c_0$, which is not a probability distribution, whatever c_0. It is thus impossible for (34.34) to be satisfied by any probability measure μ. Of course, π fails to be in Π^*_{MC} for this choice of λ; indeed,

$$\int_0^{t_*(x)} \mathrm{e}^{-\Lambda(t,x)} \mathrm{d}t = \frac{1}{\lambda(x)} = \left(\frac{q}{p}\right)^x,$$

and therefore

$$\int_E \int_0^{t_*(x)} \mathrm{e}^{-\Lambda(t,x)} \mathrm{d}t \pi(\mathrm{d}x) = \sum_{k=1}^{\infty} \left(\frac{q}{p}\right)^k \left(\frac{p}{q}\right)^k \left(1 - \frac{p}{q}\right) = \infty.$$

What is happening here is that $\lambda_k \to 0$ as $k \to \infty$, so that the process slows down for large x and takes too long to get back to the origin. Again, we do not know whether $\pi \in \Pi^*_{\mathrm{MC}}$ is a general necessary condition for the existence of a corresponding $\mu \in \Pi_{\mathrm{PDP}}$.

Returning now to general PDPs, suppose there are constants λ_1, λ_2, such that

(34.35) $$0 < \lambda_1 \leqslant \lambda(x) \leqslant \lambda_2 < \infty, \qquad \text{for all} \quad x \in E;$$

then clearly $\Pi^*_{\mathrm{PDP}} = \Pi_{\mathrm{PDP}}$ and $\Pi^*_{\mathrm{MC}} = \Pi_{\mathrm{MC}}$. Condition (34.35) is the simplest sufficient condition for this, but other conditions may also be sufficient in specific cases (for example, $\Pi^*_{\mathrm{MC}} = \Pi_{\mathrm{MC}}$ if $t_*(x)$ is bounded). Let us denote by $\pi = A\mu$ the correspondence between $\mu \in \Pi^*_{\mathrm{PDP}}$ and $\pi \in \Pi_{\mathrm{MC}}$ given by (34.23), and by $\mu = B\pi$ the correspondence between $\pi \in \Pi^*_{\mathrm{MC}}$ and $\mu \in \Pi_{\mathrm{PDP}}$ given by (34.30). We will show that these maps are inverses of each other.

(34.36) *Proposition.* Suppose (34.35) holds. Then $\mu = BA\mu$ for $\mu \in \Pi_{\mathrm{PDP}}$ and $\pi = AB\pi$ for $\pi \in \Pi_{\mathrm{MC}}$. Thus $B = A^{-1}$.

Proof. From (34.23) $\pi = A\mu$ satisfies

$$\int_E f(x)\pi(\mathrm{d}x) = \frac{1}{D_1}\int_E \lambda(v)\int_E f(y)Q(\mathrm{d}y;x)\mu(\mathrm{d}x) + \frac{1}{D_1}\int_\Gamma\int_E f(y)Q(\mathrm{d}y;x)\sigma(\mathrm{d}x)$$

for bounded measurable f, where D_1 is the denominator of (34.23) and σ is the boundary measure corresponding to $\mu\in\Pi_{\mathrm{PDP}}$. Using (34.20) we see that if $f\in\mathscr{A}\mathscr{C}$, this is equivalent to

$$\int_E f(x)\pi(\mathrm{d}x) = \frac{1}{D_1}\left(\int_E (\lambda f - \mathfrak{X}f)\mathrm{d}\mu + \int_\Gamma f\,\mathrm{d}\sigma\right). \tag{34.37}$$

Now from (34.30), $\mu' = B\pi$ is given for bounded measurable g by

$$\int_E g(x)\mu'(\mathrm{d}x) = \frac{\int_E f_g(x)\pi(\mathrm{d}x)}{\int_E f_1(x)\pi(\mathrm{d}x)}, \tag{34.38}$$

where

$$f_g(t) = \int_0^{t_*(x)} g(\phi(t,x))\,\mathrm{e}^{-\Lambda(t,x)}\mathrm{d}t$$

and we define $1(x)\equiv 1$. As in the proof of Theorem (34.31) we find that $\mathfrak{X}f_g = \lambda f_g - g$, and for $z\in\Gamma$ we have $f_g(z) = \lim_{t\downarrow 0} f_g(\phi(-t,z)) = 0$. Thus from (34.37)

$$\int_E f_g\mathrm{d}\pi = \frac{1}{D_1}\int_E g\,\mathrm{d}\mu$$

and from (34.38)

$$\int_E g(x)\mu'(\mathrm{d}x) = \int_E g\mathrm{d}\mu.$$

Thus $\mu' = BA\mu = \mu$.

For the converse, (34.30) defines $B\pi$ as

$$\int_E g(x)B\pi(\mathrm{d}x) = \frac{1}{D_2}\int_E\int_0^{t_*(x)} g(\phi(t,x))\mathrm{e}^{-\Lambda(t,x)}\mathrm{d}t\pi(\mathrm{d}x), \tag{34.39}$$

where D_2 is the normalizing constant. A simple computation using Proposition (34.13) shows that the boundary measure σ corresponding to $\mu = B\pi$ is given by

$$\int_\Gamma h(x)\sigma(\mathrm{d}x) = \frac{1}{D_2}\int_E h(\phi(t_*(x), x)\mathrm{e}^{-\Lambda(t_*(x),x)}\pi(\mathrm{d}x) \tag{34.40}$$

for bounded measurable $h:\Gamma \to \mathbb{R}$. Hence from the definition (34.23) together with (34.39) and (34.40) for $F \in \mathscr{E}$,

$$\begin{aligned}
AB\pi(F) &= \frac{1}{D_1}\left(\int_E \lambda(x)Q(F;x)B\pi(\mathrm{d}x) + \int_\Gamma Q(F;z)\sigma(\mathrm{d}x)\right) \\
&= \frac{1}{D_1 D_2}\int_E \lambda(\phi(t,x)Q(F;\phi(t,x))\mathrm{e}^{-\Lambda(t,x)}\mathrm{d}t\pi(\mathrm{d}x) \\
&\quad + \int_E Q(F;\phi(t_*(x),x)\mathrm{e}^{-\Lambda(t_*(x),x)}\pi(\mathrm{d}x) \\
&= \frac{1}{D_1 D_2}\int_E p(F,x)\pi(\mathrm{d}x) \\
&= \frac{1}{D_1 D_2}\pi(F),
\end{aligned}$$

where the last inequality follows, from the fact that $\pi \in \Pi_{\mathrm{MC}}$. Since $AB\pi$ and π are both probability measures, $D_1 D_2 = 1$ and hence $\pi(F) = AB\pi(F)$, $F \in \mathscr{E}$. This completes the proof. □

(34.3) Conditions for the existence of stationary distributions

It is not a simple matter to give general conditions under which continuous-time Markov processes, or discrete-time processes (Markov chains) on general state spaces, have stationary distributions, and we cannot survey this whole subject here. There are two main techniques: a *stochastic stability* method based on supermartingale convergence, and a *recursive* or *renewal* method applicable when the process goes through 'cycles'. Typical results for these two approaches are given without proof below, and their application to PDP problems is exemplified.

The first method, for which the reader is referred to Tweedie (1975) concerns a Markov chain $Z = (z_k)$ on a separable metric state space

$(E, \mathscr{E})$ with stationary transition function $p(A, x)$. Z is said to be *φ-irreducible* if for some non-zero σ-finite measure φ on $\mathscr{E}$ we have $\sum_{n=0}^{\infty} 2^{-n} P_x[z_n \in A] > 0$ for every $x \in E$ whenever $\varphi(A) > 0$. The transition function p is *strongly continuous* if the map $x \to p(A, x)$ is continuous for each $A \in \mathscr{E}$.

(34.41) *Theorem* (Tweedie, 1975, Theorem 4.2). Let Z be a φ-irreducible Markov chain with strongly continuous transition function. A stationary distribution for Z exists if there is a compact set $C \subset E$, a measurable function $g: E \to \mathbb{R}_+$ and numbers $0 < \varepsilon, B < \infty$, such that:

1. $\varphi(C) > 0$;
2. $\int_E g(y) p(\mathrm{d}y, x) \leqslant g(x) - \varepsilon, \quad x \notin C$; and
3. $\int_E g(y) p(\mathrm{d}y, x) \leqslant B, \qquad x \in C.$

Recall that the process $Y_k = g(z_k)$ is a supermartingale if, for all $x \in E$, $g(x) \geqslant \mathbb{E}_x g(z_1)$ and this implies that the expectation is decreasing: $\mathbb{E}_x Y_{k+1} \leqslant \mathbb{E}_x Y_k$. Condition 2 says – very roughly – that $g(z_k)$ is a supermartingale outside some compact set, whose expectation is uniformly decreasing. An obvious consequence of Theorem (34.41) is the following.

(34.42) *Corollary.* If Z is φ-irreducible, the transition function p is strongly continuous and the state space E is compact then a stationary distribution exists.

(34.43) *Example.* The birth-death process (34.3)
Here $E = \mathbb{Z}_+$ and it is evident that the process is φ-irreducible with φ the counting measure $\varphi(\{k\}) = 1, k \in \mathbb{Z}_+$. If we take $g(x) = x$, then for $x \geqslant 1$,

$$\mathbb{E}_x g(x_1) = p(x+1) + (1-p)(x-1) = x + (2p-1).$$

Thus the conditions of the theorem are met with $C = (0)$, $\varepsilon = 1 - 2p$ as long as $p < \frac{1}{2}$.

For PDPs, the following sufficient conditions for strong continuity of $p(A, x)$ follow directly from the definition (34.12).

(34.44) *Proposition.* $p(A, x)$ defined by (34.12) is strongly continuous if

1. $t_*: E \to]0, \infty]$ is continuous;
2. $\lambda(\phi(t, y)) \to \lambda(\phi(t, x))$ as $y \to x$ for almost every $t \in [0, t_*(x)[$;
3. for every $A \in \mathscr{E}$

$$Q(A; \phi(t, y)) \to Q(A; \phi(t, x)) \text{ as } y \to x \text{ for almost every } t \in [0, t_*(x)[,$$

$$Q(A; y) \to Q(A; z) \quad \text{as} \quad y \to z, \quad \text{for all } z \in \Gamma.$$

(34.45) *Example.* The capacity expansion model (21.13)
This model will be as described in section (21.13) except that we will suppose that the demand rate μ is shortage-dependent, i.e. that the jump rate is $\lambda(v, \zeta) = \mu_v$ for some given numbers μ_v. We will also suppose that a 'threshold' construction policy is applied: investment takes place at maximum rate k whenever $v \geqslant N$ and not at all whenever $v < N$. The state space for the process $x_t = (v_t, \zeta_t)$ is then $\mathbb{Z} \times [0, p]$, as shown in Fig. 2.9. It is important to note that $E_1 = E_2 \cup E_3$ is an invariant set, where $E_2 = \{x = (v, \eta) \in E : v \geqslant N\}$ and $E_3 = \{(v, 0) \in E : v = N-K, N-K+1, \ldots, N-1\}$. This means that if $x \in E_1$, then $P_x[x_t \in E_1$ for all $t \geqslant 0] = 1$. If $x = (v, \zeta) \notin E_1$, then $x_t \in E_1$ for all $t \geqslant T_{N-v}$ so that $x_t \in E_1$ eventually if $\mu_v \geqslant \delta_1 > 0$ for $v < N$. Thus any stationary distribution must have support in E_1.

(34.46) *Proposition.* Suppose $\mu_v \geqslant \delta_1 > 0$ for $v < N$ and that

$$\limsup_{v \to \infty} \mu_v < \frac{Kk}{p}. \tag{34.47}$$

Then there exists a stationary distribution.

Remark. Note that Kk/p is the increase in capacity per unit time at maximum build rate while the left-hand side of (34.47) is (almost) an upper bound for the mean rate of demand increase.

Proof. In view of (34.47) there exists $\varepsilon' > 0$ and $M > N$ such that

$$\mu_v \leqslant \delta k := \frac{1}{p}(K - \varepsilon')k, \qquad \text{for } v > M,$$

and we can assume that $M - N > K$. Now define

$$g(v, \zeta) = \begin{cases} v + \delta(p - \zeta), & v > N \\ 2N - v, & v \leqslant N. \end{cases}$$

Then for $x = (v, \zeta)$ with $v > M$ we have, defining $a := (p - \zeta/k) = t_*(x)$,

$$\begin{aligned}
\mathbb{E}_x[g(z_1) - g(z)] &= \int_0^a \mu_v e^{-\mu_v t}(1 - \delta kt)dt + (\delta\zeta - K)e^{-\mu_v a} \\
&= \left(1 - \frac{\delta k}{\mu_v}\right)(1 - e^{-\mu_v a}) + (\delta p - K)e^{-\mu_v a} \\
&\leqslant (\delta p - K)e^{-\mu_v a} \quad (\text{since } \delta k \geqslant \mu_v) \\
&= -\varepsilon' e^{-\mu_v (p-\zeta)/k} \\
&\leqslant -\varepsilon' e^{-\mu^* p/k},
\end{aligned}$$

where μ^* is an upper bound for $(\mu_v, v > M)$. On the other hand if $v < N$ then $\mathbb{E}_x[g(z_1) - g(x)] = -1$. Thus the conditions of Theorem (34.41) are met with this choice of g and with $C = \{(v, \zeta) : N \leqslant v \leqslant M\}$ and a stationary distribution exists.

The second approach to the existence of stationary distributions is based on the following idea (Asmussen, 1987, Chapter 5). Let (x_t) be a PDP, let $y \in E$ be some fixed point in the state space and denote $\tau_y = \inf\{t > 0 : x_t = y\}$. Suppose that $P_x[\tau_y < \infty] = 1$ for all $x \in E$, i.e. whatever the starting point, the process hits y in finite time. If we denote $T_0^* = 0$ and $T_k^* = \inf\{t > T_{k-1}^* : x_t = y\}$ for $k \geqslant 1$ then it is clear from the construction of the process that $S_k^* := T_k{}^* - T_{k-1}^*$ is a sequence of i.i.d. random variables and that the processes $\{x_t^k, k = 1, 2\}$ are independent, where

$$x_t^k = \begin{cases} x_t + T_{k-1} & 0 \leqslant t < S_k^* \\ \Delta & t \geqslant S_k^*, \end{cases}$$

Δ being an adjoined cemetery state. In this case (x_t) is said to be a *regenerative process* and the times T_k^* are the *regeneration points.* Of course, the regeneration points form a renewal process, the so-called *imbedded renewal process.* The same applies if there is some point $y \in \Gamma$ such that $P_x[\sigma_y < \infty] = 1$, where $\sigma_y = \inf\{t : x_{t-} = y\}$. In this case the components x_t^k are copies of the process (x_t), where (x_t) has initial distribution π given by $\pi(A) = Q(A; y)$, $A \in \mathscr{E}$. The main result is now as follows.

Theorem 3.10 (Asmussen, 1987, Theorem V.1.2). Suppose $\mathbb{E}_y[\tau_y] = \mu_1 < \infty$ for some $y \in E$. Then the process x_t has a unique stationary distribution μ, given by

$$\int_E f(x)\mu(\mathrm{d}x) = \frac{1}{\mu_1}\mathbb{E}_y \int_0^{\tau_y} f(x_s)\mathrm{d}s. \tag{34.49}$$

Alternatively, suppose there exists $y \in \Gamma$ such that $\mathbb{E}^\pi[\sigma_y] = M_2 < \infty$, where π is the probability measure $\pi(A) = Q(A; y)$ and $\mathbb{E}^\pi$ is the expectation for the process with initial distribution π. Then there is a unique stationary distribution μ given by

$$\int_E f(x)\mu(\mathrm{d}x) = \frac{1}{\mu_2}\mathbb{E}^\pi \int_0^{\sigma_y} f(x_s)\mathrm{d}s.$$

This result gives an alternative approach to showing the existence of a stationary distribution in the capacity expansion model.

(34.50) *Example.* Capacity expansion (Example (34.45)) continued. We will suppose here that the demand arrival rate is a constant $\mu < b := Kk/p$. It is not hard to extend the result to the case of shortage-dependent rates μ_v satisfying (34.47). We claim that the point $y = (N, 0)$ is a recurrence point, specifically that $\mathbb{E}_x[\tau_y] < \infty$ for all $x \in E$. To show this, define $\tau_1 = \inf\{t : x_t \in \{(N,0), (N-1,0), \ldots, (N-K), 0\}$. It suffices to show that $\mathbb{E}_x[\tau_1] < \infty$, since $\tau_y = \tau_1 + \sigma$, where σ is the time taken to 'climb up' to $(N, 0)$ after τ_1; thus the expected value of σ is at most K/μ. By the same kind of reasoning it suffices to consider points $x = (v, \zeta)$ such that $v \geqslant N$ and, further, it suffices to consider $\zeta = 0$. If the initial capacity is c_0 and the initial demand $d_0 = v + c_0$, then for $t \leqslant \tau_1$ the capacity is precisely

$$c_t = c_0 + K[kt/p] \geqslant c_0 - K + bt,$$

where $[kt/p]$ denotes the integer part of kt/p, while the demand is $d_t = c_0 + v + N_t$, where N_t is a Poisson process with rate μ. Thus

$$v_t = d_t - c_t \leqslant N_t + v + K - bt = v_t', \quad t \leqslant \tau_1,$$

and hence $\tau_1 \leqslant \tau_2 = \inf\{t : v_t' \leqslant N\}$. By the strong law of large numbers $v'_{T_n}/n \to (1 - b/\mu) < 0$ a.s. as $n \to \infty$, which implies that $\tau_2 < \infty$ a.s. Further, τ_2 is just the length of the first 'busy cycle' for an $M/G/1$ queue in which the service time requirement is 1 with probability 1. We show in Lemma (34.51) below that τ_2 has finite expectation when $b > \mu$, and hence, invoking Theorem (34.48), that there is a unique stationary distribution under this condition, given by (34.49).

(34.51) *Lemma.* Let (N_t) be a Poisson process with rate μ. For $b, x > 0$ let $\tau = \inf\{t: N_t \leqslant bt - x\}$. If $b > \mu$ then $\mathbb{E}\tau < \infty$.

Proof. For $\alpha, c \in \mathbb{R}$, let $x_t = \alpha N_t - ct$. Then x_t is a PDP with generator

$$\mathfrak{A}f(x) = -cf'(x) + \lambda(f(x+\alpha) - f(x)).$$

In particular, if $f(x) = \mathrm{e}^x$, then $\mathfrak{A}f(x) = (\lambda(\mathrm{e}^\alpha - 1) - c)\mathrm{e}^x$. Choosing $c = \lambda(\mathrm{e}^\alpha - 1)$ we have $\mathfrak{A}f(x) = 0$ and conclude that the process

$$M_t = \exp(\alpha N_t - \lambda(\mathrm{e}^\alpha - 1)t)$$

is a martingale. (This can also be checked directly, using the independent increments property of N_t.) With τ defined as above, it is easy to see that, with probability one, $N_\tau = b\tau - x$. Now M_t is a positive supermartingale, so applying the optional sampling theorem (13.7) we have

$$1 = \mathbb{E}M_0 \geqslant \mathbb{E}M_\tau = \mathbb{E}\exp([\alpha b - \lambda(\mathrm{e}^\alpha - 1)]\tau - \alpha x)$$

which we can write as

$$\mathbb{E}\,\mathrm{e}^{\beta(\alpha)\tau} \leqslant \mathrm{e}^{\alpha x},$$

where $\beta(\alpha) = \alpha b - \lambda(\mathrm{e}^\alpha - 1)$. Now $\beta(0) = 0$ and $\beta'(0) = b - \lambda$. Thus if $b > \lambda$ there exists $\alpha_0 > 0$ such that $\beta_0 := \beta(\alpha_0) > 0$. Using the inequality $1 + \beta x \leqslant \mathrm{e}^{\beta x}$ for $\beta > 0$ we then have

$$\mathbb{E}(1 + \beta_0\tau) \leqslant \mathbb{E}\,\mathrm{e}^{\beta_0\tau} \leqslant \mathrm{e}^{\alpha_0 x}$$

so that

$$\mathbb{E}\tau \leqslant \frac{1}{\beta_0}(\mathrm{e}^{\alpha_0 x} - 1). \qquad \square$$

35 Notes and references

31, 32 These sections develop material originally presented in Davis (1984b). There are strong analogies with stochastic calculus based on Brownian motion (Karatzas and Shreve, 1988, =KS). For example the differential formula (31.4) is the Ito formula (KS, Theorem 3.3.3), while (32.4) and (32.5) are akin to the Feynman–Kac formula (KS, Theorem 5.7.6). The transformations introduced in section are standard in Markov process theory; see for example Williams (1979). On the other hand, the recursive computations of section 32.2 are of course entirely special to PDPs.

33 (33.1) *renewal theory*: this example was discussed in Davis (1984b). Further information on renewal theory can be found in many places, for example Cox (1962) or Asmussen (1987).

(33.2) *risk theory*: this section is from Dassios and Embrechts (1989).

(33.18) *permanent health insurance*: the model is due to Waters *et al.* (1991) but the analysis here is new.

34 For geneal information on stationary distributions, see Ethier and Kurtz (1986) or Asmussen (1987). Proposition (34.7) is from Ethier and Kurtz (1986, Proposition 4.9.2). Section (34.2) is from Costa (1990), to whom all the results are due. The stochastic stability result (Theorem 34.41) is due to Tweedie (1975). Recurrence and regeneration are discussed by Asmussen (1987). Kaspi and Mandelbaum (1992) develop a much more sophisticated approach based on Harris recurrence and apply it to a queueing network system modelled as a PDP. A great deal more information about recurrence, stability and stationary distributions for general Markov processes both in discrete and continuous time can now be found in the papers and book of Meyn and Tweedie (1992, 1993). Unfortunately it has not been possible, in the time available, to make use of these results here.

CHAPTER 4

Control theory

In Chapter 3 we discussed calculation of expectations and distributions for a given PDP model. In this chapter and the next we consider *parametrized families* of PDPs, with the general objective of computing *minimum* expected values of functionals representing costs or penalties in control problems. This chapter concerns 'continuous control' in which the local characteristics $\mathfrak{X}$, λ, Q depend on an additional control parameter. This is then selected at each instant of time in order to steer the process in favourable directions or to influence the jump behaviour in favourable ways. The capacity expansion problem introduced in (21.13) is a typical example, and this is discussed further in (42.13) below. Chapter 5 deals with *impulse control*, in which instead of acting at every moment the decision maker takes occasional 'interventions'. The repair shop model (21.15) is a case in point.

In continuous control it is important to isolate exactly what the class of admissible decision strategies or control policies should be. This question is addressed in section 41. In section 42 we give a simple formulation of dynamic programming for PDPs which is adequate to solve some problems, but is too restrictive to qualify as a general theory; the development of one is the goal of the remainder of the chapter. In fact, this theory is an intriguing mixture of 'deterministic' and 'stochastic' features. We show in section 44 that the control problem can be formulated as a discrete-stage Markov decision model (DMDM) in which the stages are the jump times T_k, $k = 1,2,\ldots$, of the process and at each stage the 'decision' is the control function that solves a deterministic optimal control problem. We need to topologize these control functions in an appropriate way; this is the subject of section 43, where relaxed controls and the Young topology are introduced. Sections 45 and 46 return to dynamic programming and show that a weak form of the Bellman equation gives a necessary and sufficient condition for optimality.

Finally, section 47 summarizes some information on the DMDM as required for section 44.

41 Feedback control of PDPs

In this chapter the state space of the PDP will be taken as a subset E of $\mathbb{R}^d$ (described in more detail in Assumptions (41.1)), rather than a family of such subsets as in the general definition of section 24. This is simply for notational convenience; see Remark (24.9). A vector field $\mathfrak{X}$ is then identified with a Lipschitz function $g:\mathbb{R}^d \to \mathbb{R}^d$ such that $\mathfrak{X}f(x) = \sum_i g_i(x)\partial f/\partial x^i$ and that corresponding integral curve $\phi(t, x)$ is the unique solution of the ODE.

$$\frac{\mathrm{d}}{\mathrm{d}t}\phi(t,x) = g(\phi(t,x)), \quad \phi(0,x) = x.$$

Control arises when g and/or the other local characteristics λ, Q depend on an additional *control parameter*. It is important to distinguish at the outset between control in the *interior* of the state space E and control on the *boundary* ∂E. In applications these are usually quite different kinds of action, and we will suppose that the parameters associated with them take values in possibly different sets U_0, U_Γ, respectively. We now need different notations for the transition measure from interior points and from boundary points; these will be denoted Q, R, respectively. Our standing assumptions, in force throughout this chapter, will be as follows. Here and throughout, $\mathscr{P}(E)$ will denote the set of probability measures on a (Borel) space E with the topology of weak convergence; this means that a sequence $\mu_n \in \mathscr{P}(E)$ converges to $\mu_n \in \mathscr{P}(E)$ if and only if $\mu_n(f) \to \mu(f)$ for all $f \in C_b(E)$, where $\mu(f) := \int_E f(x)\mu(\mathrm{d}x)$.

(41.1) *Assumptions*

(a) The state space is $\hat{E} = E \cup \{\Delta\}$, where $E = \{x \in \mathbb{R}^d : \psi(x) < 0\}$ for some $\psi \in C^1(\mathbb{R}^d)$ such that $\|\nabla\psi(x)\| \geqslant 1$ on $\partial E = \{x : \psi(x) = 0\}$. Here $\{\Delta\}$ is an isolated point. (Note that E is not necessarily bounded or connected.)

(b) The control parameter sets U_0, U_Γ are compact metrizable spaces. (In most applications they are compact subsets of Euclidean spaces, but the Euclidean structure plays no role in the theory.)

(c) The functions $g:\hat{E} \times U_0 \to \mathbb{R}^d$ and $\lambda:\hat{E} \times U_0 \to \mathbb{R}_+$ are bounded and Lipschitz continuous on E, uniformly in U_0; $\lambda(\Delta, u) = g(\Delta, u) = 0$ for all $u \in U_0$. The bound and Lipschitz constant will be denoted M and L, respectively.

(d) $Q: E \times U_0 \to \mathscr{P}(\hat{E})$ and $R: \partial E \times U_\Gamma \to \mathscr{P}(E)$ are continuous functions such that for all $\theta \in C_b(\hat{E})$ the maps $(x, u_0) \to \int_{\hat{E}} \theta(y) Q(dy;\ x, u_0)$ $(x \in E, u_0 \in U_0)$ and $(x, u_\Gamma) \to \int_E \theta(y) R(dy; x, u_\Gamma)$ $(x \in \partial E,\ u_\Gamma \in U_\Gamma)$ are Lipschitz continuous in x, uniformly in u_0 and u_Γ, respectively.

(e) There exists $\delta > 0$ such that $\lambda(x, u_0) \geqslant \delta$ and $Q(\{\Delta\};\ x, u_0) = \delta/\lambda(x, u_0)$, for all $(x, u_0) \in E \times U_0$.

The last assumption means that killing at rate δ is included in the definition of the process. Finite time horizon problems are included by the standard state augmentation procedure described in section 31, and in this case Assumption (41.1(e)) is not required. As in section 24, it will generally be unnecessary to specify R on all of the boundary ∂E; we will define the *active boundary* Γ in section 42 below.

The natural class of controls in Markovian optimization problems is that of *feedback policies*, a feedback policy v being in the present context a pair of measurable functions $v_0: E \to U_0$ and $v_\Gamma: \partial E \to U_\Gamma$. These are described as a *policy* because they prescribe a rule of action: if the state is x, apply control $v_0(x)$; if the boundary is hit at z, apply boundary control $v_\Gamma(z)$. We now wish to construct a PDP corresponding to a control policy (v_0, v_Γ). In principle, this is simple: $v = (v_0, v_\Gamma)$ defines a set of local characteristics $(g^v, \lambda^v, Q^v, R^v)$ by the recipe

$$g^v(x) = g(x, v_0(x)), \quad \lambda^v(x) = \lambda(x, v_0(x)),$$
$$Q^v(dy; x) = Q(dy; x, v_0(x)), \quad R^v(dy; z) = R(dy; z, v_\Gamma(z)),$$

and we construct a PDP having these local characteristics as in section 24, giving a family of sample space measures P_x^v. The problem now is to choose a policy v to minimize a given *cost function*, which we shall usually assume to be of the form

$$(41.2) \qquad J_x(v) = \mathbb{E}_x^v \left\{ \int_0^\infty l(x_t, v_0(x_t))dt + \int_0^\infty c(x_{t-}, v_\Gamma(x_{t-}))dp^*(t) \right\},$$

where $l: \hat{E} \times E_0 \to \mathbb{R}_+$ and $c: \partial E \times U_\Gamma \to \mathbb{R}_0$ are bounded non-negative functions. A policy $\hat{v}$ minimizing $J_x(v)$ for all $x \in E$ over all admissible policies v is *optimal*. This formulation, however, runs into an immediate obstruction: between jumps times, the trajectory x_t of the system must satisfy the ordinary differential equation

$$\frac{\mathrm{d}}{\mathrm{d}t}x_t = g^v(x_t) = g(x_t, v_0(x_t)). \tag{41.3}$$

In order to guarantee existence and uniqueness of a solution, the function $x \to g^v(x)$ must be at least locally Lipschitz continuous, and this puts a severe restriction on the class of control policies v_0 that can be considered; for example it rules out discontinuous functions v_0, which arise naturally in many applications where 'bang-bang' control is optimal. In particular problems it may be the case that, due to special structure, (41.3) is uniquely solvable for an adequately large class of functions v_0 even if the local Lipschitz property fails; we will outline a treatment of such problems in section 43 below. But for a general theory, feedback policies as described above are inadequate.

Fortunately, there is a very satisfactory solution to this problem, due to Vermes (1985) who introduced the class of *piecewise open-loop controls* which we describe next. It involves defining a new three-component process $\hat{x}_t = (x_t, \eta_t, \tau_t)$ on $E' = \hat{E} \times \hat{E} \times \mathbb{R}_+$, where x_t is the trajectory of the system, η_t is the value of (x_s) at the last jump time T_k before and $\tau_t = t - T_k$ is the time since the last jump.

Boundary controls are, as before, feedback functions $v_\Gamma : \partial E \to U_\Gamma$ and the control applied is $v(x_{t-})$ when $x_{t-} \in \partial E$: An *interior control* u is a measurable function $u: E^0 \times \mathbb{R}_+ \to U_0$ and the control applied at (x_t, η_t, τ_t) is $u(\eta_t, \tau_t)$; it depends only on the last two components, not on the first. Corresponding to a pair (u, v_Γ), local characteristics are defined as follows: with $\hat{x} := (x, \eta, \tau)$

$$\hat{g}^u(\hat{x}) = \begin{bmatrix} g(x, u(\eta, \tau)) \\ 0 \\ 1 \end{bmatrix}$$

$$\hat{\lambda}^u(\hat{x}) = \lambda(x, u(\eta, \tau)),$$

$$\hat{Q}^u(A \times A \times \{0\}; \hat{x}) = Q(A; x, u(\eta, \tau))$$

$$\hat{R}^{v_\Gamma}(A \times A \times \{0\}; \hat{x}) = R(A; x, v_\Gamma(x)), \qquad x \in \partial E.$$

We now construct a PDP with these local characteristics using the method of section 24. That this is possible is due to *Carathéodory's theorem* in ordinary differential equations (Warga, 1972, Theorems II.4.1, II.4.2) of which the following is a special case.

(41.4) *Theorem.* Suppose $f:\mathbb{R}^d \times \mathbb{R}_+ \to \mathbb{R}^d$ is such that $f(x,t)$ is Lipschitz continuous in x, uniformly in t, and for each $x \in \mathbb{R}^d$ the function $t \to f(x,t)$ is measurable. Then for $y_0 \in \mathbb{R}^d$ there is a unique absolutely continuous function y such that for all $t \geqslant 0$

$$y(t) = y_0 + \int_0^t f(y(s), s)\mathrm{d}s.$$

The point here is that f need only be *measurable* in t: no smoothness is required.

Returning to the construction of the PDP, note that for fixed $\eta \in \hat{E}$ the function $f(x,t) = g(x, u(\eta, t))$ satisfies the conditions of Theorem (41.4) if $g(x,u)$ is Lipschitz in x and measurable in u and $u(\eta, t)$ is a measurable function. It follows that the equation

$$\frac{\mathrm{d}}{\mathrm{d}t}\hat{x}_t = \hat{g}^u(\hat{x}_t), \quad x_0 = (x, x, 0)$$

or, in more explicit form,

$$\begin{aligned} \dot{x}_t &= g(x_t, u(\eta_t, \tau_t)), & x_0 &= x \\ \dot{\eta}_t &= 0, & \eta_0 &= x \\ \dot{\tau}_t &= 1, & \tau_0 &= 0, \end{aligned}$$

has a unique solution $(x_t, \eta_t, \tau_t) = (x_t, x, t)$. We can thus take the class of interior controls to be the set of all measurable functions $u: \hat{E} \times \mathbb{R}_+ \to U_0$ and construct solutions in the above manner. The remaining parts of the PDP construction pose no new problems and are carried out exactly as in section 24. Generally, it is not necessary to include the additional components (η_t, τ_t) explicitly in a controlled PDP model: the above construction is clearly equivalent to choosing, at each jump time T_k, an open loop control depending on the initial condition x_{T_k}, to be used until the next jump time, i.e. a function $v_0(t)$ such that $x_t = \phi(t - T_k, x_{T_k})$ for $t \in [T_k, T_{k+1}[$, where $\phi(t,x)$ satisfies

$$\frac{\mathrm{d}}{\mathrm{d}t}\phi(t,x) = g(\phi(t,x), v_0(t)), \quad \phi(0,x) = x.$$

The control policy should be thought of as a rule specifying how this choice depends on the initial state x.

The situation is in fact directly analogous to that obtaining in deterministic optimal control theory where one seeks to minimize a cost, say $\Phi(x(T))$, over controls $u(\cdot)$ for a differential system in the form

$$\frac{d}{dt}x(t) = f(x(t), u(t)), \quad x(0) = x_0, \quad t \in [0, T]. \tag{41.5}$$

The usual class of controls is that of open loop controls, i.e. measurable functions $t \to u(t)$. There is no extra generality in using feedback controls $(x(t), t) \to \psi(x(t), t)$ because for any ψ for which the feedback equation

$$\frac{d}{dt}x^{\psi}(t) = f(x^{\psi}(t)), \psi(x^{\psi}(t), t)), \quad x^{\psi}(0) = x_0$$

is solvable we can define an open loop control $u(t) = \psi(x^{\psi}(t), t)$ and the solution of (41.5) is then $x^{\psi}(t)$ again. Thus the infimal cost over open loop controls is the same as that over the, at first sight more general, class of feedback controls. The initial state x_0 is not usually included in the notation, but we should really write $u(x_0, t)$ since the control chosen will clearly depend on the initial state x_0 considered. In the PDP context we must be explicit about this dependence since we are considering a sequence of choices of control starting at initial points x_{T_k} which are randomly selected.

42 Naïve dynamic programming

As pointed out above, the class of feedback policies does not provide an adequate basis for a *general* theory of PDP controls, but is adequate for the solution of many specific problems, for example the capacity expansion problem described in (21.13), which is discussed again in (42.13). In this section we will show how the ideas of dynamic programming can be used to provide strong sufficient conditions for optimality. This is useful on two counts: first, practical problems can be solved by applying them, and second, we identify the appropriate form of the Bellman equation, which will reappear in section 45 later in the chapter in a more general context. We assume that the reader has at least some acquaintance with dynamic programming; introductory accounts in the same vein as this book can be found in Fleming and Rishel (1975), Davis (1977) or Rishel (1990). Several ideas but few specific results from these sources will be used.

The combined system local characteristics g, λ, Q, R are supposed to satisfy Assumption (41.1) and we denote by $\mathscr{U}'_0$ the set of all measurable functions $v_0: E \to U_0$ and by $\mathscr{U}_\Gamma$ the set of all measurable functions $v_\Gamma: \partial E \to U_\Gamma$. Now consider the equation

$$\frac{\mathrm{d}}{\mathrm{d}t}\phi(t) = g(\phi(t), v_0(\phi(t))), \quad \phi(0) = x \in E \tag{42.1}$$

for $v_0 \in \mathscr{U}'_0$. We denote by $\mathscr{U}_0$ the subset of $\mathscr{U}'_0$ such that for $v_0 \in \mathscr{U}_0$ equation (42.1) has, for each $x \in E$, a unique solution on the interval $[0, t_*(x)[$, where $t_*(x) = \inf\{t: \phi(t) \in \partial E\}$. Thus $\mathscr{U}_0$ consists of those control functions for which the controlled PDP can be constructed in the direct way outlined in section 41. We denote $\mathscr{U}_{\mathrm{F}} := \mathscr{U}_0 \times \mathscr{U}_\Gamma$ (F for 'feedback'). The objective is to choose a control $v \in \mathscr{U}_{\mathrm{F}}$ so as to minimize the cost

$$J_x(v) = \mathbb{E}_x^v \left\{ \int_0^\infty l(x_t, v_0(x_t))\mathrm{d}t + \int_0^\infty c(x_{t-}, v_\Gamma(x_{t-}))\mathrm{d}p^*(t) \right\}. \tag{42.2}$$

As pointed out in section 31, several apparently more general forms of cost function are actually special cases of (42.2). We make two further assumptions.

(42.3) *Assumptions.* (a) $l: E \times U_0 \to \mathbb{R}_+$ and $c: \partial E \times \mathscr{U}_\Gamma \to \mathbb{R}_+$ are non-negative, bounded measurable functions.

(b) For any $x \in E$, $v \in \mathscr{U}_\Gamma$ and $t > 0$, $\mathbb{E}_x^v N_t < \infty$, where $N_t = \sum_i I_{(t \geq T_i)}$ and T_i are the jump times of the process with control v. In particular, $T_i \to \infty$ P_x^v a.s.

Since by Assumption (42.1(e)) killing is included in the definition of the process, these assumptions imply that $J_v(x) < \infty$ for $v \in \mathscr{U}_{\mathrm{F}}$.

For $u_0 \in U_0$, $u_\Gamma \in U_\Gamma$ and $f \in C^1(\bar{E})$ we denote

$$\mathfrak{A}^{u_0} f(x) = g(x, u_0) \cdot \nabla f(x) + \lambda(x) \int_E (f(y) - f(x)) Q(\mathrm{d}y; x, u_0), \quad x \in E \tag{42.4}$$

$$\mathfrak{C}^{u_\Gamma} f(x) = \int_E (f(y) - f(x)) R(\mathrm{d}y; x, u_\Gamma), \quad x \in \partial E. \tag{42.5}$$

By slight abuse of notation, $\mathfrak{A}^v, \mathfrak{C}^v$ will denote (42.4) and (42.5) with $v_0(x)$ and $v_\Gamma(x)$ replacing u_0 and v_Γ, respectively, for $v=(v_0, v_\Gamma)\in\mathscr{U}_{\mathrm{F}}$. Thus $\mathfrak{A}^v$ is the generator of the controlled process with control v, and $\mathfrak{C}^v f=0$ if $f\in\mathscr{D}(\mathfrak{A}^v)$.

The ideas of dynamic programming (Fleming and Rishel, 1975) lead to consideration of the so-called (Hamilton–Jacobi–) Bellman equation, to be solved for a function V on $\bar{E}$. In the present context, because of the boundary conditions, this is actually the following pair of equations:

(42.6) $$\min_{u_0\in U_0}\{\mathfrak{A}^{u_0}V(x)+l(x,u_0)\}=0, \quad x\in E$$

(42.7) $$\min_{u_\Gamma\in U_\Gamma}\{\mathfrak{C}^{u_\Gamma}V(x)+c(x,u_\Gamma)\}=0, \quad x\in\partial E.$$

Introduction of these equations will be justified by showing that if a solution V of (42.6) and (42.7) having suitable properties can be found, then $V(x)$ is the minimum value of $J_x(v)$ over all feedback controls $v\in\mathscr{U}_{\mathrm{F}}$.

(42.8) *Theorem.* Suppose Assumptions (41.1) and (42.3) are satisfied and that

1. V is a piecewise C^1 solution of (42.6) and (42.7).
2. For any $v\in\mathscr{U}_{\mathrm{F}}$, $t>0$, $\mathbb{E}^u_x\sum_i I_{(t\geqslant T_i)}|V(x_{T_i})-V(x_{T_i-})|<\infty$.
3. For any $v\in\mathscr{U}_{\mathrm{F}}$, $\mathbb{E}^v_x V(x_t)\to 0$ as $t\to\infty$.
4. There exists $\hat{v}\in\mathscr{U}_{\mathrm{F}}$ such that at each $x\in E$ (respectively $x\in\partial E$) the value $\hat{v}_0(x)[\hat{v}_\Gamma(x)]$ achieves the minimum in (42.6) [(42.7)].

Then $\hat{v}$ is optimal in $\mathscr{U}_{\mathrm{F}}$ and $V(x)=J_x(\hat{v})$.

(42.9) *Remark.* Assumptions (41.1), together with condition 1 of Theorem (42.8), ensure that the minima in (42.6) and (42.7) are achieved at each x by control values $\hat{v}_0(x)$, $\hat{v}_\Gamma(x)$, and standard measurable selection theorems will guarantee that the maps $x\to\hat{v}_0(x)$, $x\to\hat{v}_\Gamma(x)$ are measurable. The content of condition 4 in the theorem is that $\hat{v}\in\mathscr{U}_{\mathrm{F}}$, i.e. the system equations are solvable when the minimizing value $\hat{v}_0(x)$ is 'plugged in'.

Proof. Suppose V is a solution to (42.6) and (42.7) satisfying the conditions of Theorem (42.8), let $v\in\mathscr{U}_{\mathrm{F}}$ be an arbitrary feedback control and let (x_t) denote the corresponding controlled process.

Then by the PDP Differential formula (31.4), for any $t > 0$

$$(42.10) \qquad V(x_t) - V(x) = \int_0^t \mathfrak{A}^v V(x_s)\mathrm{d}s + \int_0^t \int_E \mathfrak{B}^v V(y,s) q(\mathrm{d}y, \mathrm{d}s) + \int_0^t \mathfrak{C}^v V(x_{s-})\mathrm{d}p^*(s).$$

($V \in \mathscr{D}_0(\mathfrak{A}^v)$ in view of condition 1 of Theorem (42.8) and Assumption (42.3(b)).) Now from (42.6) and (42.7) $\mathfrak{A}^v(x) \geqslant -l(x, u_0(x))$ and $\mathfrak{C}^v V(x) \geqslant -c(x, u_\Gamma(x))$. By condition 2 of Theorem (42.8), the stochastic integral term is a martingale. Thus, taking expectations in (42.10) we obtain

$$V(x) \leqslant \mathbb{E}_x^v \left\{ \int_0^t l(x_s, u_0(x_s))\mathrm{d}s + \int_0^t c(x_{s-}, u_\Gamma(x_{s-}))\mathrm{d}p^*(s) \right\} + \mathbb{E}_x^v V(x_t).$$

Now let $t \to \infty$ and invoke condition 2 of Theorem 4.2 to conclude that

$$(42.11) \qquad V(x) \leqslant J_x(v).$$

Finally, let $\hat{v}$ be the control policy referred to in condition 4 of Theorem (42.8). Then the same argument as above applies, but equality replaces inequality throughout. Hence

$$(42.12) \qquad V(x) = J_x(\hat{v}).$$

Now (42.11) and (42.12) say that $\hat{v}$ is optimal. □

The merit of Theorem (42.8) is that it identifies, by a straightforward argument, the Bellman equations (42.6) and (42.7) as the appropriate objects of study. Its deficiency, on the other hand, is that it involves so many *ad hoc* conditions not directly related to the data of the problem. There is no general way of putting these conditions in more readily verifiable form: they simply have to be examined on a case-by-case basis. In the remainder of this section we re-examine the capacity expansion problem described in Chapter 2, which is one problem that is amenable to solution by a direct attack on the Bellman equation.

(42.13) *The capacity expansion problem*

The model for this problem is described in (21.13). The state space for the PDP model is $E = \mathbb{Z} \times [0, p]$ and a feedback control $u \in \mathscr{U}_\mathrm{F}$ is a function $u: E \to [0, \kappa]$. In the following, we will take $\kappa = 1$ (this just amounts to a choice of units.) It is clear from section (21.13) that the generator of the controlled PDP is

(42.14) $$\mathfrak{A}^u V_\nu(\zeta) = u(\nu,\zeta) V'_\nu(\zeta) + \mu(V_{\nu+1}(\zeta) - V_\nu(\zeta)),$$

where we write $V_\nu(\zeta) = V(\nu,\zeta)$ and $V'_\nu(\zeta) = \mathrm{d}V_\nu(\zeta)/\mathrm{d}\zeta$, with the boundary condition

(42.15) $$V_\nu(p) = V_{\nu-K}(0), \quad \nu \in \mathbb{Z}.$$

As things stand the state space has an infinite number of components, and to make the problem tractable we must introduce some form of truncation. The easiest way to do this is to introduce an extra ingredient into the problem specification: let us agree that if the shortage level ν ever reaches some prescribed large value M then the contractors are fired and have to pay a fixed penalty P. The total expected cost to be minimized is now

(42.16) $$J_x(u) = \mathbb{E}_x\left\{\int_0^\tau \mathrm{e}^{-\delta t}(u(x_t) + h(\nu_t))\mathrm{d}t + P\mathrm{e}^{-\delta\tau}\right\},$$

where $\tau = \inf\{t : \nu_t = M\}$. It is clear that if M is large enough this modification will have a negligible effect on the problem. The Bellman equation associated with the modified problem is

(42.17) $$\min_{u\in[0,1]} \{\mathfrak{A}^u V_\nu(\zeta) - \delta V_\nu(\zeta) + u + h_\nu\} = 0$$

together with the boundary conditions (42.15) and

(42.18) $$V_M(\zeta) = P.$$

By Theorem (42.8), any solution of (42.15) and (42.16)–(42.17) satisfying conditions 1–3 of Theorem (42.8) is the value function. Strictly speaking, the conditions of the theorem are not met unless the function h is bounded; however, we will see later that the values of $h(\nu)$ for large negative ν are irrelevant to the problem. The term involving u in (42.17) is $u[V'_\nu(\zeta) + 1]$ and therefore the minimizing value of u is

(42.19) $$u^0(\nu,\zeta) = \begin{cases} 1 & \text{if } V'_\nu(\zeta) < -1 \\ 0 & \text{if } V'_\nu(\zeta) \geq -1. \end{cases}$$

u^0 is therefore *bang-bang*: one either builds at maximum rate or not at all, and the state space divides into two regions, the 'build' region $\mathscr{B} = \{x : u^0 = 1\}$ and the 'wait' region $\mathscr{W} = \{x : u^0 = 0\}$. It is easy to conjecture roughly what these regions look like; indeed, it must be

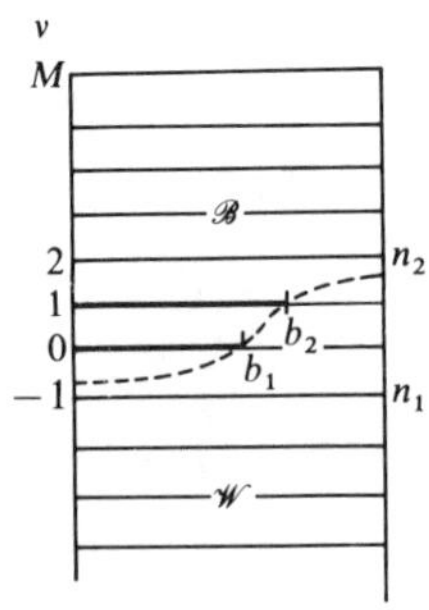

Fig. 4.1. *An FRD policy for capacity expansion.*

the case that if $(v,\zeta)\in\mathscr{B}$ then $(v',\zeta)\in\mathscr{B}$ for $v'>v$ (greater shortage, everything else the same), while if $(v,\zeta)\in\mathscr{W}$ then $(v',\zeta)\in\mathscr{W}$ for $v'<v$. Thus the build region 'sits on top of' the wait region (Fig. 4.1).

What is not so obvious, but turns out to be true, is that, optimally, the boundary between $\mathscr{B}$ and $\mathscr{W}$ constitutes what we will call an FRD (follow realized demand) policy: there are integers $n_1<n_2$ and numbers $0<b_1<b_2<\cdots<b_{n_2-n_1-1}<p$ such that $(v,\zeta)\in\mathscr{B}$ if $v\geqslant n_2$ or $n_1<v<n_2$ and $\zeta<b_{v-n_1}$. Figure 4.1 shows an FRD policy with $n_1=-1$, $n_2=2$; thus $\mathscr{B}$ consists of $(\{0\}\times[0,b_1[)\cup(\{1\}\times[0,b_2[)$ together with all (v,ζ) with $v\geqslant 2$. We have also taken $K=4$. Starting at $(0,0)$ the controlled process will, if no demand increments materialize, hit $(0,b_1)\in\mathscr{W}$ at time b_1. It waits there until the next demand increment takes it to $(1,b_1)$, when building is resumed; and so on. This explains the term FRD. Note that $n_2=2$ is the lowest level of shortage at which a project is completed, and therefore the set $\mathscr{I}=\mathscr{B}\cup\{(-1,0),(-2,0)\}$ is an *invariant set*: the process never leaves $\mathscr{I}$ once it is entered. Further, $\mathscr{I}$ is entered starting from any point in $\mathscr{W}$, since in $\mathscr{W}$ the process simply jumps vertically upwards in unit steps at Poisson times.

Let us now discuss how to compute the expected cost for a given FRD policy, without any concern for optimality at this stage. This is most easily described in terms of an example, so we will take the FRD policy of Fig. 4.1, and the upper limit $M=20$. In view of the formula (42.14) for the generator, $v_v(\zeta):=J_{(v,\zeta)}(u)$ satisfies the boundary conditions (42.15), (42.18) and

(42.20) $$v_v'(\zeta)+\mu(v_{v+1}(\zeta)-v_v(\zeta))-\delta v_v(\zeta)+h_v=0,\quad (v,\zeta)\in\mathscr{B},$$

(42.21) $$\mu(v_{v+1}(\zeta)-v_v(\zeta))-\delta v_v(\zeta)+h_v=0,\quad (v,\zeta)\in\mathscr{W}.$$

Let us consider solving (42.20), ignoring for a moment the boundary with $\mathscr{W}$. Since $v_{20} = P$, equations (42.20) with $\nu = 19,18,17,16$ are a closed system of linear differential equations, with unique solution given a four-vector $\boldsymbol{X}$ of initial conditions, $\boldsymbol{X}^i = v_{20-i}(0)$, $i = 1,\ldots,4$. We now use the boundary condition (42.15) which says that

$$v_{16-i}(0) = v_{20-i}(p), \quad i = 1,\ldots,4.$$

This provides the initial condition for the next quartet of equations determining $v_{15}(\zeta),\ldots,v_{12}(\zeta)$. Hence all v_ν are determined recursively and are affine functions of $\boldsymbol{X}$. Thus the seven-vector $\boldsymbol{Y} = [v_2(b_2), v_1(b_2), v_1(b_1), v_0(b_1), v_0(0), v_{-1}(0), v_{-2}(0)]^T$ is given by

$$\boldsymbol{Y} = A\boldsymbol{X} + \boldsymbol{b}, \tag{42.22}$$

for some computable 7×4 matrix A and seven-vector $\boldsymbol{b}$. We now use the fact that the points $(1, b_2), (0, b_1), (-1, 0), (-2, 0)$ are in $\mathscr{W}$, and therefore (42.21) holds, and that the functions $\zeta \to v_\nu(\zeta)$ are continuous at these points (Theorem 32.2). Thus from (42.21) we obtain

$$Y_2 = \frac{\mu}{\alpha} Y_1 + \frac{1}{\alpha} h_1$$

$$Y_4 = \frac{\mu}{\alpha} Y_3 + \frac{1}{\alpha} h_0$$

$$Y_6 = \frac{\mu}{\alpha} Y_5 + \frac{1}{\alpha} h_{-1}$$

$$Y_7 = \frac{\mu}{\alpha} Y_6 + \frac{1}{\alpha} h_{-2},$$

where $\alpha = \mu + \delta$, which we can write as $C\boldsymbol{Y} = \boldsymbol{d}$, where C is 4×7 and $\boldsymbol{d}$ is a four-vector. From (42.22) we now have $C\boldsymbol{Y} = CA\boldsymbol{X} + C\boldsymbol{b} = \boldsymbol{d}$ and hence

$$\boldsymbol{X} = (CA)^{-1}(\boldsymbol{d} - C\boldsymbol{b}).$$

The 4×4 matrix CA is guaranteed to be non-singular since, by Theorem (32.10), equations (42.15), (42.18), (42.20) and (42.21) are known to have a unique solution. Once the initial condition $\boldsymbol{X}$ is determined, the solution is easily computed at all points in $\mathscr{B}$, and then at all points in $\mathscr{W}$ by using (42.21). (This is why no truncation of the state space is required in the negative ν direction.)

This FRD policy is optimal if (42.19) is satisfied, i.e. $v'_v < -1$ in $\mathscr{B}$ and $v'_v \geqslant -1$ in $\mathscr{W}$. Consider for example a point $(1, c)$, where $c > b_2$. Then $(1, c) \in \mathscr{W}$ while $(2, c) \in \mathscr{B}$ and from (42.21) we see that $v'_1(c) = (\mu/\alpha)v'_2(c)$. To satisfy (42.19) we must have $v'_2(c) \in [-\alpha/\mu, -1[$. Similarly, $v'_1(c) \in [-\alpha/\mu, -1[$ for $c \in [b_1, b_2[$ and $v'_0(c) \in [-\alpha/\mu, -1[$ for $c \in [0, b_1[$. If v satisfies all these conditions then it satisfies the Bellman equation and this policy is optimal. It is quite easy to write a computer program that searches over FRD policies for one satisfying the Bellman equation. This was done in Davis, Dempster, Sethi and Vermes (1987b), where numerical results are reported for the simplest case $K = 1$. We found that a *simple FRD policy*, i.e. one with $n_2 = n_1 + 1$ is optimal when there is no penalty for excess capacity, while general FRD policies are optimal when $h(v)$ and $h(-v)$ increase monotonically in $v > 0$. This was true in all the cases studied but no theorem is available stating that it is generally true.

A standard technique for solving Bellman equations is the method of *policy improvement* (see, for example, Theorem VI.6.1 of Fleming and Rishel, 1975). Define $F(u, v) = \mathfrak{A}^u v - \delta v + h + u$ so that the Bellman equation (42.17) reads

$$\lim_{u \in [0,1]} F(u, v) = 0.$$

Choose an arbitrary feedback policy u_0 and define sequences $u_k \in \mathscr{U}_F$ and functions v_k for $k = 0, 1, \ldots,$, recursively by

$$\begin{cases} F(u_k, v_k) = 0 \\ F(u_{k+1}, v_k) = \min_{u \in [0,1]} F(u, v_k) \quad (\leqslant 0), \end{cases} \tag{42.23}$$

where in all cases it is understood that v_k satisfies the boundary conditions (42.15), (42.18). Then

$$0 \geqslant F(u_{k+1}, v_k) - F(u_{k+1}, v_{k+1}) = \mathfrak{A}^{u_{k+1}}(v_k - v_{k+1}) - \delta(v_k - v_{k+1})$$

and it follows easily that

$$J_x(u_{k+1}) \leqslant J_x(u_k).$$

The policy improvement algorithm thus generates a sequence of controls with decreasing cost. In this case all u_k are bang-bang, i.e. define build and wait regions $\mathscr{B}_k$ and $\mathscr{W}_k$. Numerical work (Skiadas, 1986) indicates that these rapidly converge to regions $\mathscr{B}, \mathscr{W}$ satisfying the Bellman equation but, again, no general proof of convergence is available, and this is an interesting area for future research.

43 Relaxed controls

We now return to consideration of control using piecewise open loop policies as discussed in section 41. At each jump time T_k we choose a control function $u(\cdot)$, depending on the state x_{T_k}, to be used until the next jump time. To make further progress with a general theory, we have to introduce a topology on the class of control functions, ideally in such a way that:

1. the class of controls is compact; and
2. the solution trajectories depend continuously on the controls.

These are conflicting requirements. To see this in general terms, consider a map $f: X \to (Y, \mathscr{S})$, where X is a set and $(Y, \mathscr{S})$ a topological space ($\mathscr{S}$ is the class of open sets in Y). If $\mathscr{T}_1$ and $\mathscr{T}_2$ are two topologies on X then $\mathscr{T}_2$ is *stronger* than $\mathscr{T}_1$ if $\mathscr{T}_1 \subset \mathscr{T}_2$, i.e. every 1-open set is 2-open. The function f is i-continuous if $f^{-1}(\mathscr{S}) \subset \mathscr{T}_i$, $i = 1,2$. Clearly, then, f is 2-continuous if it is 1-continuous, so that a stronger topology makes it easier for a given function to be continuous. On the other hand, X is *i-compact* if every cover by open sets of $\mathscr{T}_i$ has a finite sub-cover; thus 2-compactness implies 1-compactness since every 1-cover is *a fortiori* a 2-cover. So a stronger topology makes it *more difficult* for a space to be compact. For given X, f and $(Y, \mathscr{S})$ it is a juggling act to find a topology which is weak enough to make X compact but strong enough to make f continuous. To carry out this act for the control problem, we first introduce the notion of *relaxed controls*. Intuitively, the idea is that at each time t, instead of choosing a fixed control value $u(t) \in U_0$ we randomize over the set of possible values U_0, i.e. choose a value according to some probability distribution $v_t(\mathrm{d}u)$. The mean value of the velocity vector under this randomization is $\int_{U_0} g(x, u) v_t(\mathrm{d}u)$ and the corresponding trajectory is $\phi(t)$ satisfying

$$\text{(43.1)} \qquad \frac{\mathrm{d}}{\mathrm{d}t}\phi(t) = \int_{U_0} g(\phi(t), u) v_t(\mathrm{d}u), \quad \phi(0) = x.$$

Here g is the family of controlled vector fields as defined in section 41. It is convenient to assume that g is defined on all of $\mathbb{R}^d \times U_0$ (not just for $x \in E$) and satisfies Assumption (41.1(c)) with $\mathbb{R}^d$ replacing $\hat{E}$; this enables us to define solution trajectories $\phi(t)$ for arbitrary t without worrying (at this stage) whether $t \leqslant t_*(x)$.

(43.2) *Definition.* The set $\mathscr{R}$ of *relaxed controls* is the set of measurable functions $v:\mathbb{R}_+ \to \mathscr{P}(U_0)$.

Since g is bounded and continuous, the function

$$\tilde{g}(x,t) = \int_{U_0} g(x,u)v_t(\mathrm{d}u)$$

satisfies the Carathéodory conditions (Theorem (41.4)) for any $v \in \mathscr{R}$, so (43.1) has a unique solution. Relaxed controls are a generalization of ordinary controls: if $u(\cdot)$ is a measurable U_0-valued function then $v_t(\mathrm{d}u) := \delta_{u(t)}(\mathrm{d}u)$ is in $\mathscr{R}$, where δ_u is the Dirac measure at $u \in U_0$, and the right-hand side of (43.1) is just $g(\phi(t), u(t))$ in this case. There is one case in which $\mathscr{R}$ is effectively no larger than the set of ordinary controls: this is when the set $g(x, U_0) = \{g(x,u): u \in U_0\}$ is convex. Then for each t, $\int_{U_0}(\phi(t), u)v_t(\mathrm{d}u) \in g(\phi(t), U_0)$ and hence there is some $u(t) \in U_0$ such that $\int_{U_0} g(\phi(t), u)v_t(\mathrm{d}u) = g(\phi(t), u(t))$. A *measurable selection theorem* (Warga, 1972, Theorem I.7.7) assures that $u(t)$ can be chosen in such a way that the function $t \to u(t)$ is measurable. Then $u(\cdot)$ is an ordinary control and clearly $\phi(t) = \tilde{\phi}(t)$ for all $t \geqslant 0$, where $\tilde{\phi}(t)$ is the solution of $(\mathrm{d}/\mathrm{d}t)\tilde{\phi}(t) = g(\tilde{\phi}(t), u(t))$, $\tilde{\phi}(0) = x$. In general, going to relaxed controls has the effect of convexifying the velocity set: we are able to choose any velocity in $\mathrm{co}\{g(x, U_0)\}$ rather than just those in $g(x, U_0)$. Of course, this strictly enlarges the set of solution trajectories that can be generated; for example if $d = 1$, $\dot{\phi}(t) = u(t)$, $\phi(0) = x$ and $U_0 = \{-1, +1\}$ then the trajectory $\phi(t) \equiv x$ is generated by the relaxed control $u_t(\mathrm{d}u) = \frac{1}{2}\delta_{\{-1\}}(\mathrm{d}u) + \frac{1}{2}\delta_{\{+1\}}(\mathrm{d}u)$ but is not generated by any ordinary control. There are, however, theorems which assert that under reasonable conditions any 'relaxed' trajectory can be uniformly approximated by 'ordinary' trajectories. In our example, for instance, the sequence $u^n(t)$ given by $u^n(t) = 1$, $t \in [2(k-1)/n, (2k-1)/n[$ and $u^n(t) = -1$, $t \in [(2k-1)/n, 2k/n[$, $k = 1, 2, \ldots$, gives a solution trajectory ϕ^n satisfying $|\phi^n(t) - x| \leqslant 1/n$, so that (u^n) 'approximates' v_t. We shall not make use of general results of this sort in this book, but they are important in the interpretation of the results proved here.

Let us now introduce a topology on $\mathscr{R}$. Let $X = L_1(\mathbb{R}_+; C(U_0))$; thus X consists of functions $f(t,u)$ that are measurable in t, continuous in u and satisfy

$$\| f \| := \int_0^\infty \max_{u \in U_0} | f(t,u)| \mathrm{d}t < \infty.$$

X is a Banach space under this norm; its dual space is $X^* = L_\infty(\mathbb{R}_+; C^*(U_0))$. $C^*(U_0)$ (the dual space of $C(U_0)$) consists of the set of signed measures[1] on U_0 under the total variation norm: any signed measure v can be minimally decomposed into the difference $v^+ - v^-$ of two positive measures v^+ and v^-, and $\|v\|_{C^*} = v^+(U_0) + v^-(U_0)$. Thus X^* consists of measurable functions $v: \mathbb{R}_+ \to C^*(U_0)$ such that

$$\|v\|_* = \operatorname*{ess\,sup}_{t \in \mathbb{R}_+} \|v_t\|_{C^*} < \infty.$$

The *weak* topology* on X^* is the topology which is equivalent to the following notion of convergence: for $v^n, v \in X^*$, $v^n \to v$ if and only if $(f, v^n) \to (f, v)$ for all $f \in X$, where (f, v) denotes the pairing between $f \in X$ and $v \in X^*$. The *unit ball* in X^* is the set $B_1 = \{v \in X^*: \|v\|_* \leqslant 1\}$. It is a standard result in functional analysis (Alaoglu's Theorem, Dunford and Schwartz, 1958, Theorem V.4.2) that B_1 *is compact in the weak* topology*. Now the set $\mathscr{R}$ is a subset of B_1; in fact $v \in \mathscr{R}$ is a measurable function $v: \mathbb{R}_+ \to \mathscr{P}(U_0) \subset C^*(U_0)$ and $\|v\|_* = 1$.

(43.3) *Proposition.* $\mathscr{R}$ is a closed subset of B_1, and hence compact.

Proof. Take a sequence $(v^n) \subset \mathscr{R}$ and suppose $v^n \to v \in B_1$ in the weak* topology. (Certainly the limit is in B_1 since B_1 is closed). First, v_t must be a positive measure for almost all t. If not, there exists $f \in X$ such that $f \geqslant 0$ but $(f, v) < 0$. (To produce such an f, consider continuous functions approximating the indicator function of the support of v^-.) But this leads to a contradiction since $(f, v^n) \to (f, v) < 0$, while $(f, v^n) \geqslant 0$ because v^n is a positive measure. Now let us show that, for almost all t, v_t has total mass 1. Take $f(t, u) = \mathrm{e}^{-\alpha t}$ for some $\alpha > 0$; then certainly $f \in X$, so

$$\int_0^\infty \mathrm{e}^{-\alpha t}\,\mathrm{d}t = (f, v^n) \to (f, v) = \int_0^\infty \mathrm{e}^{-\alpha t} v_t(U_0)\,\mathrm{d}t.$$

Thus

$$\int_0^\infty \mathrm{e}^{-\alpha t}(1 - v_t(U_0))\,\mathrm{d}t = 0, \quad \text{for all } \alpha > 0.$$

By the uniqueness theorem for Laplace transforms, this implies $v_t(U_0) = 1$ a.e. (i.e. except on a set of Lebesgue measure zero) and

[1] A signed measure is a countably additive set function with no requirement of non-negativity.

hence that $v \in \mathscr{R}$. Thus $\mathscr{R}$ is closed; compactness follows from part 3 of Theorem (11.3).

(43.4) *Definition.* The *Young topology* $\mathscr{Y}$ on $\mathscr{R}$ is the relative weak* topology of $\mathscr{R}$ considered as a subset of B_1.

Thus $(\mathscr{R}, \mathscr{Y})$ is a compact space. We now have to show that trajectories are continuous functions of controls.

(43.5) *Theorem.* (Warga, 1972, p. 325). Let g satisfy Assumption (41.1(c)) (with $\mathbb{R}^d$ replacing $\hat{E}$), take $r \in \mathscr{R}$ and let $\phi(\cdot)$ be the unique solution of the equation

$$\frac{\mathrm{d}}{\mathrm{d}t}\phi(t) = \int_{U_0} g(\phi(t), u) r_t(\mathrm{d}u), \qquad \phi(0) = x \in \mathbb{R}^d. \tag{43.6}$$

Then the map $(x, r) \to \phi(\cdot)$ is continuous from[2] $\mathbb{R}^d \times \mathscr{R} \to C([0, T]; \mathbb{R}^d)$ for any $T \in \mathbb{R}_+$.

Proof. Let $x^n \to x$, $r^n \to r$ and ϕ^n, ϕ be the corresponding solutions of (43.6). Then writing (43.6) in integral form,

$$\begin{aligned} &|\phi^n(t) - \phi(t)| \\ &\quad \leqslant |x^n - x| + \left| \int_0^t \int_{U_0} g(\phi^n, u) r_s^n(\mathrm{d}u)\mathrm{d}s - \int_0^t \int_{U_0} g(\phi^n, u) r_s(\mathrm{d}u)\mathrm{d}s \right| \\ &\quad \leqslant |x^n - x| + \int_0^t \int_{U_0} |g(\phi^n, u) - g(\phi, u)| r^n(\mathrm{d}u)\mathrm{d}s + |y^n(t)| \end{aligned}$$

where

$$y^n(t) = \int_0^t \int_{U_0} g(\phi, u)(r_s^n(\mathrm{d}u) - r_s(\mathrm{d}u))\mathrm{d}s.$$

Thus, using the Lipschitz continuity of g and the Gronwall–Bellman inequality, we obtain

$$\begin{cases} |\phi^n(t) - \phi(t)| \leqslant |x^n - x| + K \displaystyle\int_0^t |\phi^n(s) - \phi(s)|\mathrm{d}s + |y^n(t)| \\ \qquad\qquad \leqslant |x^n - x| + |y^n(t)| + \mathrm{e}^{Kt} \displaystyle\int_0^t |y^n(s)|\mathrm{d}s. \end{cases} \tag{43.7}$$

[2] $C([0, T]; \mathbb{R}^d)$ is a Banach space under the *uniform norm* $\|y\| = \max_{t \in [0,T]} |y(t)|$.

Now we can write $y^n(t)$ as $y^n(t) = (f_t, r^n - r)$, where $f_t(s, u) = g(\phi(s, u))I_{(s \leqslant t)} \in X$; hence $y^n(t) \to 0$ as $n \to \infty$ for each $t \in \mathbb{R}_+$. Since $s \to g(\phi(s), u)$ is bounded, it is easily seen that $y^n(t)$ is equicontinuous on any interval $[0, T]$, and it follows from the Ascoli theorem (Royden, 1988, Theorem 9.30) that $\max_{0 \leqslant t \leqslant T} |y^n(t)| \to 0$. In view of (43.7), this implies that

$$\max_{0 \leqslant t \leqslant T} |\phi^n(t) - \phi(t)| \to 0. \qquad \square$$

44 Control via discrete-time dynamic programming

Having defined a suitable class of control functions, namely the set $\mathscr{R}$ of relaxed controls, we are now in a position to give a general formulation of the control problem. We continue to suppose that the data (ψ, g, λ, Q, R) of the problem satisfy Assumption (41.1); some further assumptions will be imposed below.

For a relaxed control $r \in \mathscr{R}$, let $\phi^r(t, x)$ be the corresponding solution trajectory starting at $x \in E$, and let t_* be the time at which this trajectory hits the boundary (or $+\infty$ if it never does.) Now let

$$\Gamma = \{z \in \partial E : z = \phi^r(t_*, x) \text{ for some } x \in E, r \in \mathscr{R}\}.$$

This is the *active boundary*: the set of points from which boundary jumps might possibly be made. Note that since λ is assumed to be bounded, there is no distinction between Γ and Γ^* as there was in section 24.

The class of piecewise open-loop controls $\mathscr{U}_{\mathrm{P}}$ is defined as the set of measurable functions $v : E \to \mathscr{R} \times U_\Gamma$ ('measurable' means that if $v(x) = (v^0(x), v^\Gamma(x))$ then the map $v^f : E \to \mathbb{R} \times U_\Gamma$ defined by $v^f(x) = ((f, v^0(x)), v^\Gamma(x))$ is measurable for each $f \in X$). The PDP (x_t) corresponding to $v \in \mathscr{U}_{\mathrm{P}}$ is constructed as in section 24 with slight modifications because of the use of relaxed controls. Starting at $x \in E$, we let $x_t = \phi(t)$ for $t < T_1$, where $(\phi(t), \chi(t))$ is the unique solution of the system of equations

$$\text{(44.1)} \qquad \frac{\mathrm{d}}{\mathrm{d}t}\phi(t) = \int_{U_0} g(\phi(t), u) v_t^0(\mathrm{d}u, x), \quad \phi(0) = x$$

$$\text{(44.2)} \qquad \frac{\mathrm{d}}{\mathrm{d}t}\chi(t) = -\chi(t) \int_{U_0} \lambda(\phi(t), u) v_t^0(\mathrm{d}u, x), \quad \chi(0) = 1.$$

The boundary hitting time is $t_* = \inf\{t, \phi(t) \in \Gamma)\}$ and the distribution function of the first jump time T_1 is

(44.3) $$P_x^v(T_1 > t) = I_{(t<t_*)}\chi(t).$$

The conditional distribution of x_{T_1}, given T_1, is

(44.4) $$P_x^v[x_{T_1} \in A \mid T_1] = I_{(t<t_*)} \int_{U_0} Q(A:\phi(T_1), u) v_t^0(\mathrm{d}u, x) + I_{(t=t_*)} R(A; \phi(t_*), v^\Gamma(x)).$$

The process now restarts from x_{T_1} according to the same recipe.

It is important to note that this is *not* boundary feedback control. The boundary action $v^\Gamma(x)$ is a function of the *starting point* of the trajectory, not the boundary hitting point. The latter is $\phi(t_*)$, determined by v^0, and this is the point at which action $v_\Gamma(x) \in U_\Gamma$ is taken. There is nothing in the theory excluding the possibility that there might be a $v \in \mathscr{U}_\mathrm{P}$ and starting points $x, x' \in E$ with corresponding trajectories ϕ, ϕ' under v such that $\phi(t_*) = \phi'(t'_*)$ but $v_\Gamma(x') \neq v_\Gamma(x')$, i.e. two trajectories starting at different points arrive at the same boundary point and then take different boundary actions. So $\mathscr{U}_\mathrm{P}$ is not equivalent to boundary feedback control; but we will show that the sort of behaviour just described will never be optimal.

We need to impose some conditions. The first of these (Assumption 44.4(a)) is the usual one to the effect that jump times cannot accumulate. Conditions 2 and 3 relate to the behaviour of the vector field g at the boundary of the state space. The *outward normal* $v(x)$ at $x \in \partial E = \{x : \psi(x) = 0\}$ is defined by $v(x) = \nabla\psi(x)/\|\nabla\psi(x)\|$.

(44.5) *Assumption* (a) For any $x \in E$, $v \in \mathscr{U}_\mathrm{P}$ and $t \in \mathbb{R}_+$ we have $\mathbb{E}_x^v \sum_i I_{(t \geqslant T_i)} < \infty$.

(b) There exists $\alpha > 0$ such that $g(x, u_0) \cdot v(x) \geqslant \alpha$ for all $x \in \Gamma, u_0 \in U_0$.

(c) There is an extension of g to all of $\mathbb{R}^d$ such that if $\phi(t)$ is any solution trajectory as in (44.1) starting at $\phi(0) = x \in \Gamma$, then $d(\phi(t), E) \geqslant \alpha t$.

Here d is the distance function $d(x, E) = \inf_{y \in E} |x - y|$. Assumption (44.5(b)) implies (since g is continuous) that trajectories close to the boundary must proceed towards it at a uniform rate. This is an unpleasantly strong condition, but it seems that something close to it is necessary if we are to get the continuity result of Theorem

(44.9) below. It was originally introduced by Vermes (1985). Given Assumption (44.5(b)), Assumption (44.5(c)) is pretty harmless. It is needed for technical reasons in the proof of Theorem (44.9).

The cost corresponding to $v \in \mathscr{U}_P$ is

(44.6)
$$J_x(v) = \mathbb{E}_x^v \left\{ \int_0^\infty \int_{U_0} l(x_t, u) v^0_{t-\sigma(t)}(\mathrm{d}u, x_{\sigma(t)}) \mathrm{d}t + \int_0^\infty c(x_{t-}, v_\Gamma(x_{\sigma(t)})) \mathrm{d}p^*(t) \right\},$$

where $\sigma(t)$ is the time of the last jump *strictly before t*. The first form of control problem, precisely stated, is now to minimize $J_x(v)$ given by (44.6) over the class $\mathscr{U}_P$ of piecewise open-loop controls. This class is, however, not quite the natural one because, as mentioned above, the boundary component is not feedback from the boundary as one would expect it to be. We therefore introduce a further class of controls, denoted $\mathscr{U}$, where each element $u \in \mathscr{U}$ is a pair (u_0, u_Γ) of measurable functions $u_0 : E \to \mathscr{R}$ and $u_\Gamma : \Gamma \to U_\Gamma$. Note that $\mathscr{U}$ can be regarded as a subset of $\mathscr{U}_P$. Indeed, for $u \in \mathscr{U}$ let $\phi(t)$ be the trajectory corresponding to $u_0(x)$ and define $v_\Gamma(x) = u_\Gamma(\phi(t_*))$ if $t_* < \infty$, $v_\Gamma(x) = u' \in U_\Gamma$ if $t_* = \infty$, where u' is an arbitrary element of U_Γ. Then $v(x) := (u_0(x), v_\Gamma(x))$ defines a policy in $\mathscr{U}_P$. As pointed out earlier, the inclusion $\mathscr{U} \subset \mathscr{U}_P$ is strict, in that there may be policies in $\mathscr{U}_P$ which are not equivalent to boundary feedback control. The cost for a policy $u \in \mathscr{U}$ is

$$J_x(u) = \mathbb{E}_x^u \left\{ \int_0^\infty \int_{U_0} l(x_t, v) u_0(t - \sigma(t))(\mathrm{d}v, x_{\sigma(t)}) \mathrm{d}t + \int_0^\infty c(x_{t-}, u_\Gamma(x_{t-})) \mathrm{d}p^*(t) \right\}.$$

It will be shown in Theorem (44.17) below that there is an optimal policy $\hat{v} \in \mathscr{U}_P$, and then in Theorem (44.22) that we can construct an element $\hat{u} \in \mathscr{U}$ such that $J_x(\hat{u}) = J_x(\hat{v})$. Thus $\hat{u}$ is optimal in the larger class $\mathscr{U}_P$.

We next show that minimizing $J_x(v)$ over $v \in \mathscr{U}_P$ is equivalent to optimizing a discrete-stage Markov decision model (DMDM). A description of the DMDM is given in section 47 at the end of this chapter, together with a summary of the results we shall use here.

Define functions $h: E \times \mathscr{R} \times U_\Gamma \to \mathbb{R}$ and $q: E \times \mathscr{R} \times U_\Gamma \to \mathscr{P}(E)$, as follows

$$
(44.7)\quad
\begin{cases}
h(x, r, u_\Gamma) = \displaystyle\int_0^{t_*} \chi(t) \int_{U_0} l(\phi(t), u) r_t(\mathrm{d}u)\mathrm{d}t + c(\phi(t_*), u_\Gamma)\chi(t_*). \\[2ex]
q(A; x, r, u_\Gamma) = \displaystyle\int_0^{t_*} \chi(t) \int_{U_0} Q(A; \phi(t), u)\lambda(\phi(t), u) r_t(\mathrm{d}u)\mathrm{d}t \\
\qquad\qquad + R(A; \phi(t_*), u_\Gamma)\chi(t_*),
\end{cases}
$$

where $\phi(t), \chi(t)$ are given by (44.1) and (44.2) with r_t replacing v_t^0. Given a control $v \in \mathscr{U}_P$, a discrete-time Markov process $\{Z_k, k = 0, 1, \ldots\}$ can be constructed having q as its transition function, i.e. satisfying

$$
(44.8)\qquad P^v[Z_{k+1} \in A \mid Z_k = z] = q(A; z, v(z)).
$$

(44.9) *Theorem.* For $v \in \mathscr{U}_P$,

$$
(44.10)\qquad J_x(v) = \mathbb{E}_x \left\{ \sum_{k=0}^{\infty} h(Z_k, v(Z_k)) \right\},
$$

where h and q are defined by (44.7) and Z_k is the Markov chain having transition function q as in (44.8).

Proof. For $x \in E, r \in \mathscr{R}, u_\Gamma \in U_\Gamma$, construct the process $x(t)$ for $t \leqslant T_1$ as in (44.1)–(44.4) above, with r replacing $v^0(x)$ and u_Γ replacing $v^\Gamma(x)$. It is then evident that

$$
q(A; z, r, u_\Gamma) = P_z^{(r, u_\Gamma)}[x_{T_1} \in A],
$$

i.e. that q is the transition function of the Markov chain $\{x_{T_k}, k = 0, 1, \ldots\}$, which we can therefore identify with Z_k. Thus it suffices to show that for $v \in \mathscr{U}_P$,

$$
J_x(v) = \mathbb{E}_x^v \left\{ \sum_{k=0}^{\infty} h(x_{T_k}, v(x_{T_k})) \right\}.
$$

Let $\mathfrak{F}_t$ denote the natural filtration of the controlled process. Then using the strong Markov property, which holds for the augmented process introduced in section 41 by a slight extension of Theorem

(25.5), we have (with $T_0 := 0$)

$$\begin{aligned} J_x(v) &= \mathbb{E}_x^v\left\{\mathbb{E}_x^v\left[\sum_{k=0}^{\infty}\int_{T_k}^{T_{k+1}}\int_{U_0} l(x_t,u)v_t^0(\mathrm{d}u, x_{T_k})\mathrm{d}t\right.\right. \\ &\quad \left.\left. + \int_{]T_k,T_{k+1}]} c(x_{t-}, u_\Gamma(x_{T_k}))\mathrm{d}p^*(t)\middle| \mathfrak{F}_{T_k}\right]\right\} \\ &= \mathbb{E}_x^v\left\{\sum_{k=0}^{\infty}\mathbb{E}_x^v[\cdots | x_{T_k}]\right\} \\ &= \mathbb{E}_x^v\left\{\sum_{k=0}^{\infty} h(x_{T_k}, v(x_{T_k}))\right\}. \qquad \square \end{aligned}$$

This completes the identification of the PDP control problem with a discrete-time Markov decision problem, namely that of minimizing the cost (44.10) over stationary policies $v: E \to \mathscr{R} \times U_\Gamma$, where the controlled transition measure of the process is $q(A; z, r, u_\Gamma)$ given by (44.7).

(44.11) *Theorem.* Suppose Assumptions (41.1) and (44.5) hold. Then the functions $h: E \times \mathscr{R} \times U_\Gamma \to \mathbb{R}$ and $q: E \times \mathbb{R} \times U_\Gamma \to \mathscr{P}(E)$ defined by (44.7) are continuous.

Proof. We first express h, q in a more convenient form. For $(x, r, U_\Gamma) \in E \times \mathscr{R} \times U_\Gamma$ let $\xi(t) \in E \times \mathbb{R}^3 =: \tilde{E}$ be the trajectory defined as follows, where we write $\phi(t) = (\xi^1(t), \ldots, \xi^d(t))$ so that $\xi(t) = (\phi(t), \xi^{d+1}(t), \xi^{d+2}(t) + \xi^{d+3}(t))$. In the fourth equation, θ is an arbitrary element of $C_b(E)$.

$$\frac{\mathrm{d}}{\mathrm{d}t}\phi(t) = \int_{U_0} g(\phi(t), u)r_t(\mathrm{d}u), \qquad \phi(0) = x$$

$$\frac{\mathrm{d}}{\mathrm{d}t}\xi^{d+1}(t) = -\xi^{d+1}(t)\int_{U_0}\lambda(\phi(t), u)r_t(\mathrm{d}u), \qquad \xi^{d+1}(0) = 1$$

$$\frac{\mathrm{d}}{\mathrm{d}t}\xi^{d+2}(t) = \xi^{d+1}(t)\int_{U_0} l(\phi(t), u)r_t(\mathrm{d}u), \qquad \xi^{d+2}(0) = 0$$

$$\frac{\mathrm{d}}{\mathrm{d}t}\xi^{d+3}(t) = \xi^{d+1}(t)\int_{U_0}\int_E \theta(y)Q(\mathrm{d}y; \phi(t), u)\lambda(\phi(t), u)r_t(\mathrm{d}u), \quad \xi^{d+3}(0) = 0.$$

This is a system of equations in the form

$$\text{(44.12)} \qquad \frac{\mathrm{d}}{\mathrm{d}t}\xi(t) = \int_{U_0} \hat{g}(\xi(t), u) r_t(\mathrm{d}u), \quad \xi_0 = (x, 1, 0, 0)$$

and by Theorem (43.5) the map $r \mapsto \xi(\cdot)$ is continuous. ($\hat{g}^{d+1}(t)$ is not bounded as a function of ξ^{d+1}, but this does not matter since $0 < \xi^{d+1}(t) \leqslant 1$ for any $r \in \mathscr{R}$.) We see from (44.7) and Theorem (44.9) that

$$\text{(44.13)} \qquad h(x, r, u_\Gamma) = \xi^{d+2}(t_*) + \xi^{d+1}(t_*) c(\phi(t_*), u_\Gamma)$$

$$\text{(44.14)} \quad \int_E \theta(y) q(\mathrm{d}y; x, r, u_\Gamma) = \xi^{d+3}(t_*) + \xi^{d+1}(t_*) \int_E \theta(y) R(\mathrm{d}y; \phi(t_*), u_\Gamma),$$

where, as before, $t_* = \inf\{t : \phi(t) \in \Gamma\}$. Thus $h(x, r, u_\Gamma)$ and $\int_E \theta(y) q(\mathrm{d}y;\, x, r, u_\Gamma)$ are continuous functions of $(u_\Gamma, \xi(t))$. To prove the theorem, we therefore have to show that $\xi(t_*)$ depends continuously on (x, r).

Note first that the state space for $\xi(t)$, namely $\tilde{E}$, is given by $\tilde{E} := \{\xi \in \mathbb{R}^{d+3} : \hat{\psi}(\xi) < 0\}$, where $\hat{\psi}(\xi) = \psi(\xi^1, \ldots, \xi^d)$, so that the new active boundary is $\tilde{\Gamma} = \Gamma \times \mathbb{R}^3$ and Assumption (44.5(b)) implies that $\hat{g}(\xi, u_0) \cdot \hat{v}(\xi) \geqslant \alpha$ for all $\xi \in \tilde{\Gamma}$, where $\hat{v}(\xi) = \nabla\hat{\psi}(\xi)/\|\nabla\hat{\psi}(\xi)\|$ is the outward normal at $\xi \in \tilde{\Gamma}$.

Take $x_n \to x, r_n \to r$ and let $\xi_n(t), \xi(t)$ be the corresponding solutions of (44.12) with exit times, t_*^n, t_*.

Case 1: $t_* = \infty$. For $\varepsilon > 0$, let $T_\varepsilon(t)$ denote the ε-tube around $\{\xi(s), 0 \leqslant s \leqslant t\}$, i.e.

$$T_\varepsilon(t) = \{y \in \mathbb{R}^{d+3} : |\xi(s) - y| < \varepsilon \quad \text{for some } s \in [0, t]\}$$

$$= \bigcup_{0 \leqslant s \leqslant t} B_\varepsilon(\xi(t)),$$

where $B_\varepsilon(\xi)$ denotes the ε-ball around ξ. If $t_* = \infty$ then $\xi(s) \in \tilde{E}$ (an open set) for all $s \in \mathbb{R}_+$, and hence for any $t \in \mathbb{R}_+$ there exists $\varepsilon > 0$ such that $T_\varepsilon(t) \subset E$. Since $\xi^n(s) \to \xi(s)$ uniformly on $[0, t]$, $\{\xi_n(s), 0 \leqslant s \leqslant t\} \subset T_\varepsilon(s)$ for large n, and it follows that $t_*^n \to \infty$ as $n \to \infty$. Since $\xi^{d+1}(t) \leqslant \mathrm{e}^{-\delta t}$ the second terms on the right of (44.13) and (44.14) converge to 0 as $n \to \infty$, and for the same reason $\xi_n^{d+2}(t) \to \xi_n^{d+2}(\infty)$ and $\xi_n^{d+3}(t) \to \xi_n^{d+3}(\infty)$ uniformly in n as $t \to \infty$. Let $\beta(n, t)$, $\beta(t)$ denote and $\xi_n^{d+2}(t)$ or $\xi_n^{d+3}(t)$ and $\xi^{d+2}(t)$ or $\xi^{d+3}(t)$, respectively. Then for any t,

$$|\beta(n,\infty)-\beta(\infty)| \leqslant |\beta(n,\infty)-\beta(n,t)| + |\beta(n,t)-\beta(t)| + |\beta(t)-\beta(\infty)|.$$

For given $\varepsilon > 0$ we can choose t_0 (independently of n) so that the first and third terms on the right are each less than $\varepsilon/3$ for $t > t_0$ and then n so that the central term is less than $\varepsilon/3$. Thus $\beta(n,\infty) \to \beta(\infty)$. Now

$$|\beta(\infty)-\beta(n,t_*^n)| \leqslant |\beta(\infty)-\beta(n,\infty)| + |\beta(n,\infty)-\beta(n,t_*^n)|.$$

Thus $\beta(n,t_*^n) \to \beta(\infty)$ since $t_*^n \to \infty$ and $\beta(n,t) \to \beta(n,\infty)$ uniformly in n as $t \to \infty$. This completes the proof for the case $t_* = \infty$.

Case 2: $t_* < \infty$. Recall that controls $r_t \in \mathscr{R}$ are defined for all $t \in \mathscr{R}_+$. The corresponding trajectory $\xi(t)$ hits the boundary at $z := \xi(t_*)$ and then continues into $\tilde{E}^c$ (Assumption 44.5(b)). By Assumption (44.5(a)), we can choose ε_1 such that $\hat{g}(\xi, u_0) \cdot \hat{v}(z) \geqslant \alpha/2$ for all $u_0 \in U_0$ and $\xi \in B_{\varepsilon_1}(z)$. Introducing a translation and a local change of coordinates (possible because $\psi(x)$ is assumed C^1), we can assume that $z = 0$ and that, for some $\varepsilon_2 \leqslant \varepsilon_1$, $\hat{E} \cap B_{\varepsilon_2}(0) = \{\xi : \xi^1 < 0\} \cap B_{\varepsilon_2}(0)$ (call this $B^-_{\varepsilon_2}(0)$) so that $\Gamma \cap B_{\varepsilon_2}(0) = \{\xi : \xi^1 = 0)\} \cap B_{\varepsilon_2}(0)$ (call this Γ_{ε_2}). Thus $\hat{v}(z) = (1, 0, \ldots, 0)$ and $\hat{g}^1(\xi, u_0) \geqslant \beta$ for all $\xi \in B^-_{\varepsilon_2}$ and $u_0 \in U_0$, for some $\beta > 0$. (β could be smaller than $\alpha/2$ because of the coordinate change.) In $B^-_{\varepsilon_2}(0)$, the trajectory $\xi(t)$ has velocity at least β in the 1-direction and at most M in all other directions, where M is a bound on $|\hat{g}(\xi, u_0)|$ (Assumption 44.1(c)), and $\xi(t_*) = 0$. The trajectory is therefore contained in the cone $-C$, where $C = \{\xi : \xi^1 \geqslant 0, |\xi^k| \leqslant M\xi^1/\beta, k = 2, \ldots, d+3\}$. It follows easily, using also Assumption (44.5(c)), that for some $\varepsilon_3 > 0$, $T_{\varepsilon_3}(\infty) \cap \Gamma \subset \Gamma_{\varepsilon_2}$. ($T_{\varepsilon_3}(\infty)$ is the ε_3-tube around the whole trajectory.)

Now consider neighbouring trajectories $\xi_n(t)$. Since these converge uniformly to $\xi(t)$ on any interval $[0, T]$, there exists n_0 such that, for $n > n_0$, $\xi_n(t_*) \in B_{\varepsilon_3}(0)$ and $\{\xi_n(s), 0 \leqslant s \leqslant t_*\} \subset T_{\varepsilon_3}(t_*)$. If, for sufficiently large n, $t_*^n < t_*$ then, because of Assumption (44.5(b)), $\xi_n(t_*) \in B^+_{\varepsilon_3}(0) := B_{\varepsilon_3}(0) \backslash B^-_{\varepsilon_3}(0)$, and it must be the case that $\xi_n(t_*^n) \in \Gamma_{\varepsilon_2}$. If on the other hand $t_*^n \geqslant t_*$ infinitely often then we only need consider subsequences, still labelled $\xi_n(t)$, for which $t_*^n \geqslant t_*$ for all n. In this case, since $\xi_n(t_*) \to \xi(t_*) = 0$, we can choose n_1 such that for $n \geqslant n_1$, $\xi_n(t_*) \in B^-_{\varepsilon_4}(0)$, where $\varepsilon_4 = \beta\varepsilon_3/2M$. Now, as argued before, the trajectory segment $\{\xi_n(s) : t_* \leqslant s \leqslant t_*^n\}$ must lie inside the cone $C' = \xi_n(t_*) + C$ and it can be checked that with this choice of ε_4,

$(y+C)\cap\Gamma\subset B_{\varepsilon_3}(0)\cap\Gamma$ for any $y\in B^-_{\varepsilon_4}(0)$. Hence $|\xi_n(t^n_*)-\xi_n(t_*)|<\varepsilon_3$. Thus in all cases $\xi_n(t^n_*)\to\xi(t_*)$, showing that $\xi(t_*)$ is a continuous function of (x,r). This completes the proof. □

(44.15) *Example*

The following example shows that some condition along the lines of Assumption (44.5(b)), mandating uniform motion towards the boundary once in a neighbourhood of it, cannot be dispensed with if the continuity property of Theorem (44.11) is to hold. Take $E=U_0=[-1,1]^2$ and the dynamic equation

$$\frac{\mathrm{d}}{\mathrm{d}t}\phi(t)=u(t),\quad \phi(0)=\begin{bmatrix}0\\0\end{bmatrix}.$$

Consider the control

$$u(t)=\begin{cases}a_1 & 0\leqslant t<1\\ a_2 & 1\leqslant t<3\\ 0 & t\geqslant 3,\end{cases}$$

where $a_1^T=[1,0]$ and $a_2^T=[-1,1]$, and let $u_n(t)=u(t+1/n)$. Certainly $u_n\to u$ in the Young topology: this requires that

$$\int_0^3 f(t,u_n(t))\mathrm{d}t\to\int_0^3 f(t,u(t))\mathrm{d}t \tag{44.16}$$

for all measurable functions f which are continuous in u and satisfy

$$\int_0^3 \max_u |f(t,u)|\mathrm{d}t<\infty.$$

But here $u_n(t)\to u(t)$ for almost every t, and (44.16) holds by dominated convergence. However $t_*=1$, whereas $t^n_*=2-1/n$, so $\phi_n(t^n_*)$ does not converge to $\phi(t_*)$. We see that even in the simplest cases continuity is lost if the trajectories are able to 'back off' from the boundary.

We are now in a position to state the main result of this section.

(44.17) *Theorem.* Under Assumptions (41.1) and (44.5) there is an optimal policy $\hat{v}$ in $\mathscr{U}_P$ for the discrete-time Markov decision problem of minimizing $J_x(v)$ given by (44.10) over controlled discrete-time processes with transition measure q given by Theorem (44.9). Further, let

$$V(x) = \min_{v \in \mathscr{U}_{\mathrm{P}}} J_x(v). \tag{44.18}$$

Then V is bounded and continuous, and satisfies

$$V(x) = \inf_{(r, v_\Gamma) \in \mathscr{R} \times U_\Gamma} \mathbb{E}[h(x, (r, v_\Gamma)) + V(Z_1)], \quad x \in E. \tag{44.19}$$

The control $(r, u_\Gamma) = \hat{v}(x)$ achieves the infimum in (44.19).

Proof. This is an application of general results in the theory of discrete-stage Markov decision models (DMDM) summarized below in section 47. The four ingredients of a DMDM are sets S, U and functions q, h. In this application, $S = \hat{E} = E \cup \{\Delta\}$ is the PDP state space, $U = \mathscr{R} \times U_\Gamma$ and q, h are the functions given by (44.7). In view of the standing assumptions and Theorem (44.11) this model satisfies the LSC conditions (47.3). The class $\mathscr{U}_{\mathrm{P}}$ coincides with the class Π_{SM} of stationary Markov policies for the DMDM in section 47. Hence by Theorem (47.8) an optimal element $\hat{v} \in \mathscr{U}_{\mathrm{P}}$ exists, the value function V of (44.18) is lower semicontinuous (l.s.c.) and (44.19) is satisfied. (In fact, $\hat{v}$ is optimal in the wider class Π of possibly non-stationary, non-Markovian policies.)

The remaining claim in the Theorem is that V is continuous; for this it suffices, by Proposition (47.2), part 3, to show that V is upper semicontinuous (u.s.c.), since lower semicontinuity is already established. The model as it stands does not satisfy the conditions of the u.s.c. model of Proposition (47.7) since the function h is not non-positive. We can, however, produce a modified model that does satisfy these conditions. In this new model, the cost functions l, c are replaced by $\tilde{l} = l - b$ and $\tilde{c} = c - b$ where b is an upper bound for l and c. The cost in the new model using control $v \in \mathscr{U}_{\mathrm{P}}$ is

$$\begin{aligned}
\tilde{J}_x(v) &= \mathbb{E}_x^v \left\{ \int_0^\infty \int_{U_0} (l(x_t, u) - b) v^0_{t-\sigma(t)}(\mathrm{d}u, x_{\sigma(t)}) \mathrm{d}t \right. \\
&\qquad \left. + \int_0^\infty (c(x_{t-}, v_\Gamma(x_{\sigma(t)})) - b) \mathrm{d}p^*(t) \right\}, \\
&= J_x(v) - b\mathbb{E}_x^v \left\{ \int_0^\infty \mathrm{e}^{-\delta t} \mathrm{d}t + \int_0^\infty \mathrm{d}p_t^* \right\} \\
&= J_x(v) - b\left\{ \frac{1}{\delta} + \mathbb{E}_x^v p_\infty^* \right\},
\end{aligned}$$

where we have used the fact that killing at rate δ is included in the definition of the process. The new model satisfies the conditions of the u.s.c. model described in section 47 and hence $\tilde{J}_x(v)$ is u.s.c. by Theorem (47.9). It is not hard to show that under our standing assumptions the function $x \mapsto \mathbb{E}_x^v p_\infty^*$ is continuous; thus $J_x(v)$ is u.s.c for each $v \in \mathscr{U}_P$. For each $x \in E$, let v_n^x be a sequence in $\mathscr{U}_P$ such that $J_x(v_n^x) \downarrow V(x)$ as $n \to \infty$, and define

$$V_n(x) = \inf_{z \in E} J_x(v_n^z).$$

Then V_n is u.s.c., since $\{x : V_n(x) \geq c\} = \bigcap_{z \in E} \{x : J_x(v_n^z) \geq c\}$ (Definition 47.1), and $V_n(x) \downarrow V(x)$. It follows that V is u.s.c. □

(44.20) *Remark.* For a measurable function $\psi : E \to \mathbb{R}_+$, $r \in \mathscr{R}$ and $v_\Gamma \in U_\Gamma$, define

$$G_{(r,v_\Gamma)}\psi(x) = \mathbb{E}_x \left\{ \int_0^{T_1} \int_{U_0} l(x_t, u) r_t(\mathrm{d}u) + I_{(T_1 = t_*)} c(x_{T_1 -}, v_\Gamma) + \psi(x_{T_1}) \right\}$$

and

$$G\psi(x) = \inf_{(r, v_\Gamma) \in \mathscr{R} \times U_\Gamma} G_{(r,v_\Gamma)}\psi(x), \tag{44.21}$$

where x_t is the trajectory of the controlled process on $[0, T_1]$. In particular, $x_t = \phi(t)$ for $t < T_1$, where $\phi(t)$ satisfies

$$\frac{\mathrm{d}}{\mathrm{d}t}\phi(t) = \int_{U_0} g(\phi(t), u) r_t(\mathrm{d}u), \quad \phi(0) = x.$$

G is the operator introduced in section 47 below, specialized to the present application, and (44.19) states that V satisfies $V = GV$, i.e. V is a fixed point of G.

Theorem (44.17) asserts the existence of an optimal policy in $\mathscr{U}_P$. We now show that the optimal control can be expressed in terms of feedback from the boundary, i.e. that there is an optimal control $\hat{u} \in \mathscr{U} \subset \mathscr{U}_P$.

(44.22) *Theorem.* Under the conditions of Theorem (44.17) there exists a Borel measurable function $\hat{u}_\Gamma : \Gamma \to U_\Gamma$ such that for all $x \in \Gamma$

$$\begin{aligned} c(x, \hat{u}_\Gamma(x)) + \int_E V(y) R(\mathrm{d}y; x, \hat{u}_\Gamma(x)) \\ = \min_{u \in U_\Gamma} \left\{ c(x,u) + \int_E V(y) R(\mathrm{d}y; x, u) \right\}. \end{aligned} \tag{44.23}$$

Let $\hat{v} = (\hat{v}_0, \hat{v}_\Gamma)$ be an optimal policy in $\mathscr{U}_P$. Then $\hat{u} \in \mathscr{U}$ defined by $\hat{u} = (\hat{v}_0, \hat{u}_\Gamma)$ is also optimal in $\mathscr{U}_P$.

Proof. Since $V \in C_b(E)$, Assumptions (41.1) imply that the function in braces on the right of (44.23) is continuous in (x, u), and U_Γ is compact. The measurable selection theorem given by Bertsekas and Shreve (1978, Proposition 7.33, p. 153) now ensures that there is a *Borel selector* $\hat{u}_\Gamma$ satisfying Theorem (44.23).

Let $\hat{v} \in \mathscr{U}_P$ be optimal and choose $x \in E$ such that $P_x^{\hat{v}}[T_1 = t_*(x)] > 0$. Let $z := \phi(t_*)$ be the boundary hitting point and suppose $u' := \hat{v}(x)$ does not achieve the minimum on the right-hand side of (44.23) at z. Form a new non-stationary policy $\pi = (\mu_0, \mu, \mu, \ldots)$ (section 47), where $\mu_0(x) = (\hat{v}_0(x), \hat{v}_\Gamma(z))$ and $\mu = \hat{v}$. Let

$$J = \mathbb{E}_x \left\{ \int_0^{T_1} \int_{U_0} l(\phi(t), u) \hat{v}_t^0(\mathrm{d}u)\mathrm{d}t \right.$$

$$\left. + I_{(T_1 < t_*)} \int_E \int_{u_0} V(y) Q(\mathrm{d}y; \phi(T_1, u)) \hat{v}_{T_1}^0(\mathrm{d}u). \right.$$

Then by the strong Markov property, the cost of policy π is

$$J_x(\pi) = J + P_x[T_1 = t_*] \left\{ c(z, \hat{u}_\Gamma(z)) + \int_E V(y) R(\mathrm{d}y; z, \hat{u}_\Gamma(z)) \right\}$$

$$< J + P_x[T_1 = t_*] \left\{ c(z, u') + \int_E V(y) R(\mathrm{d}y;\ z, u') \right\} = J_x(\hat{v}).$$

Thus π has lower cost than $\hat{v}$. But the discrete-time decision theory results imply that $\hat{v}$ is also optimal in the class of non-stationary policies, so this is impossible and $\hat{v}(x)$ must minimize the right side of (44.23). Thus $\hat{v}$ and $\hat{u}$ are equivalent policies. □

We now want to understand the relation between the discrete-time formulation developed in this section and the continuous-time dynamic programming of section 42. The key to this is the 'imbedding' result (44.19) of Theorem (44.17), which expresses the value function V as (admittedly in disguised form) the solution of a deterministic optimal control problem. We sketch out the ideas briefly here and develop them in detail in the next two sections.

As in the proof of (44.19), we can express the fixed-point property

$G = GV$ (Remark 44.20), in an obvious compressed notation, as

(44.24)

$$V(x) = \inf_{r \in \mathscr{R}} \mathbb{E}_x \left\{ \int_0^{T_1} l \mathrm{d}s + I_{(T_1 < t*)} \int_E V \mathrm{d}Q + I_{(T_1 = t*)} \left(c + \int_E V \mathrm{d}R \right) \right\}$$

(44.25)

$$= \inf_{r \in \mathscr{R}} \int_0^{t*} \chi(t) f_0^V(\phi(t), r(t)) \mathrm{d}t + F^V(\phi(t_*)) \chi(t_*),$$

where

$$(44.26) \quad f_0^V(x, r) = \int_{U_0} (l(x, u) + \int_E V(y) Q(\mathrm{d}y; x, u) \lambda(x, u)) r(\mathrm{d}u), x \in E$$

and

$$(44.27) \qquad F^V(x) = c(x, \hat{u}_\Gamma(x)) + \int_E V(y) R(\mathrm{d}y; x, \hat{u}_\Gamma(x)), \quad x \in \Gamma.$$

Here $\phi(t)$ is the trajectory corresponding to $r \in \mathscr{R}$ starting at $x \in E$ and we have reverted to the notation

$$\chi(t) = \exp \left\{ - \int_0^t \int_{U_0} \lambda(\phi(t), u) r_t(\mathrm{d}u) \mathrm{d}t \right\}.$$

Thus the distribution of T_1 is $P_x[T_1 > t] = I_{(t < t*)} \chi(t)$ and (44.25) is obtained by evaluating the expectation in (44.24) using this distribution. Finally, we have taken boundary optimization out of the problem by using the optimal boundary feedback function $\hat{u}_\Gamma$ in the definition of F^V. We thus see that $V(x)$ is the value function for the deterministic optimization problem (44.25) over controls $r \in \mathscr{R}$, subject to the dynamics

$$(44.28) \qquad \frac{\mathrm{d}}{\mathrm{d}t} \phi(t) = f(\phi(t), r(t)) \quad \phi(0) = x,$$

where, here and below, we shall write

$$f(x, r) = \int_{U_0} g(x, u) r(\mathrm{d}u)$$

for $r \in \mathscr{P}(U_0)$. This is of course still an implicit characterization of V, since V also appears in the coefficients, as the notation f_0^V, F^V indicates.

Let us fix V (it is defined by (44.18)) and consider the optimal control problem (44.25) and (44.28) independently. Let $\bar{V}(x)$ denote the value of the infimum in (44.25). Then a standard dynamic programming argument (Fleming and Rishel's (1975) verification theorem, Theorem IV. 4.4) tells us that if a function $W: E \to \mathbb{R}$ satisfies the *Bellman equation*

$$\text{(44.29)} \qquad \min_{r \in \mathcal{R}} \{\nabla W(x) \cdot f(x,r) + f_0^V(x,r) - \lambda(x,r) W(x)\} = 0, \quad x \in E,$$

$$\text{(44.30)} \qquad W(x) = F^V(x), \quad x \in \Gamma,$$

where $\lambda(x,r) := \int_{U_0} \lambda(x,u) r(\mathrm{d}u)$, and if $r = \hat{u}(x)$ achieves the minimum then $W = \bar{V}$ and $\hat{u}$ is optimal. But then $W = V$, so writing f_0^V and F^V as defined in (44.26) and (44.27) explicitly in (44.29) and (44.30) we see that (44.29) is equivalent to

$$\min_{r \in \mathcal{R}} \Bigg\{ \nabla V(x) \cdot f(x,r) + \int_{U_0} \bigg(l(x,u) + \int_E V(y) Q(\mathrm{d}y; x,u) \lambda(x,u) \bigg) r(\mathrm{d}u) - \lambda(x,r) V(x) \Bigg\} = 0,$$

i.e.

$$\min_{r \in \mathcal{R}} \Bigg\{ \int_{U_0} \bigg(\nabla V(x) \cdot g(x,u) + \lambda(x,r) \int_E (V(y) - V(x)) Q(\mathrm{d}y; x,u) + l(x,u) \bigg) r(\mathrm{d}u) \Bigg\} = 0,$$

or

$$\text{(44.31)} \qquad \min_{r \in \mathcal{R}} \int_{U_0} \{\mathfrak{A}^u V(x) + l(x,u)\} r(\mathrm{d}u) = 0,$$

where $\mathfrak{A}^u$ is the generator of the PDP with control parameter $u \in U_0$. This is – apart from the use of relaxed controls – precisely the stochastic Bellman equation (42.6) arising in the 'naïve' continuous-time stochastic control approach of section 42. (The boundary condition (42.7) corresponds to (44.23) in this section.)

The problem with this approach is – as always – that V has to be C^1 for (44.29) to be well-defined and the conditions of the verification theorem to be met, and we do not know *a priori* that this is the case. However, we can show, under one extra condition, that V is actually Lipschitz continuous, and we can use F.H. Clarke's (1983) theory of 'non-smooth analysis' (a calculus for locally Lipschitz functions) to state a somewhat weaker form of (44.29) which turns out to be a *necessary and sufficient* condition for optimality.

In the next section we introduce some ideas from non-smooth analysis and study a deterministic problem similar to (44.25) and (44.28). Then in section 46 we return to the PDP problem and establish the exact relation between deterministic and stochastic dynamic programming.

45 Non-smooth analysis and deterministic optimal control

We first introduce the Clarke generalized gradient and some of its properties. Only those properties strictly needed will be mentioned, and the reader can refer to Clarke (1983) for the whole story.

Let $f:\mathbb{R}^d \to \mathbb{R}$ be a locally Lipschitz continuous function. For $x, y \in \mathbb{R}^d$ the *upper generalized directional derivative* of f at x in direction y is

$$f^0(x;y) := \limsup_{\substack{z \to x \\ \varepsilon \to 0}} \frac{f(z+\varepsilon y) - f(z)}{\varepsilon}.$$

If we replace 'lim sup' by 'lim inf' we obtain the *lower* directional derivative f_0.

(45.1) *Definition.* The *Clarke generalized gradient* of f at x is the subset of $\mathbb{R}^d$ defined by

$$\partial f(x) = \{\zeta \in \mathbb{R}^d : f^0(x;y) \geqslant \zeta \cdot y \text{ for all } y \in \mathbb{R}^d\}.$$

It is clear that if f is continuously differentiable at x then $\partial\phi(x) = \{\nabla f(x))\}$.

(45.2) *Proposition*

1. ∂f is a non-empty, convex, compact subset of $\mathbb{R}^d$ and $|\zeta| \leqslant L$ for all $\zeta \in \partial f(x)$, where L is a Lipschitz constant for f in any neighbourhood of x.

2. $$f^0(x;y) = \max_{\zeta \in \partial f(x)} \zeta \cdot y, \quad f_0(x;y) = \min_{\zeta \in \partial f(x)} \zeta \cdot y.$$

3. The function $x \to \partial f(x)$ is closed as a set-valued mapping, i.e. if $x_n \to x$ and $\zeta_n \to \zeta$, where $\zeta_n \in \partial f(x_n)$, then $\zeta \in \partial f(x)$.
4. $$\partial f(x) = \text{co}\{\lim \nabla f(x_i) : x_i \to x\}.$$

   Recall that a locally Lipschitz function is differentiable almost everywhere. Here x_i is any sequence of points converging to x such that $\nabla f(x_i)$ exists at x_i, and 'co$\{\cdots\}$' denotes the closed convex hull, i.e. the smallest closed convex set containing the elements in the braces.
5. (Chain rule.) Let $f = g \circ h$, where $h: \mathbb{R}^d \to \mathbb{R}^d$ and $g: \mathbb{R}^n \to \mathbb{R}$. Denote $h(x) = (h_1(x), \ldots, h_n(x))$. If g, h_i are locally Lipschitz then so is f, and

$$\partial f(x) \subset \text{co}\left\{ \sum_{i=1}^{n} \alpha_i \cdot \zeta_i : \zeta_i \in \partial h_i(x), \alpha = (\alpha_1, \ldots, \alpha_n) \in \partial g(h(x)) \right\}.$$

(45.3) *Definition.* A locally Lipschitz function f is *regular at* x if the ordinary directional derivative

$$f'(x; y) := \lim_{\varepsilon \downarrow 0} \frac{f(z + \varepsilon y) - f(z)}{\varepsilon}$$

exists for all y, and $f^0(x; y) = f'(x, y)$. f is *regular* if it is regular at all $x \in \mathbb{R}^d$.

Now let us turn to the optimal control problem of minimizing the cost

$$J_x(u) = \int_0^{t_*} \chi(t) f_0(x(t), u(t)) \mathrm{d}t + \chi(t_*) F(x(t_*)) \tag{45.4}$$

over trajectories of the differential system

$$\frac{\mathrm{d}}{\mathrm{d}t} x(t) = f(x(t), u(t)), x(0) = x \in E. \tag{45.5}$$

As usual,

$$\chi(t) = \exp\left(- \int_0^t \lambda(x(s), u(s)) \mathrm{d}s \right)$$

and $t_* = \inf\{t : x(t) \in \partial E\}$. We shall suppose that Assumptions (41.1) hold *mutatis mutandis*. In particular, $u(t)$ takes values in some compact space U, the functions f_0, f, λ, F are bounded and Lipschitz continuous and $\lambda(x, u) \geqslant \delta > 0$. Thus $J_x(u)$ is well defined even when

$t_* = \infty$. We say that $(x(\cdot), u(\cdot))$ is an *admissible pair* if $u:\mathbb{R}_+ \to U$ is a measurable function and $x(\cdot)$ is the solution of (45.5) starting at some $x \in E$. The active boundary is $\Gamma := \{x \in \partial E : x = x(t_*)$ for some admissible pair with $t_* < \infty\}$.

We first show that existence of a locally Lipschitz solution of the Bellman equation (i.e. the exact analogue of (44.29)), formulated in terms of generalized gradients, provides a sufficient condition for optimality.

(45.6) *Theorem*

1. Suppose W is locally Lipschitz and satisfies the generalized Bellman equation

$$\min_{\substack{\zeta \in \partial W(x) \\ u \in U}} \{\zeta \cdot f(x,u) - \lambda(x,u) W(x) + f_0(x,y)\} = 0, \qquad x \in E \tag{45.7}$$

$$W(x) = F(x) \quad x \in \Gamma. \tag{45.8}$$

Suppose also that $\lim_{t \to \infty} W(x(t)) = 0$ for all admissible pairs (x,u) such that $t_* \to \infty$. Then $W(x) \leqslant J_x(u)$ for all admissible controls u.

2. Suppose in addition that W is regular and that for some admissible pair $(\hat{x}, \hat{u})$ there exists for almost all t an element $\hat{\zeta}(t) \in \partial W(\hat{x}(t))$ such that

$$\zeta(t) \cdot f(\hat{x}(t), \hat{u}(t)) - \lambda(\hat{x}(t), \hat{u}(t)) W(\hat{x}(t)) + f_0(\hat{x}(t), \hat{u}(t)) = 0. \tag{45.9}$$

Then $(\hat{x}, \hat{u})$ is optimal and $W(x) = J_x(\hat{u})$.

Proof

1. Let (x,u) be any admissible pair with $x(0) = x$. Then $\theta(t) := W(x(t))$ is a composition of Lipschitz maps and hence Lipschitz. By the chain rule (part 5 of Proposition (45.2))

$$\dot{\theta}(t) \in \text{co}\{\zeta(t) \cdot f(x(t), u(t)) : \zeta(t) \in \partial W(x(t))\}.$$

In view of (45.7), for any $\zeta(t) \in \partial W(x(t))$,

$$\zeta(t) \cdot f(x(t), u(t)) - \lambda(x(t), u(t)) W(x(t)) + f_0(x(t), u(t)) \geqslant 0,$$

and hence, since $\chi(t) > 0$,

$$\chi(t)\dot{\theta}(t) - \chi(t)\lambda(x(t), u(t)) W(x(t)) \geqslant -\chi(t) f_0(x(t), u(t)).$$

However, the left hand side is just $\mathrm{d}(\chi(t) W(x(t)))/\mathrm{d}t$, so integrating from 0 to t_* and using the boundary condition (45.8) and the fact that $\chi(0)=1$ gives

$$W(x) \leqslant J_x(u).$$

2. Let $(\hat{x}, \hat{u})$ have the stated properties and W be a regular solution of (45.7) and (45.8). Let t be such that $\dot{\theta}(t)$ exists. Then

$$\begin{aligned}
\dot{\theta}(t) = \frac{\mathrm{d}}{\mathrm{d}t} W(\hat{x}(t)) &= \lim_{\varepsilon \to 0} \frac{W(\hat{x}(t-\varepsilon)) - W(\hat{x}(t))}{-\varepsilon} \\
&= \lim_{\varepsilon \to 0} \frac{W(\hat{x}(t) - \varepsilon\dot{\hat{x}}(t)) - W(\hat{x}(t))}{-\varepsilon} \text{(since } W \text{ is Lipschitz)} \\
&= -W^0(\hat{x}(t); -\dot{\hat{x}}(t)) \quad \text{(using regularity)} \\
&\leqslant -\zeta(t)\cdot(-\dot{\hat{x}}(t)) \quad \forall \zeta(t) \in \partial W(\hat{x}(t)) \\
&\qquad \text{(part 2 of proposition 45.2)} \\
&= \zeta(t)\cdot f(\hat{x}(t), \hat{u}(t)).
\end{aligned}$$

If (45.9) holds for some $\hat{\zeta}(t) \in \partial W(\hat{x}(t))$ then we have

$$\frac{\mathrm{d}}{\mathrm{d}t} W(\hat{x}(t)) \leqslant \hat{\zeta}(t)\cdot f(\hat{x}(t), \hat{u}(t)) = \lambda(\hat{x}(t), \hat{u}(t)) W(\hat{x}(t)) - f_0(\hat{x}(t), \hat{u}(t)).$$

Thus, in abbreviated notation

$$\frac{\mathrm{d}}{\mathrm{d}t}(\chi(t) W(\hat{x}(t))) = \chi \frac{\mathrm{d}W}{\mathrm{d}t} - \lambda\chi W \leqslant -\chi f_0, \tag{45.10}$$

and integrating from 0 to t_* we obtain (since (45.10) holds for almost every t)

$$W(x) \geqslant J_x(\hat{u}).$$

But, from 1, $W(x) \leqslant J_x(u)$, so that $W(x) = J_x(\hat{u})$, i.e. $(\hat{x}, \hat{u})$ is optimal. □

Let us now define

$$H(x) = \inf_{(x,u)\text{admissible}} J_x(u). \tag{45.11}$$

We wish to show that if H is locally Lipschitz continuous then it does satisfy the generalized Bellman equation (45.7) and (45.8), which therefore provides a *necessary*, as well as *sufficient*, condition for optimality. For this purpose it is convenient to start by considering a control problem in *Mayer* form, namely that of minimizing a function

$\Phi(x(t_*))$ over admissible pairs (x, u). When $t_* = \infty$ we take the cost as $\lim_{t\to\infty} \Phi(x(t))$ and assume that this limit exists. Define

$$\tilde{H}(x) = \inf_{u(\cdot)} \Phi(x(t_*)).$$

The following simple result is given by Fleming and Rishel (1975, Theorems IV.3.1 and IV.3.2).

(45.12) *Lemma*

1. For any admissible pair (x, u), the function $t \to \tilde{H}(x(t))$ is non-decreasing.
2. If $(\hat{x}, \hat{u})$ is optimal then $t \to \tilde{H}(\hat{x}(t))$ is constant.

Proof

1. Suppose that $x(0) = x$ and $(x(\cdot), u(\cdot))$ is an admissible pair. Then $\tilde{H}(x(t)) = \inf \Phi(\check{x}(t_*))$, where the infimum is taken over all admissible pairs $(\check{x}, \check{u})$ such that $\check{x}(0) = x$ and $\check{x}(s) = x(s), \check{u}(s) = u(s)$ for $s \in [0, t]$. Since this set is decreasing as t increases, $\tilde{H}(x(t))$ is non-decreasing.
2. Suppose $(\hat{x}, \hat{u})$ is optimal starting from x, and let t be any time less than t_*. Since $(\hat{x}, \hat{u})$ is optimal, $\tilde{H}(x) = \Phi(\hat{x}_{t_*})$. On the other hand, $(\hat{x}, \hat{u})$ restricted to $[t, t_*[$ is an admissible pair starting at $\hat{x}(t)$ so $\tilde{H}(\hat{x}(t)) \leqslant \Phi(\hat{x}(t_*))$. Thus $\tilde{H}(\hat{x}(t)) \leqslant \tilde{H}(x)$, which implies that $\tilde{H}(\hat{x}(t)) = \tilde{H}(x)$ in view of part 1. □

(45.13) *Theorem.* Suppose $\tilde{H}$ is locally Lipschitz. Then $\tilde{H}$ satisfies the Bellman equation in the form

$$\beta \cdot f(x, u) \geqslant 0 \qquad \text{for all } \beta \in \partial \tilde{H}(x),\ u \in U. \tag{45.14}$$

If for each initial point $x \in E$ there exists an optimal admissible pair $(\hat{x}, \hat{u})$ then $\tilde{H}$ satisfies

$$\min_{\substack{\beta \in \partial H(x) \\ u \in U}} \beta \cdot f(x, u) = 0, \qquad x \in E,$$

$$\tilde{H}(x) = \Phi(x) \quad x \in \Gamma,$$

and

$$\min_{\beta \in \partial \tilde{H}(x(t))} \beta \cdot f(\hat{x}(t), \hat{u}(t)) = 0, \quad \text{a.s. } t \in [0, t_*[.$$

Proof

If $\tilde{H}$ is differentiable at $y \in E$ then, from Lemma (45.12), part 1, $\nabla \tilde{H}(y) \cdot f(y,u) \geqslant 0$. Now $\partial \tilde{H}(y) = \mathrm{co}\{\lim \nabla \tilde{H}(y_i), y_i \to y\}$, i.e. any $\beta \in \partial \tilde{H}(y)$ is a convex combination of $\beta_k = \lim_{i\to\infty} \nabla \tilde{H}(y_i^k)$ for sequences $y_i^k \to y$ as $i \to \infty$. Since f is continuous,

$$\beta_k \cdot f(y,u) = \lim_i \nabla \tilde{H}(y_i^k) \cdot f(y_i^k, u) \geqslant 0$$

and hence $\beta \cdot f(y,u) \geqslant 0$. This is (45.14).

For the second part, we have to show that $\hat{\beta} \cdot f(y,\hat{u}) = 0$ holds for some $\hat{\beta} \in \partial \tilde{H}(y), \hat{u} \in U$. From Lemma (45.12), part 2, $t \to \tilde{H}(\hat{x}(t))$ is constant when $(\hat{x}, \hat{u})$ is an optimal admissible pair starting at $x(0) = y$, so

$$0 = \frac{\mathrm{d}}{\mathrm{d}t} \tilde{H}(\hat{z}(t)) \subset \mathrm{co}\{\partial \tilde{H}(\hat{x}(t) f(\hat{x}(t), u(t))\}$$

by the chain rule. Thus there exists $\beta(t) \in \partial \tilde{H}(\hat{x}(t))$ such that $\beta(t) \cdot f(\hat{x}(t), u(t)) = 0$. From Proposition (45.2), part 1, we know that $\| \beta_t(t) \| \leqslant L$ for $0 \leqslant t \leqslant t_0$, where L is a Lipschitz consistent for $\tilde{H}$ in a neighbourhood of y containing the trajectory $\{x(t), 0 \leqslant t \leqslant t_0\}$. We also have $\hat{u}(t) \in U$, compact. Taking a sequence $t_n \downarrow 0$, we can extract subsequences, still labelled t_n, such that $\beta(t_n) \to \hat{\beta}$, $\hat{u}(t_n) \to \hat{u}$ for some $\beta \in \mathbb{R}^d$ and $u \in U$. By Proposition (45.2), part 3, (stating that $\partial \tilde{H}$ is closed as a set-valued map) $\hat{\beta} \in \partial \tilde{H}(y)$, and since f is continuous we have $\hat{\beta} \cdot f(y, \hat{u}) = 0$. This completes the proof □

To make the connection between the Mayer problem and our original problem (45.4) and (45.5), we introduce extra state components x^{d+1}, x^{d+2}, as was done in the proof of Theorem (44.9). The extended state $\bar{x} = (x^1, \ldots, x^{d+2})$ satisfies (writing x for $(x^1, \ldots, x^d)$)

$$\begin{aligned} \dot{x}(t) &= f(x(t), u(t)), \\ \dot{x}^{d+1}(t) &= -x^{d+1}(t)\lambda(x(t), u(t)) \\ \dot{x}^{d+2}(t) &= x^{d+1}(t) f_0(x(t), u(t)). \end{aligned}$$

If we define $\Phi(\bar{x}) = x^{d+2} + x^{d+1} F(x)$, then clearly

$$J_x(u) = \Phi(\bar{x}(t_*)),$$

where the initial conditions are $x(0) = x, x^{d+1}(0) = 1, x^{d+2}(0) = 0$. If we allow arbitrary initial conditions x^{d+1} and x^{d+2} and take the infimum over admissible controls then we obtain

$$\tilde{H}(\bar{x}) = x^{d+2} + x^{d+1}H(x), \quad \bar{x} = (x, x^{d+1}, x^{d+2})$$

for $x^{d+1} > 0$. Thus if $H(x)$ is locally Lipschitz continuous, so is $\tilde{H}(\bar{x})$, and the generalized gradients are related as follows:

$$\partial\tilde{H}(\bar{x}) = (x^{d+1}\partial H(x)) \times \{H(x)\} \times \{1\}. \tag{45.15}$$

(45.16) *Theorem.* Define H by (45.11) and suppose that H is locally Lipschitz. Then H satisfies the generalized Bellman equation

$$\min_{\substack{\zeta\in\partial H(x)\\ u\in U}} \{\zeta\cdot f(x,u) - \lambda(x,u)H(x) + f_0(x,u)\} \geqslant 0, \qquad x\in E, \tag{45.17}$$

$$H(x) = F(x), \quad x\in\Gamma.$$

If for every initial point $x\in E$ there exists an optimal admissible pair $(\hat{x}, \hat{u})$, then equality holds in (45.17) and

$$\min_{\zeta\in\partial H(\hat{x}(t))} \{\zeta\cdot f(\hat{x}(t), \hat{u}(t)) - \lambda(\hat{x}(t)), \hat{u}(t))H(\hat{x}(t) + f_0(\hat{x}(t), \hat{u}(t))\} = 0.$$

$$\text{a.e. } t\in[0, t_*[.$$

Proof. We have only to transcribe the results of Theorem (45.13) with the present setting. In view of (45.15) any $\beta\in\partial\tilde{H}(\bar{x})$ takes the form $\beta = (x^{d+1}\zeta, H(x), 1)$ for some $\zeta\in\partial H(x)$ and hence (45.14) becomes

$$x^{d+1}\zeta\cdot f(x,u) - x^{d+1}\lambda(x,u)H(x) + x^{d+1}f_0(x,u) \geqslant 0, \;\; \zeta\in\partial H(x), \;\; u\in U,$$

which is equivalent to (45.17) on $\{x : x^{d+1} > 0\}$. The remaining statements follow from Theorem (45.13), using the fact that $x^{d+1}(t) > 0$ for all t and admissible $u(\cdot)$ if $x^{d+1}(0) > 0$. □

46 Necessary and sufficient conditions for optimality: the generalized Bellman equation

We now return to the PDP control problem of Section 44. Recall that this is to minimize the cost

$$J_x(u) = \mathbb{E}_x\left\{\int_0^\infty \int_{U_0} l(x_t, v)u_0(t-\sigma(t), x_{\sigma(t)})(\mathrm{d}v)\mathrm{d}t + \int_0^\infty c(x_{t-}, u_\Gamma(x_{t-}))\,\mathrm{d}p^*(t)\right\}$$

over the set of controls $\mathscr{U}$, where each $u \in \mathscr{U}$ is a pair $u = (u_0, u_\Gamma)$ of measurable functions $u_0: E \to \mathscr{R}$ and $u_\Gamma: \Gamma \to U_\Gamma$, and $\sigma(t)$ denotes the last jump time of the PDP (x_t) strictly before t. We denote by V the value function $V(x) = \inf_{u \in \mathscr{U}} J_x(u)$. We showed in Theorems (44.17) and (44.22) that under Assumptions (41.4) and (44.5) there exists an optimal control $\hat{u} \in \mathscr{U}$ for this problem. It was also shown that $V(x) = J_x(\hat{u})$ is continuous. We now want to characterize V terms of the *generalized Bellman equation*, which is as follows, to be solved for a function $W: E \to \mathbb{R}$:

$$\text{(46.1)} \qquad \min_{r \in \mathscr{R}, \zeta \in \partial W(x)} \int_{U_0} \Big\{ \zeta \cdot g(x, u) + l(x, u) + \lambda(x, u) \int_E (W(y) - W(x)) Q(\mathrm{d}y; x, u) \Big\} r(\mathrm{d}u) = 0, \quad x \in E,$$

$$\text{(46.2)} \qquad \min_{u_\Gamma \in U_\Gamma} \int_E (W(y) - W(x)) R(\mathrm{d}y; x, u_\Gamma) + c(x, u_\Gamma) = 0, \quad x \in \Gamma.$$

This is precisely the Bellman equation introduced at (44.31) or in the naïve dynamic programming argument of Section 42 ((42.6) and (42.7)) except that relaxed controls are used and the directional derivative of W has been replaced by the minimizing element of the generalized gradient set. Of course, (46.1) and (46.2) only make sense for functions W that are locally Lipschitz continuous. We will show the following:

1. Under an additional assumption, (46.4) below, the value function V is locally Lipschitz continuous and satisfies (46.1) and (46.2).
2. If W is any regular locally Lipschitz continuous function satisfying (46.1) and (46.2) then W is fixed point of the operator G defined at (44.21), i.e. W satisfies $W = GW$. We will state a condition, Condition (46.16) below, under which the value function V is the *unique* fixed point of G.

Thus, if additional assumptions (46.4) and Condition (46.16) hold, a regular function W is the value function for the control problem *if and only if* W satisfies (46.1) and (46.2). Further, optimal controls are characterized in terms of minimizing elements in (46.1) and (46.2) as the theorems later in this section describe.

(46.3) *Theorem.* Suppose Assumptions (41.1) and (44.5) hold, together with the following condition:

(46.4) $$\delta > \sup_{\substack{y,z\in E \\ u_0\in U_0}} \frac{(y-z)\cdot(g(y,u_0)-g(z,u_0))}{|y-z|^2}.$$

Then V is Lipschitz continuous.

(46.5) *Remark.* Let α denote the right-hand side of (46.4). The condition is nugatory if $\alpha < 0$. Denote $\beta = \frac{1}{2}(\delta + \alpha)$; then (46.4) implies that $\beta < \delta$ and that for all $y, z \in E$, $u_0 \in U_0$

(46.6) $$(y-z)\cdot(g(y,u_0)-g(z,u_0)) \leqslant \beta|y-z|^2.$$

Condition (46.4) cetrainly holds if $\delta > L$, where L is a Lipschitz constant for g.

(46.7) *Example*
The following example shows that condition (46.4) cannot generally be dispensed with if V is to be Lipschitz continuous. Recall that deterministic systems are a special case of PDPs and consider a control-free scalar system with dynamics

$$\frac{\mathrm{d}}{\mathrm{d}t}x(t) = bx(t), \quad x(0) = x \in \mathbb{R}_+$$

with $b > 0$. Then

$$V(x) = \int_0^\infty \mathrm{e}^{-\delta t} l(x(t))\,\mathrm{d}t.$$

Take $l(x) = x \wedge 1$; then l is bounded and Lipschitz continuous on $\mathbb{R}_+$ but a straightforward calculation shows that

$$V(x) = \begin{cases} \dfrac{1}{b-\delta}\left(\dfrac{b}{\delta}x^{\delta/b} - x\right), & 0 \leqslant x \leqslant 1 \\ \dfrac{1}{\delta}, & x \geqslant 1. \end{cases}$$

The condition (46.4) is satisfied if and only if $\delta > b$ and we note that V is in fact not Lipschitz continuous on $\mathbb{R}_+$ if this condition is violated. For example if $\delta = \frac{1}{2}b$ then

$$V(x) = \frac{1}{\delta}(2x^{1/2} - x), \quad x \leqslant 1.$$

Proof of Theorem (46.3). From Theorem (44.17) we know that (see (44.25))

$$V(x) = \inf_{r \in \mathscr{R}} J^V(x, r),$$

where

$$J^V(x, r) = \int_0^{t_*} \chi(t) f_0^V(\phi(t), r(t)) \mathrm{d}t + F^V(\phi(t_*))\chi(t_*)$$

and f_0^V, F^V are given by (44.26) and (44.27). Since $V \in C_b(E)$, Assumption (41.1) ensures f_0^V, F^V are Lipschitz continuous, uniformly in r. We will show that $J^V(x, r)$ is Lipschitz continuous with constant not depending on r, which implies that V is Lipschitz with the same constant. Take a control $r \in \mathscr{R}$ and two starting points, x and z. There are several cases to consider: $t_*(x) = t_*(z) = \infty$; $t_*(x) < \infty$, $t_*(z) = \infty$; $t_*(x)$, $t_*(z) < \infty$. We shall analyse only the first of these; the remainder follow by similar arguments (Ye, 1990). Now with $t_*(x) = t_*(z) = \infty$ we have $\chi(t_*) = 0$, and with an obvious compressed notation,

$$\begin{aligned} |J^V(x, r) - J^V(z, r)| &= \left| \int_0^\infty \chi_t(x) f_0^V(x) \mathrm{d}t - \int_0^\infty \chi_t(z) f_0^V(z) \mathrm{d}t \right| \\ &\leqslant \left| \int_0^\infty \chi_t(x)(f_0^V(x) \mathrm{d}t - f_0^V(z)) \mathrm{d}t \right| \\ &\quad + \left| \int_0^\infty (\chi_t(x) - \chi_t(z)) f_0^V(z) \mathrm{d}t \right| \\ &\leqslant \int_0^\infty K \mathrm{e}^{-\delta t} |\phi_t(x) - \phi_t(z)| \, \mathrm{d}t + M \int_0^\infty |\chi_t(x) - \chi_t(z)| \, \mathrm{d}t \end{aligned}$$

where K and M are a Lipschitz constant and a bound for f_0^V, respectively, and we recall that $\chi(t) \leqslant \mathrm{e}^{-\delta t}$ since $\lambda \geqslant \delta$. Now using (46.6) we have

$$\frac{\mathrm{d}}{\mathrm{d}t} |\phi_t(x) - \phi_t(z)|^2 = 2(\phi_t(x) - \phi_t(z)) \cdot (g(x) - g(z)) < 2\beta |\phi_t(x) - \phi_t(z)|^2$$

and hence

$$|\phi_t(x) - \phi_t(z)| < |x - z| \mathrm{e}^{\beta t}.$$

Also $\chi_0(x) = 1$ and

$$\frac{d}{dt}(\chi_t(x) - \chi_t(z)) = -\chi_t(x)\lambda_t(x) + \chi_t(z)\lambda_t(z)$$

$$= -\lambda_t(x)(\chi_t(x) - \chi_t(z) + \chi_t(z)(\lambda_t(z) - \lambda_t(x))$$

so that

$$|\chi_t(x) - \chi_t(z)| = \left| \int_0^t \exp\left(-\int_s^t \lambda_u(x) du \right) \chi_s(z)(\lambda_s(z) - \lambda_s(x)) ds \right|$$

$$\leqslant \int_0^t e^{-\delta(t-s)} e^{-\delta s} |\lambda_s(z) - \lambda_s(x)| \, ds$$

$$\leqslant e^{-\delta t} \int_0^t K |\phi_s(x) - \phi_s(z)| \, ds$$

$$\leqslant K e^{-\delta t} |x - z| \int_0^t e^{\beta s} \, ds$$

$$\leqslant \frac{K}{\beta} |x - z| \, e^{-(\delta - \beta)t}.$$

Thus

$$|J^V(x, r) - J^V(z, r)| \leqslant \int_0^\infty K e^{-\delta t} |x - z| \, e^{\beta t} \, dt + M \int_0^\infty \frac{K}{\beta} |x - z| \, e^{-(\delta - \beta)t} \, dt$$

$$\leqslant \frac{K}{(\delta - \beta)} \left(1 + \frac{M}{\beta} \right) |x - z|.$$

This completes the proof. □

(46.8) *Theorem.* Suppose Assumptions (41.1), (44.5) and (46.4) hold. Then the value function V satisfies the generalized Bellman equation (46.1) and (46.2).

Proof. We know from (44.25) that V satisfies

$$V(x) = \min_{r \in \mathscr{R}} \int_0^{t_*} \chi(t) f_0^V(\phi(t), r(t)) \, dt + F^V(\phi(t_*)) \chi(t_*), \tag{46.9}$$

where f_0^V and F^V are defined by (44.26) and (44.27), $\phi(t)$ is the deterministic trajectory corresponding to $r\in\mathscr{R}$ and

$$\chi(t)=\exp\left(-\int_0^t\int_{U_0}\lambda(\phi(t),u)r_t(\mathrm{d}u)\mathrm{d}t\right).$$

We now apply Theorem (45.13) to this deterministic optimal control problem. In order to do this, we must first put the problem in Mayer form. Define an augmented state process $z(t)\in\mathbb{R}^{d+2}$ by taking $\bar{z}(t):=(z^1(t),\ldots,z^d(t))=\phi(t), z^{d+1}(t)=\chi(t)$ and

$$z^{d+2}(t)=\int_0^t\chi(s)f_0^V(\phi(s),r(s))\mathrm{d}s.$$

Then $z(t)$ satisfies

(46.10) $$\frac{\mathrm{d}}{\mathrm{d}t}z(t)=f(z(t),r(t)),\quad z^T(0)=(x,1,0),$$

where

$$f^T=(\bar{f}^T,f^{d+1},f^{d+2}),$$

with

$$\bar{f}(z,r)=\int_{U_0}g(\bar{z},u)r(\mathrm{d}u)$$

$$f^{d+1}(z,r)=-z^{d+1}\int_{U_0}\lambda(\bar{z},u)r(\mathrm{d}u)$$

$$f^{d+2}(z,r)=z^{d+2}\int_{U_0}f_0^V(\bar{z},u)r(\mathrm{d}u).$$

The cost for a control $r\in\mathscr{R}$ is then

(46.11) $$J_z(r)=z^{d+2}(t_*)+z^{d+1}(t_*)F^V(\bar{z}(t_*))=:\Phi(z).$$

Here we are allowing general starting points for z^{d+1} and z^{d+2} with $z^{d+1}>0$; we recover the cost for the original problem (46.9) starting at $\phi(0)=x$ by taking $z=(x,1,0)$ as in (46.10). Clearly the value function $\tilde{H}$ of the control problem (46.11) is related to V by

$$\tilde{H}(z)=z^{d+2}+z^{d+1}V(\bar{z}).$$

Control problem (46.11) now satisfies the conditions of Theorem (45.13); in particular, $\tilde{H}$ is locally Lipschitz continuous. Its generalized gradient is related to that of V by

$$\partial\tilde{H}(z) = (z^{d+1}\partial V(\bar{z})) \times \{V(\bar{z})\} \times \{1\}.$$

(Note that at points where V is differentiable we have $\nabla\tilde{H}^T(z) = (z^{d+1}\nabla V^T(\bar{z}), V(\bar{z}), 1)$.) Applying Theorem (45.13) (in particular, taking $U := \mathscr{P}(U_0)$) we find that $\tilde{H}$ satisfies

(46.12) $$\min_{\substack{\beta\in\partial\tilde{H}(z)\\ r\in\mathscr{P}(U_0)}} \beta\cdot f(z,r) = 0 \quad \text{in} \quad E\times\mathbb{R}^2.$$

Now any $\beta\in\partial\tilde{H}$ takes the form $\beta^T = (z^{d+1}\zeta^T, V(\bar{z}), 1)$ for some $\zeta\in\partial V(\bar{z})$, and hence (46.12) becomes

$$\min_{\substack{\zeta\in\partial V(\bar{z})\\ r\in\mathscr{P}(U_0)}} z^{d+1}\zeta\cdot\bar{f}(\bar{z},r) - V(\bar{z})f^{d+1}(z,r) + f^{d+2}(z,r) = 0 \quad \text{in} \quad E\times\mathbb{R}^2$$

or

$$\min_{\substack{\zeta\in\partial V(\bar{z})\\ r\in\mathscr{P}(U_0)}} z^{d+1}\int_{U_0}\Big\{\zeta\cdot g(x,u) - V(x)\lambda(x,u) + l(x,u)$$

$$+ \lambda(x,u)\int_E V(y)Q(\mathrm{d}y;x,u)\Big\}r(\mathrm{d}u) = 0, \quad x\in E.$$

Since $z^{d+1} > 0$, this is equivalent to (46.1). We know from Theorem (44.22) that V satisfies (46.2) at all $x\in\Gamma$. This completes the proof. □

We now turn to sufficient conditions. The main result is as follows.

(46.13) *Theorem.* Suppose Assumptions (41.1) and (44.5) hold and that $W: E\cup\Gamma\to\mathbb{R}$ is a regular locally Lipschitz continuous function satisfying the generalized Bellman equation (46.1) and (46.2). Suppose further that there is an admissible control $\hat{u} = (\hat{u}_0, \hat{u}_\Gamma)\in\mathscr{U}$ with the properties 1 and 2 below. Then W is a fixed point of the discrete-time dynamic programming operator G of (44.21).

To state properties 1. and 2., fix $x\in E$, denote $\hat{r}_t = \hat{u}_0(x,t)$, $\hat{u}_\Gamma = \hat{u}_\Gamma(x)$, and let $\hat{\phi}(t)$ be the deterministic trajectory satisfying

(46.14) $$\frac{\mathrm{d}}{\mathrm{d}t}\hat{\phi}(t) = \int_{U_0} g(\hat{\phi}(t),u)\hat{r}_t(\mathrm{d}u).$$

Now let $\hat{u}_\Gamma = \hat{u}_\Gamma(\hat{\phi}(t_*))$, where t_* is the boundary hitting time of $\hat{\phi}(t)$. The properties in question are:

1. For almost every t, there exists an element $\hat{\zeta}(t) \in \partial W(\hat{\phi}(t))$ such that

$$\int_{U_0} \Big\{ \hat{\zeta}(t) \cdot g(\hat{\phi}(t), u) + l(\hat{\phi}(t), u)$$
$$+ \lambda(\hat{\phi}(t), u) \int_E (W(y) - W(\hat{\phi}(t))) Q(\mathrm{d}y; \hat{\phi}(t), u) \Big\} \hat{r}_t(\mathrm{d}u) = 0.$$

2. The minimum in (46.2) is achieved at $u_\Gamma = \hat{u}_\Gamma$ when x is replaced by $\hat{\phi}(t_*)$.

Proof. This is an application of Theorem (45.6). In that theorem, we take $U = \mathscr{P}(U_0)$,

$$f(z, r) = \int_{U_0} g(z, u) r(\mathrm{d}u),$$

$$f_0(z, r) = \int_{U_0} \Big\{ l(z, u) + \lambda(z, u) \int_E W(y) Q(\mathrm{d}y; z, u) \Big\} r(\mathrm{d}u), \quad z \in E$$

and

$$F(z) = C(z, \hat{u}_\Gamma) + \int_E W(y) R(\mathrm{d}y; \hat{u}_\Gamma), z \in \Gamma.$$

We also define

$$\hat{\chi}(t) = \exp\Big\{ - \int_0^t \int_{U_0} \lambda(\hat{\phi}(s), u) \hat{r}_s(\mathrm{d}u)\, \mathrm{d}s \Big\},$$

where $\hat{\phi}(s)$ is given by (46.14). We then conclude from Theorem (45.6) that

$$W(x) = \int_0^{t_*} \hat{\chi}(t) f_0(\hat{\phi}(t), \hat{r}_t)\, \mathrm{d}t + \hat{\chi}(t_*) F(\hat{\phi}(t_*))$$
$$= \int_0^{t_*} \hat{\chi}(t) \int_{U_0} \Big\{ l(\hat{\phi}(t), u) + \lambda(\hat{\phi}(t), u) \int_E W(y) Q(\mathrm{d}y; \hat{\phi}(t), u) \Big\} \hat{r}_t(\mathrm{d}u)$$
$$+ \hat{\chi}(t_*) \Big(c(\hat{\phi}(t_*), \hat{u}_\Gamma) + \int_E W(y) R(\mathrm{d}y; \hat{\phi}(t_*), \hat{u}_\Gamma) \Big)$$
$$= \mathbb{E}_x \Big\{ \int_0^{T_1} \int_{U_0} l(x_t, u) \hat{r}_t(\mathrm{d}u) \mathrm{d}t + c(x_{T_1-}, \hat{u}_\Gamma(x_{T_1-}) I_{(T_1 = t_*)} + W(x_{T_1}) \Big\}$$
$$= G_{\hat{u}} W(x),$$

where x_t denotes the PDP corresponding to control $\hat{u}\in\mathscr{U}$, and we have used the notation of Remark (44.20). It also follows from Theorem (45.6) by a similar analysis that $W(x)\leqslant G_u W(x)$ for an arbitrary control $u\in\mathscr{U}$; thus

$$W(x)=\min_{u\in\mathscr{U}} G_u W(x) = GW(x). \qquad \square$$

(46.15) *Corollary.* Suppose W and $\hat{u}$ are as in Theorem (46.13) and G has a unique fixed point. Then $W(x)=V(x), x\in E$ and $\hat{u}$ is optimal, where V is the value function.

Proof. It was shown in Section 44 that V is a fixed point of G and that any control u satisfying $V=G_u V$ is optimal. If G has a unique fixed point then W must coincide with V and the result follows. □

As discussed in Section 47 below, a general condition under which G has a unique fixed point is that G be an *n-stage contraction* operator: $\| G^n\psi_1 - G^n\psi_2\| \leqslant \rho \|\psi_1-\psi_2\|$ for some integer n and $\rho\in]0,1[$, where

$$\|\psi\| = \sup_{x\in E}|\psi(x)|$$

(Proposition (47.8)).

In general, even under Assumption (41.1) (i.e. the existence of a discount rate $\delta>0$), G is not a one-stage contraction if $t_*(x)<\infty$ for some x. To see this, consider an uncontrolled, unkilled PDP x_t with cost rate $l=0$; then $G\psi(x)=\mathbb{E}_x \exp(-\delta T_1)\psi(x_{T_1})$. If we take $\psi_1(x)=1$ and $\psi_2(x)=0$, then $|G(\psi_1-\psi_2)(x)|=\mathbb{E}_x\exp(-\delta T_1)$, and $\sup_x \mathbb{E}_x\exp(-\delta T_1)=1$ since $T_1\leqslant t_*(x)$ and $t_*(\phi(t,x))\to 0$ as $t\to t_*(x)$. However G is a two-stage contraction under the following very reasonable condition, satisfied in virtually every application. This condition is a modification of the one introduced in part 2 of Proposition (24.6). We suppose that Assumption (41.1) holds, take $\varepsilon>0$ and define $A_\varepsilon=\{x\in E: \inf_{u\in\mathscr{U}} t_*^u(x)\geqslant\varepsilon\}$.

(46.16) *Condition.* There exists $\varepsilon>0$ such that $R(A_\varepsilon; z, u_\Gamma)=1$ for all $z\in\Gamma$ and $u_\Gamma\in U_\Gamma$.

Informally this just says that jumps from the boundary are always to points whose distances from the boundary (as measured by the boundary hitting time) are uniformly bounded away from zero.

(46.17) *Proposition.* If Assumption (41.1) and Condition (46.6) hold, then G is a two-stage contraction mapping.

Proof. It suffices to show that there exists $\rho \in]0, 1[$ such that, for any $u \in \mathscr{U}$, $\| G_u^2 \psi_1 - G_u^2 \psi_2 \| \leqslant \rho \| \psi_1 - \psi_2 \|$. Indeed, if this is the case then, for any $x \in E$, $G_u^2 \psi_1(x) \leqslant G_u^2 \psi_2(x) + \rho \| \psi_1 - \psi_2 \|$. Taking the infimum over u on both sides we obtain $G^2 \psi_1(x) \leqslant G^2 \psi_2(x) + \rho \| \psi_1 - \psi_2 \|$. The same inequality holds with ψ_1 and ψ_2 interchanged, so that $|G^2 \psi_1(x) - G^2 \psi_2(x)| \leqslant \rho \| \psi_1 - \psi_2 \|$. The result follows.

To show that G_u^2 is a contraction it is enough, since G_u^2 is an affine operator, to show that $\mathbb{E}_x^u \exp(-\delta T_2) \leqslant \rho < 1$ for some constant $\rho < 1$ not depending on $u \in \mathscr{U}$ or $x \in E$, where T_2 is the second jump time of the controlled process. By Assumption (41.1(c)), $\lambda(x, u) \leqslant M < \infty$. Fix $u \in \mathscr{U}$ and consider first $x \in E_\varepsilon = \{x \in E : t_*(x) \geqslant \varepsilon'\}$ for some $\varepsilon' > 0$. Recall that in the construction of the PDP, the first jump time T_1 is obtained by transformation of a $U[0, 1]$-distributed random variable U. If we apply a similar transformation to the same U, but replacing $\lambda(x, u)$ by M, we obtain a random variable $\tilde{T}_1$ such that $\tilde{T}_1 \leqslant T_1$ a.s. Thus $\hat{T}_1 := \tilde{T}_1 \wedge \varepsilon' \leqslant T_1 < T_2$ a.s. and hence $\mathbb{E}_x \exp(-\delta T_2) \leqslant \mathbb{E}_x \exp(-\delta \hat{T}_1) \leqslant \mathbb{E}_x \exp(-\delta \hat{T}_1)$. However the survivor function of $\hat{T}_1$ is $I_{(t < \varepsilon')} \exp(-Mt)$, and a simple calculation shows that

$$\mathbb{E}_x \mathrm{e}^{-\delta \hat{T}_1} =: \rho(\varepsilon') = \frac{M + \delta \, \mathrm{e}^{-(M+\delta)\varepsilon'}}{M + \delta} < 1.$$

If $x \in (F_{\varepsilon'})^c$ we have $T_2 \geqslant S_2 := T_2 - T_1$ and

$$\begin{aligned} \mathbb{E}_x \mathrm{e}^{-\delta T_2} &\leqslant \mathbb{E}_x \mathrm{e}^{-\delta S_2} \\ &= \mathbb{E}_x(I_{(T_1 < t_*)} \mathrm{e}^{-\delta S_2}) + \mathbb{E}_x(I_{(T_1 = t_*)} \mathrm{e}^{-\delta S_2}) \\ &\leqslant P_x(T_1 < t_*) + \mathbb{E}_x(I_{(T_1 = t_*)} \mathbb{E}_x[\mathrm{e}^{-\delta S_2} | \mathfrak{F}_{T_1}]) \\ &\leqslant (1 - \mathrm{e}^{-M\varepsilon'}) + \rho(\varepsilon). \end{aligned}$$

In the last inequality we have used the facts that (for the first term) $T_1 \geqslant \hat{T}_1$ and (for the second) that S_2 is constructed in a similar manner to T_1, but restarting the process at x_{T_1}, and by Condition (46.16), $P_x(x_{T_1} \in E_\varepsilon) = 1$. Now choose ε' such that $\mathrm{e}^{-M\varepsilon'} > \frac{1}{2}(1 + \rho(\varepsilon))$. Then

$$\mathbb{E}_x \mathrm{e}^{-\delta T_2} \leqslant \tfrac{1}{2}(1 + \rho(\varepsilon))$$

for $x \in (E_{\varepsilon'})^c$. Hence for any $x \in E$ we have

$$\mathbb{E}_x \mathrm{e}^{-\delta T_2} \leqslant \rho := \max\{\rho(\varepsilon'), \tfrac{1}{2}(1 + \rho(\varepsilon))\} < 1.$$

This completes the proof. □

We can now combine all of the preceding results to give a necessary and sufficient condition for optimality.

(46.18) *Theorem.* Suppose Assumptions (41.1), (44.5), (46.4) and Condition (46.16) hold. Then

1. An optimal control $\hat{u} \in \mathcal{U}$ exists.
2. The value function V is locally Lipschitz continuous and satisfies the generalized Bellman equation (46.1) and (46.2)
3. If W is any regular locally Lipschitz continuous solution of (46.1) and (46.2) and Conditions 1 and 2 of Theorem (46.13) hold then W coincides with V.
4. A control $u = (u_0, u_\Gamma)$ is optimal if and only if it satisfies

$$\min_{\zeta \in \partial V(\phi(t))} \int_{U_0} \Big\{ \zeta \cdot g(\phi(t), u) + l(\phi(t), u)$$
$$+ \lambda(\phi(t), u) \int_E (V(y) - V(\phi(t))) Q(\mathrm{d}y; \phi(t), u) \Big\} u_0(x, t, \mathrm{d}u) = 0$$

for each $x \in E$ and almost all $t < t_*^u(x)$, where

$$\frac{\mathrm{d}}{\mathrm{d}t}\phi(t) = \int_{U_0} g(\phi(t), u) u_0(x, t, \mathrm{d}u), \quad \phi(0) = x$$

and

$$V(x) = c(x, u_\Gamma(x)) + \int_E V(y) R(\mathrm{d}y; x, u_\Gamma(x)), \quad x \in \Gamma.$$

(46.19) *Remark.* The only outstanding point is whether V is automatically regular under the conditions we have imposed. If this were the case, we would be able to state simply that V is the unique regular solution of the Bellman equation, whereas as things stand we are not able to guarantee that any regular solution exists.

Computation of optimal controls based on the characterizations of optimality given here is a subject in which not much systematic research has yet been done. The only method that has been explored is naïve dynamic programming, i.e. a direct solution of the Bellman equation as in the capacity expansion problem (42.13). The best general approach seems likely to be an iterative one along the lines of the iterative method for uncontrolled processes discussed in section 32.2. Essentially, this would involve computing $G^k\psi(x) = G(G^{k-1}\psi)(x)$ for some function ψ such as $\psi \equiv 0$, where G is the discrete-time dynamic programming operator defined at (44.21). We know that $G^k\psi \to V$ as $k \to \infty$, and hence solution of the generalized Bellman equation (46.1) and (46.2) is 'reduced' to the solution of a sequence of deterministic optimal control problems. This is not straightforward, as $G\psi(x)$ has to be computed for many different starting points x. A somewhat similar situation arises in the *receding horizon* method of nonlinear control system design (Michalska and Mayne, 1991), and it may be that some of the techniques developed there will be relevant to the PDP control problem too.

47 Discrete-stage Markov decision models

Our strategy in Section 44 was to reformulate the controlled PDP as a discrete-stage Markov decision model (DMDM). This section summarizes some information about DMDMs as required for this application. The material is taken from Bertsekas and Shreve (1978), referred to here as B&S. The reader can – and should – refer to B&S for full details. First, we need to recall some properties of semicontinuous functions (B&S, pp. 146–7).

(47.1) *Definition.* Let X be a metrizable space. A function $f:X\to\mathbb{R}$ is *lower semicontinuous* (l.s.c.) if for each $c\in\mathbb{R}$ the set $\{x:f(x)\leqslant c\}$ is closed; f is *upper semicontinuous* (u.s.c.) if $-f$ is l.s.c.

(47.2) *Proposition*

1. $f:X\to\mathbb{R}$ is l.s.c. if and only if $f(x)\leqslant \liminf_{n\to\infty} f(x_n)$ whenever x_n is a convergent sequence in X with limit x.
2. $f:X\to\mathbb{R}$ is l.s.c. if and only if there exists an increasing sequence f_n of continuous functions such that $f_n(x)\uparrow f(x)$ for each x as $n\to\infty$.
3. $f:X\to\mathbb{R}$ is continuous if and only if it is both l.s.c. and u.s.c.

It is clear from the definition that if f_n is a sequence of l.s.c. functions and $f_n\uparrow f$ then f is l.s.c.

We are now ready to introduce the DMDM. Its ingredients are state and control spaces S, U, a transition function $q:S\times U\to\mathscr{P}(S)$ and a cost function $h:S\times U\to\mathbb{R}$. In the l.s.c. model the assumptions are as follows: these are more restrictive than the ones given by B&S (Definition 8.7, p. 208) but suffice for our purposes.

(47.3) *l.s.c. conditions*

1. S is a Borel space;
2. U is a compact Borel space;

3. the transition function q is continuous;
4. the cost function h is non-negative and lower semicontinuous.

A control policy (*non-randomized*, in the terminology of B&S) is a sequence $\pi = (\mu_0, \mu_1, \ldots)$ of Borel measurable functions $\mu_0 : S \to U$, $\mu_k : S^{k+1} \times U^k \to U$, $k = 1, 2, \ldots$. The set of all policies is denoted Π. Given a policy μ and an initial state $x \in S$ a discrete-time stochastic process is defined by specifying the following family of conditional distributions:

$$\text{(47.4)} \qquad \begin{cases} P[Z_0 = x] = 1 \\ P[Z_{k+1} \in A \mid Z_0, Z_1, \ldots, Z_k] = q(A; Z_k, u_k), \quad A \in \mathscr{E} \end{cases}$$

where $u_0 := \mu_0(x)$, and $u_j := \mu_j(Z_0, \ldots, Z_j, u_0, \ldots, u_{j-1}), j = 0, \ldots, k$. This is a complete specification of the process since (47.4) defines a consistent set of finite-dimensional distributions. The *cost* corresponding to μ is

$$\text{(47.5)} \qquad J_\pi(x) = \mathbb{E}_x \sum_{k=0}^{\infty} h(Z_k, u_k),$$

with u_k as above. A policy $\hat{\pi}$ is *optimal* if $J_{\hat{\pi}}(x) \leqslant J_\pi(x)$ for all $\pi \in \Pi$ and $x \in S$. A policy is *Markov* if μ_k is parametrized only by Z_k, so that each μ_k is a function $\mu_k : S \to U$. Then we can write the cost simply as

$$J_\pi(x) = \mathbb{E}_x \sum_{k=0}^{\infty} h(Z_k, \mu_k(Z_k)).$$

A Markov policy is *stationary* if $\mu_k = \mu_0$ for all k. The sets of Markov and stationary Markov policies are denoted $\Pi_{\mathrm{M}}, \Pi_{\mathrm{SM}}$, respectively.

Consider first minimization over a finite number N of stages rather than the infinite horizon problem (47.5). Let ψ be a non-negative l.s.c. function and define

$$G\psi(x) = \inf_{u \in U} \left\{ h(x, u) + \int_E \psi(y) q(\mathrm{d}y; x, u) \right\}.$$

(47.6) *Proposition.* Under assumptions in section 4.7.1,

1. $G\psi$ is l.s.c.;
2. the infimum is achieved for each $x \in S$;

3. there is a Borel-measurable function $\mu_0 : S \to U$ such that

$$G\psi(x) = h(x, \mu_0(x)) + \int_E \psi(y) q(\mathrm{d}y; x, \mu_0(x)).$$

Evidently, μ_0 is an optimal policy for the one-stage problem of minimizing

$$J_{1,\pi}(x) = \mathbb{E}_x\{h(Z_0, u_0) + \psi(Z_1)\}$$

over $\pi \in \Pi$. In view of condition 4 of section (47.3) and part 1 of Proposition (47.6), the operator G maps the class of non-negative l.s.c. functions into itself, and G can therefore be applied recursively to define $G^n\psi = G(G^{n-1}\psi)$ for $n = 1, 2, \ldots, N$. It follows easily from the dynamic programming principle that

$$G^N\Psi(x) = \inf_{\pi \in \Pi} J_{N,\pi}(x),$$

where $J_{N,\pi}$ is the N-stage cost

$$J_{N,\pi}(x) = \mathbb{E}_x\left\{\sum_{k=0}^{N-1} h(Z_k, u_k) + \Psi(Z_N)\right\}.$$

Here, Ψ is some non-negative l.s.c. function and u_k is related to π as described above. Let $\hat{\pi} \in \Pi_{\mathrm{M}}$ be a policy satisfying

$$G^{N-k}\Psi(x) = h(x, \hat{\mu}_k(x)) + \int_E G^{N-k-1}\Psi(y) q(\mathrm{d}y; x, \hat{\mu}_k(x)).$$

Such a policy exists from part 3 of Proposition (47.6), and $\hat{\pi}$ is optimal, i.e. minimizes $J_{N,\pi}(x)$ for each $x \in S$ over all $\pi \in \Pi$. Thus Markov policies are optimal, even though more general policies are admissible.

We now want to consider the infinite-horizon case. Take $\Psi \equiv 0$ and define

$$J^*(x) = \lim_{N \to \infty} G^N\Psi(x).$$

Since h is non-negative, it is clear that $G^N\Psi$ is monotone increasing, so that the limit exists although it might be equal to $+\infty$ for some or all $x \in S$. (In the latter case, any policy is optimal.) Under the l.s.c. conditions (47.3) J^* is l.s.c., being the limit of an increasing sequence of l.s.c. functions.

(47.7) *Proposition.* $J^*(x) = GJ^*(x)$.

Proof. See B&S (Proposition 9.8, p. 225). □

Thus J^* is a fixed point of the operator G. The question now is whether J^* coincides with the minimal infinite-horizon cost, defined by

$$V(x) = \inf_{\pi \in \Pi} \mathbb{E}_x \left\{ \sum_{k=0}^{\infty} h(Z_k, u_k) \right\}.$$

The main result we need is as follows (B&S, Corollary 9.17.2, p. 235).

(47.8) *Theorem.* Suppose the l.s.c. conditions (47.3) are satisfied. Then $V = J^*$, V is l.s.c., and there exists a stationary policy $\pi^\circ \in \Pi_{SM}$ that is optimal in Π, i.e. satisfies $J_{\pi^\circ}(x) = V(x), x \in S$.

We also need a result for u.s.c. models. These are models satisfying the conditions (47.3) except that (4) is replaced by

4′. The cost function h is non-positive and upper semi-continuous.

(47.9) *Theorem.* (B&S, Proposition 9.21, p. 241). Under the first 3 conditions of (47.3) and 4′, J^* is upper semi-continuous.

Proposition (47.7) and Theorem (47.8) show that the value function V is a fixed point of the operator G. We need conditions under which V is the *unique* fixed point. For this we use the theory of contraction mappings. If h is bounded then G is a nonlinear operator on the space $B(S)$ of bounded Borel-measurable functions. G is a *contraction* if, for some $\rho \in]0, 1[$, $\| G\psi_2 - G\psi_1 \| < \rho \| \psi_2 - \psi_1 \|$ for all $\psi_1, \psi_2 \in B(S)$, where, as usual, $\| \psi \| = \sup_{x \in E} |\psi(x)|$. It is a standard result (B&S, p. 55) that under this condition G has a unique fixed point ψ^* and that $\| G^n \psi - \psi^* \| \to 0$ as $n \to \infty$ for any $\psi \in B(S)$. If we define

$$G_u \psi(x) = h(x, u) + \int_E \psi(y) q(\mathrm{d}y; x, u),$$

then G is a contraction if, for some $\rho \in]0, 1[$, $\| G_u \psi_2 - G_u \psi_1 \| < \rho \| \psi_2 - \psi_1 \|$ for any $u \in U$, as is easily shown.

In the applications of this chapter, G is typically not a contraction but under reasonable conditions it is an *N-stage contraction*, i.e. G^N is a contraction for some $N > 1$ (in fact, $N = 2$ in our application). Then G^N has a unique fixed point ψ°, and ψ° is also the unique fixed point of G. Indeed, we know that for any $\psi \in B(S)$, $G^{kN} \psi \to \psi^\circ$ as $k \to \infty$. Taking $\psi = G\psi^\circ$ we have $\psi^\circ = \lim_k G^{kN} G \psi^\circ = \lim_k G G^{kN} \psi^\circ =$

$G\psi^\circ$. Thus ψ° is a fixed point of G; it is unique because obviously any fixed point of G is a fixed point of G^N. We thus have the following result.

(47.10) *Proposition.* Suppose G is an operator on $B(S)$ satisfying the N-stage contraction condition for some $N \geqslant 1$. Then G has a unique fixed point.

48 Notes and references

41 The first paper on control of PDPs is due to Vermes (1985) using methods of convex analysis and duality introduced in deterministic control theory by Vinter and Lewis (1978). In particular, Vermes is responsible for identifying piecewise open-loop controls as the appropriate ones in PDP control.
42 Dynamic programming is of course the standard technique for solving stochastic control problems. Fleming and Rishel (1975) cover the theory for controlled stochastic differential equations, which is the best developed part of the subject in continuous time. An introductory account, which also discusses PDPs, is the article by Rishel (1990). The capacity expansion problem in section (42.13) was the first substantial control problem to be solved by PDP methods, by Davis, Dempster, Sethi and Vermes (1987).
43 Relaxed controls were originally introduced in the calculus of variations by L.C. Young. A comprehensive account is given by Warga (1972).
44 This section is mainly from Davis (1986) with some improvements from Dempster and Ye (1992) and Ye (1990).
45, 46 These sections follow Dempster and Ye (1992) and Ye (1990) closely, except that the use of the two-stage contraction condition is new. A summary of their results is given in Dempster (1991) while Dempster and Ye (1990) derive a stochastic maximum principle, a topic not pursued here. An earlier treatment of all of these topics, in which many similar ideas were introduced, is Ribeiro do Val (1987). An alternative treatment based on the 'viscosity solution' concept is given by Soner (1986).
47 This material is from Bertsekas and Shreve (1978). An alternative source is Dynkin and Yushkevich (1978).

CHAPTER 5

Control by intervention

In the previous chapter we described 'continuous' control of PDPs: the control acts at all times, affecting the directions of deterministic motion and/or the incremental probability of obtaining a jump of the process. Boundary control is also included: when the process hits the (fixed, given) boundary of the state space, we may have some control over where it goes. What is not directly allowed for, however, is the ability to intervene in the evolution of the process by introducing a jump at some controller-specified stopping time and moving the process at that time to some new point in the state space. Many applied problems, however, are of precisely this type: for example, inventory problems in which a sequence of restocking decisions is made, resource allocation problems involving decisions to commit funds to specific projects, or stochastic scheduling where the option to reschedule at any time is to be included.

These are all examples of what is usually known as 'impulse control', a subject whose study was initiated by Bensoussan and Lions (1975) in the context of diffusion processes and which has generated a large literature and many applications. The aim of this chapter is to develop a self-contained theory of impulse control for PDPs. The main results concern regularity of the value function, existence of optimal strategies, convergence of approximating schemes and characterization of optimality in terms of so-called *quasi-variational inequalities.* The final section, 55, gives a brief introduction to computational methods: the main ideas and results are summarized, but the reader is referred to journal articles for complete proofs. It is in fact one of the most satisfactory aspects of PDP theory as a whole that such effective computational methods are available for impulse control, particularly since impulse control probably constitutes the most important class of optimization applications.

There are several different avenues of approach to impulse control of PDPs. One could rely on general theories which are available in the literature for impulse control of right processes (Lepeltier and Marchal, 1984) of which, as we showed in section 27, the PDP is a special case. Or one could attempt a 'grand unification' in which the possibility of intervention is added to the theory of continuous control we already have (Dempster and Ye, 1993b). However, we have rejected both of these in favour of a direct approach. In relying on general theories we would be throwing away much of the special structure of PDPs and thereby weakening the available results. On the other hand incorporating interventions in a continuous control framework involves rather a lot of technical assumptions, many of which can be dispensed with if impulse control is the only control. By taking a direct approach we obtain a cleaner theory.

The simplest form of intervention is optimal stopping, in which we select a single stopping time and collect a reward which depends on the position of the process at that time. We develop optimal stopping theory in the first two sections. It has significant applications in its own right (questions such as when to buy or sell shares or Old Masters), but, perhaps more importantly, it constitutes a major step towards solving the more general problem of impulse control. It has long been recognized that the latter problem can be formulated as an optimal stopping problem with 'implicit' reward function, and thus much of the solution is already in place once the stopping problem is well understood. These ideas are developed in section 54.

In all stochastic optimization problems one can think interchangeably in terms of maximizing rewards or minimizing costs. It seems most natural to formulate optimal stopping in the former terms and impulse control in the latter. The reason for this is connected with the effects of discounting. Often the problem in optimal stopping is to choose a stopping time τ so as to extremize a function of the form

$$J_x(\tau) = \mathbb{E}_x[e^{-\delta\tau} b(x_\tau)],$$

where $\delta > 0$ and b is a given, say bounded, function. If $b \geqslant 0$ and one is seeking to maximize $J_x(\tau)$ then the presence of the discount factor encourages early stopping, since the available reward is inexorably declining as time goes on. On the other hand, *minimizing* $J_x(\tau)$ with $b \geqslant 0$ has a trivial solution, namely $\tau = \infty$, so one must allow b to take negative values to obtain a meaningful problem. In impulse control, on the other hand, the objective function often takes the form

$$J_x(\mathscr{S}) = \mathbb{E}_x^{\mathscr{S}}\left\{\int_0^\infty e^{-\delta t} l(X_t)dt + \sum_{k=1}^{\infty} e^{-\delta\tau_k} c(X_{\tau_k}, X_{\tau_k +})\right\},$$

where $\mathscr{S}$ denotes an *intervention strategy*, i.e. a choice of stopping times τ_k and points $X_{\tau_k +}$ to which the process X_t is moved at the intervention time τ_k. We interpret $c(x, y)$ as the cost of moving from x to y and look for a strategy $\mathscr{S}$ to minimize $J_x(\mathscr{S})$. The summation term now has the effect of discouraging frequent interventions, since an intervention costs less if made later. Thus the problem formulation is most naturally couched in terms of minimization in this case.

Throughout the chapter, (x_t, P_x) denotes a piecewise-deterministic Markov process satisfying the standard conditions (24.8), except that we shall take its state space to be a set $E \subset \mathbb{R}^d$ rather than a union of sets $E_v \subset \mathbb{R}^{d(v)}$ (the difference is, of course, purely notational). In particular, Assumption (24.4), stating that the jump times T_i do not accumulate, is in force; other assumptions will be added later as required. In Chapter 4, because of the possibility of boundary control, we needed to distinguish between the transition measures from interior and boundary points, which were denoted Q and R respectively. This is no longer necessary, and we revert to the common notation $Q(dy; x)$ for the transition measure for $x \in E \cup \Gamma$.

51 Optimal stopping

As in section 24, we denote by $\phi(t, x)$ the deterministic flow of the vector fields associated with the process. We define $t_*(x) = \inf\{t : \phi(t, x) \in \Gamma\}$ where $\Gamma \subset \partial E$ is the *active boundary*, and note that, because of the semigroup property of the flow ϕ, we have for $t < t_*(x)$,

$$t_*(\phi(t, x)) = t_*(x) - t.$$

It is also convenient to define

$$\Lambda(t, x) = \int_0^{t \wedge t_*(x)} \lambda(\phi(s, x))ds.$$

The distribution function of the first jump time T_1 is then given by

$$P_x(T_1 > t) = I_{(t < t_*(x))} e^{-\Lambda(t,x)}.$$

We denote

$$f(t,x) = -\frac{\mathrm{d}}{\mathrm{d}t} P_x(T_1 > t) I_{(t < t_*(x))} = \lambda(\phi(t,x)) \mathrm{e}^{-\Lambda(t,x)} I_{(t < t_*(x))}$$

and

$$s_*(x) = \inf\{t : P_x(T_1 > t) = 0\}.$$

Note that $s_*(x) \leqslant t_*(x)$, and $s_*(x) = t_*(x)$ if, for example, λ is bounded. Recall from Definition (31.1) that $\mathscr{C}$ denotes the set of real-valued functions f on $E \cup \Gamma$ such that the map $t \to f(\phi(t,x))$ is continuous on $[0, s_*(x)]$ for each $x \in E$, and that $\mathscr{C}_b$ denotes the set of bounded functions in $\mathscr{C}$. Recall also the notation $Qh(x) = \int_E h(y) Q(\mathrm{d}y; x)$ for $h \in B(E)$. We shall need the following condition.

(51.1) *Assumption.* $Qh \in \mathscr{C}_b$ for $h \in \mathscr{C}_b$.

The completed natural filtration of the PDP is denoted $\mathfrak{F}_t$, and the jump times of the process are $T_1, T_2, \ldots$.

(51.2) *Definition.* Let Σ denote the set of $\mathfrak{F}_t$-stopping times, and

$$\begin{aligned} \Sigma_f &= \{\tau \in \Sigma : P_x[\tau < \infty] = 1 \text{ for all } x \in E\}, \\ \Sigma_n &= \{\tau \in \Sigma : \tau \leqslant T_n \text{ a.s.}\}, \\ \Sigma_{f,n} &= \Sigma_f \cap \Sigma_n. \end{aligned}$$

It is shown in Theorem (A2.3) that every stopping time $\tau \in \Sigma_1$ takes the form $\tau = t \wedge T_1$ for some $t \in [0, \infty]$. Since $\Sigma_{f,1} \subset \Sigma_1$, it follows that $\tau \in \Sigma_{f,1}$ if and only if *either* $\tau = t \wedge T_1$ for some $t \in \mathbb{R}_+$ *or* $T_1 < \infty$ a.s. and $\tau = T_1$.

Let $b \in \mathscr{C}_b$ be a given *reward function*, the interpretation being that if we stop the process at a point $x \in E$ we obtain a reward $b(x)$. The *optimal stopping* problem is then to choose a stopping time $\tau \in \Sigma_f$ that maximizes the expected reward $\mathbb{E}_x b(x_\tau)$; more precisely, we define

$$v(x) := \sup_{\tau \in \Sigma_f} \mathbb{E}_x b(x_\tau)$$

and say that $\tau \in \Sigma_f$ is *optimal* if $\mathbb{E}_x b(x_\tau) = v(x)$ for all $x \in E$ or *ε-optimal* ($\varepsilon \geqslant 0$) if $\mathbb{E}_x b(x_\tau) \geqslant v(x) - \varepsilon$ for all x. The function v is the *value function* for optimal stopping. Our strategy in studying this problem will be first to consider times $\tau \in \Sigma_{f,1}$, then by induction $\tau \in \Sigma_{f,n}$ for $n > 1$, and finally to use the fact that for any $\tau \in \Sigma_f$ we have $\tau \wedge T_n \in \Sigma_{f,n}$

and $\tau \wedge T_n \to \tau$ as $n \to \infty$ since $T_n \to \infty$ a.s. In order to study the case $n = 1$ we need to introduce the following functional operators:

(51.3) $$J: B(E) \times \mathscr{C}_b \to B(\mathbb{R}_+ \times E)$$

is defined by

$$J(w, h)(t, x) = \mathbb{E}_x[w(x_{T_1})I_{(t \geqslant T_1)} + h(x_t)I_{(t < T_1)}] = Iw(t, x) + Hh(t, x)$$

where

$$Iw(t, x) = \int_0^t Qw(\phi(s, x))f(s, x)\mathrm{d}s$$

$$Hh(t, x) = I_{(t < s_*(x))}h(\phi(t, x))\mathrm{e}^{-\Lambda(t,x)} + I_{(t \geqslant s_*(x))}h(\phi(t_*(x), x)\mathrm{e}^{-\Lambda(t_*(x),x)}$$

(Note that $t \to Hh(t, x)$ is continuous on $\mathbb{R}_+$ for $h \in \mathscr{C}_b$ and that the second term in its definition is equal to 0 unless $s_*(x) = t_*(x) < \infty$.)

(51.4) $$K: B(E) \to B(E)$$

is defined by

$$\begin{aligned} Kw(x) &= \mathbb{E}_x[w(x_{T_1})] \\ &= Iw(t_*(x), x) + Qw(\phi(t_*(x), x))\mathrm{e}^{-\Lambda(t_*(x),x)}. \end{aligned}$$

In terms of these operators we have for $b \in \mathscr{C}_b$,

(51.5) $$\mathbb{E}_x b(x_{t \wedge T_1}) = I_{(t < s_*(x))}J(b, b)(t, x) + I_{(t \geqslant s_*(x))}Kb(x).$$

Next, define for $t \geqslant 0$, $w \in \mathscr{B}$, $h \in \mathscr{B}_c$,

$$J_t(w, h)(x) = \sup_{s \geqslant t} J(w, h)(s, x)$$

and

$$L(w, h)(x) = J_0(w, h)(x) \vee Kw(x).$$

In view of the remarks after Definition (51.2), the interpretation of L is as follows.

(51.6) $$L(w, h)(x) = \sup_{\tau \in \Sigma_{f,1}} \mathbb{E}_x[h(x_\tau)I_{(\tau < T_1)} + w(x_{T_1})I_{(\tau = T_1)}].$$

(51.7) *Lemma.* For $t \in \mathbb{R}_+$, $w \in B(E)$, $h \in \mathscr{C}_b$,

(51.8) $$L(w, h)(\phi(t, x))\mathrm{e}^{-\Lambda(t,x)} = (J_t(w, h)(x) \vee Kw(x)) - Iw(t, x).$$

Proof. Let $\Sigma_1^t = \{\tau \in \Sigma_1 : \tau = s \wedge T_1 \text{ for some } s \geqslant t\}$. Then

$$\begin{aligned}
&(J_t(w,h)(x) \vee Kw(x)) \\
&\quad = \sup_{\tau \in \Sigma_1^t} \mathbb{E}_x\{\mathbb{E}_x[h(x_\tau)I_{(\tau < T_1)} + w(x_{T_1})I_{(\tau = T_1)} | \mathfrak{F}_t|\} \\
&\quad = \mathbb{E}_x[w(x_{T_1})I_{(t \geqslant T_1)}] + \sup_{\tau \in \Sigma_1^t} \mathbb{E}_x[I_{(\tau < T_1)}(h(x_\tau)I_{(\tau < T_1)} + w(x_{T_1})I_{(\tau = T_1)})] \\
&\quad = Iw(t,x) + e^{-\Lambda(t,x)}L(w,h)(\phi(t,x)),
\end{aligned}$$

where we have used the strong Markov property and the fact that $x_t = \phi(t,x)$ on $(t < T_1)$ to get the final equality. □

(51.9) *Corollary.* L maps $B(E) \times \mathscr{C}_b$ into $\mathscr{C}_b$.

Proof. Multiply (51.8) by $e^{\Lambda(t,x)}$ and note that the right-hand side is then continuous in t with the appropriate limit as $t \to t_*(x)$. □

Let $b \in \mathscr{C}_b$ be the reward function in the stopping problem and define a sequence of functions $v_0, v_1, \ldots$ as follows:

(51.10)
$$\begin{aligned}
v_0 &= b \\
v_{n+1} &= L(v_n, b), \quad n = 1, 2, \ldots
\end{aligned}$$

In view of Corollary (51.9), $v_n \in \mathscr{C}_b$ for all n, and v_n is a monotone increasing sequence, i.e. $v_{n+1}(x) \geqslant v_n(x)$ for all x. To see this, note from the definition that L is monotone in both variables, i.e. $L(w', h') \geqslant L(w, h)$ whenever $w' \geqslant w, h' \geqslant h$. Now certainly $v_1 = L(b,b) \geqslant b = v_0$ (this follows from (51.6)), while if $v_n \geqslant v_{n-1}$ for some n then $v_{n+1} = L(v_n, b) \geqslant L(v_{n-1}, b) = v_n$ and the result follows by induction.

We also need, for $\varepsilon > 0$, a sequence of stopping times S_k^ε defined as follows. Let

$$r_n^\varepsilon(x) = \begin{cases} t_*(x) & \text{if } Kv_n > J_0(v_n, b)(x) \\ \inf\{s > 0 : J(v_n, b)(s,x) > J_0(v_n,b)(x) - \varepsilon\} & \text{otherwise} \end{cases}$$

$$S_1^\varepsilon = r_0^\varepsilon(x) \wedge T_1$$

$$S_{n+1}^\varepsilon = \begin{cases} r_n^{\varepsilon/2}(x), & \text{if } T_1 > r_n^{\varepsilon/2} \\ T_1 + \theta(T_1)S_n^{\varepsilon/2}, & \text{if } T_1 \leqslant r_n^{\varepsilon/2}. \end{cases}$$

Here $\theta(t)$ is the shift operator for the process, i.e. $(\theta(t)x)_s = x_{t+s}$. Thus $\theta(T_1)S_n^{\varepsilon/2}$ is $S_n^{\varepsilon/2}$ evaluated for the process restarted at x_{T_1}. One can check from the definition that $S_n^\varepsilon \leqslant T_n$ for all $n \geqslant 1$, $\varepsilon > 0$. If $t_*(x) = \infty$ then $r_n^\varepsilon(x) = \infty$ only if $P_x[T_1 < \infty] = 1$, since otherwise $T_1 \notin \Sigma_{f,1}$.

(51.11) *Theorem.* For $n = 1, 2, \ldots$ and $\varepsilon > 0$ we have $v_n \in \mathscr{C}_b$, $S_n^{\varepsilon/2} \in \Sigma_n$ and

1. $v_n(x) = \sup_{\tau \in \Sigma_n} \mathbb{E}_x b(x_\tau)$
2. $\mathbb{E}_x b(x_{S_n^\varepsilon}) \geqslant v_n(x) - \varepsilon.$

Proof. We have already noted that $v_n \in \mathscr{C}_b$ and $S_n^\varepsilon \leqslant T_n$. The theorem is therefore proved if we prove 2 and

3. $\mathbb{E}_x b(x_{\sigma \wedge T_n}) \leqslant v_n(x)$, for any $\sigma \in \Sigma_f$.

We will use an induction argument. For any $\sigma \in \Sigma_f$ there are $\mathfrak{F}_{T_n}$-measurable random variables ρ_n such that $\sigma \wedge T_n = (T_{n-1} + \rho_{n-1}) \wedge T_n$ on $\sigma > T_{n-1}$, a.s. Thus for $n = 1$,

$$\mathbb{E}_x b(x_{\sigma \wedge T_n}) \leqslant \mathbb{E}_x b(x_{\rho_0 \wedge T_1}) \leqslant L(b, b)(x) = v_1(x),$$

which is 3 for $n = 1$. If $Kb(x) > J_0(b, b)(x)$, then $r_0^\varepsilon = t_*(x)$ and

$$\mathbb{E}_x b(x_{r_0^\varepsilon \wedge T_1}) = \mathbb{E}_x b(x_{T_1}) = Kb(x) = v_1(x).$$

On the other hand if $Kb(x) \leqslant J_0(b, b)(x)$ then

$$\mathbb{E}_x b(x_{r_0^\varepsilon \wedge T_1}) = J(b, b)(r_0^\varepsilon, x) \geqslant J_0(b, b)(x) - \varepsilon = v_1(x) - \varepsilon.$$

Thus 2, as well as 3, holds for $n = 1$. Now suppose 3 holds for $n = k$. Then by the strong Markov property,

$$\text{(51.12)} \quad \mathbb{E}_x b(x_{\sigma \wedge T_{k+1}}) = \mathbb{E}_x[b(x_{\rho_0}) I_{(T_1 > \rho_0)}] + \mathbb{E}_x[I_{(T_1 \leqslant \rho_0)} \chi(x_{T_1})],$$

where $\chi(x) := \mathbb{E}_x b(x_{\theta(T_1)\sigma \wedge T_k}) \leqslant v_k(x)$ (by the induction hypothesis). Now

$$\mathbb{E}_x[I_{(T_1 \geqslant \rho_0)} v_k(x_{T_1})] = I v_k(\rho_0, x) + Q v_k(\phi(t_*(x), x)) I_{(\rho_0 \geqslant s_*(x))} e^{-\Lambda(t_*(x), x)},$$

and hence from (51.8)

$$\begin{aligned}
\mathbb{E}_x b(x_{\sigma \wedge T_{k+1}}) &= b(\phi(\rho_0, x)) I_{(\rho_0 < s_*(x))} e^{-\Lambda(\rho_*, x)} \\
&\quad + Q v_k(\phi(t_*(x), x)) I_{(\rho_0 \geqslant s_*(x))} e^{-\Lambda(t_*(x), x)} + I v_k(\rho_0, x) \\
&\leqslant J(v_k, b)(r_0, x) \vee K v_k(x) \leqslant v_{k+1}(x).
\end{aligned}$$

Thus 3 holds for all n. If 2 holds for $n=k$, then

(51.13)
$$\mathbb{E}_x(x_{S^{2\varepsilon}_{k+1}}) = b(\phi(r^\varepsilon_k(x), x))I_{(r^\varepsilon_k < s_*(x))} e^{-\Lambda(r^\varepsilon_k(x),x)} + \mathbb{E}_x[I_{(T_1 < r^\varepsilon_k)}\chi'(x_{T_1})],$$

where $\chi'(x) = \mathbb{E}_x b(x_{S^\varepsilon_k}) \geqslant v_{k(x)} - \varepsilon$ (induction hypothesis). Hence

(51.14)
$$\begin{aligned}\mathbb{E}_x[I_{(T_1 < r^\varepsilon_k)}\chi'(x_{T_1})] &\geqslant \mathbb{E}_x[I_{(T_1 < r^\varepsilon_k)} v_k(x_{T_1})] - \varepsilon \\ &= Iv_k(r^\varepsilon_k, x) + Qv_k(\phi(t_*(x), x))I_{(r^\varepsilon_k \geqslant s_*(x))} e^{-\Lambda(t_*(x),x)} - \varepsilon.\end{aligned}$$

If $Kv_k(x) > J_0(v_k, b)(x)$, then $r^\varepsilon_k(x) = t_*(x)$ and from (51.13) and (51.14),

$$\begin{aligned}\mathbb{E}_x b(x_{S^{2\varepsilon}_k}) &\geqslant Iv_k(t_*(x), x) + Qv_k(\phi(t_*(x), x))e^{-\Lambda(t_*(x),x)} - \varepsilon \\ &= Kv_k(x) - \varepsilon = v_{k+1}(x) - \varepsilon.\end{aligned}$$

On the other hand, if $Kv_k(x) \leqslant J_0(v_k, b)(x)$, then, again from (51.13) and (51.14)

$$\begin{aligned}\mathbb{E}_x b(x_{S^{2\varepsilon}_k}) &\geqslant J(v_k, b)(r^\varepsilon_k(x), x) - \varepsilon \\ &\geqslant J_0(v_k, b)(x) - 2\varepsilon = v_{k+1}(x) - 2\varepsilon.\end{aligned}$$

Thus (51.6) holds with $n = k+1$ in either case, and hence 2 holds for all n. □

Theorem (51.11) shows that the functions v_n generated recursively by (51.10) are the value functions for optimal stopping in the classes $\Sigma_{f,n}$ of finite-valued stopping times majorized by T_n, and that S^ε_n is an ε-optimal time in $\Sigma_{f,n}$. It may well be that no optimal (i.e. 0-optimal) time exists, as the following example shows.

(51.15) *Example*
Take $E = [0, 1[$, $\phi(t, x) = t + x$ for $t < t_*(x) = 1 - x$, $\lambda = 0$ and Q the uniform distribution on $[0, 1[$, i.e. $Q([0, a[; x) = a$, $0 \leqslant a \leqslant 1$. If $b(x) = x$, then $v_1(x) = 1$ but $\mathbb{E}_x b(x_{\tau \wedge T_1}) < 1$ for any $\tau \in \Sigma_1$. The reward function b achieves its maximum on the boundary, but the process cannot be stopped there ($1 \notin E$).

The functions v_n are monotone increasing and uniformly bounded (by $\sup_{x\in E}|b(x)|$, which is finite since we assume $b \in \mathscr{C}_b$). Thus the limit

(51.16) $$\hat{v}(x) = \lim_{n\to\infty} v_n(x)$$

exists and defines a function $\hat{v} \in B(E)$. We can easily show, given the result of Theorem (51.11), that $\hat{v}$ is the value function for the original stopping problem. We can also characterize $\hat{v}$ in terms of a functional equation involving the operator L.

(51.17) *Theorem*

1. $\hat{v}(x) = v(x) := \sup_{\tau \in \Sigma_f} \mathbb{E}_x b(x_\tau)$, where $\hat{v}$ is defined by (51.16)
2. $v \in \mathscr{C}_b$
3. v is the smallest solution majorizing b of the functional equation

$$w = L(w, b), \quad w \in B(E). \tag{51.18}$$

Proof

1. Since $\Sigma_f \supset \Sigma_{f,n}$ for each n,

$$v(x) = \sup_{\tau \in \Sigma_f} \mathbb{E}_x b(x_\tau) \geqslant v_n(x),$$

and hence $v(x) \geqslant \hat{v}(x) = \lim_n v_n(x)$. Conversely, take $\tau \in \Sigma_f$. Then $\tau \wedge T_n \in \Sigma_{f,n}$ and since $T_n \to \infty$ a.s. for almost every ω there is a number $n_0(\omega)$ such that $\tau(\omega) \wedge T_n(\omega) = \tau(\omega)$ for all $n > n_0(\omega)$. Hence $b(x_{\tau \wedge T_n}) \to b(x_\tau)$ a.s. as $n \to \infty$ and by bounded convergence

$$\mathbb{E}_x b(x_\tau) = \lim_{n \to \infty} \mathbb{E}_x b(x_{\tau \wedge T_n}) \leqslant \lim_n v_n(x) = \hat{v}(x).$$

Thus

$$v(x) = \sup_{\tau \in \Sigma_f} \mathbb{E}_x b(x_\tau) \leqslant \hat{v}(x),$$

so that $v(x) = \hat{v}(x)$, as claimed.

3. This is a simple consequence of monotone convergence. In view of 1 we have

$$\begin{aligned} L(v, b)(x) &= \sup_{t \geqslant 0} (Iv(t, x) + Hb(t, x)) \vee Kv(x) \\ &= \sup_{t \geqslant 0} \left(\lim_{n \to \infty} Iv_n(t, x) + Hb(t, x) \right) \vee \left(\lim_{n \to \infty} Kv_n(x) \right) \\ &= \lim_{n \to \infty} L(v_n, b)(x) \\ &= \lim_{n \to \infty} v_{n+1}(x) = v(x), \end{aligned}$$

so that v satisfies (51.18). If w satisfies (51.18) and $w \geqslant b$, then $w = L(w,b) \geqslant L(b,b) = v_1$. Similarly, if $w \geqslant v_n$, then $w = L(w,b) \geqslant L(v_n, b) = v_{n+1}$, and we conclude by induction that $w(x) \geqslant v_n(x)$ for all n and hence that $w(x) \geqslant \lim_{n\to\infty} v_n(x) = v(x)$.

2. This follows from 3 and the fact that $L: B(E) \times \mathscr{C}_b \to \mathscr{C}_b$. □

We now consider the construction of ε-optimal stopping times. Note that, since $v \geqslant b$, we can divide the state space E into two regions, $S := \{x : v(x) = b(x)\}$ and $C = \{x : v(x) > b(x)\}$. S is the *stopping set*: it contains those points x for which the optimal strategy is immediate stopping, i.e. $\tau = 0$, with reward $b(x)$. C is called the *continuation set*: the inequality $v(x) > b(x)$ indicates that there is greater reward to be obtained from some strategy $\tau > 0$ than by stopping immediately. By the Markov property, it cannot be optimal, starting at $x_0 \in C$, to stop at any time t at which $x_t \in C$. A natural conjecture, therefore, is that the best strategy is to stop immediately if $x \in S$ and otherwise to wait until x_t hits S. Since, however, the latter event may never happen, one might have to be satisfied with stopping when x_t *nearly* hits S. This motivates the definition of a family U_ε of stopping times for $\varepsilon \geqslant 0$, given as follows

(51.19) $$U_\varepsilon = \inf\{\tau \geqslant 0 : b(x_t) \geqslant v(x_t) - \varepsilon\}.$$

The following theorem shows that U_ε is indeed an ε-optimal stopping time.

(51.20) *Theorem*

1. $P_x[U_\varepsilon < \infty] = 1$ for all $x \in E, \varepsilon > 0$, i.e. $U_\varepsilon \in \Sigma_f$.
2. $\mathbb{E}_x v(x_{U_\varepsilon}) = v(x)$, $x \in E, \varepsilon > 0$.
3. $\mathbb{E}_x b(x_{U_\varepsilon}) \geqslant v(x) - \varepsilon, x \in E, \varepsilon > 0$, i.e. U_ε is ε-optimal.
4. If $U_0 \in \Sigma_f$, then 2 and 3 hold with $\varepsilon = 0$, i.e. U_0 is optimal.

Proof. Since $b \in \mathscr{C}_b$ and x_t has right-continuous sample paths, $b(x_{U_\varepsilon}) \geqslant v(x_{U_\varepsilon}) - \varepsilon$ for any $\varepsilon \geqslant 0$, from the definition of U_ε. Hence if 2 holds then $\mathbb{E}_x b(x_{U_\varepsilon}) \geqslant \mathbb{E}_x v(x_{U_\varepsilon}) - \varepsilon = v(x) - \varepsilon$; this is 3. To prove the theorem, we therefore need to show that 1 is true and that $\mathbb{E}_x v(x_{U_\varepsilon}) = v(x)$ whenever $U_\varepsilon \in \Sigma_f$.

Proof of 1. We will show that for any $x \in E$,

(51.21) $$\limsup_{t\geqslant 0} v(x_t) = \limsup_{t\geqslant 0} b(x_t), \quad P_x\text{-a.s.},$$

from which 1 follows immediately. Indeed, take $\varepsilon > 0$ and $\omega \in \Omega$ such that $U_\varepsilon(\omega) = \infty$; then $v(x_t(\omega)) > b(x_t(\omega)) + \varepsilon$ for all $t \geq 0$ and hence $\limsup_t v(x_t(\omega)) > \limsup_t v(x_t(\omega))$. Thus (5.13) implies that $P_t[\omega: U_\varepsilon(\omega) = \infty] = 0$. To show (5.13), note first that $v(x) \geq b(x)$ so $\limsup_t v(x_t) \geq \limsup_t b(x_t)$. For the converse, fix $x \in E$ and define

$$\psi(t) = \sup_{s \geq t} b(x_s), \quad \chi(t) = \mathbb{E}_x[\psi(t) \mid \mathfrak{F}_t].$$

By the Markov property, $\chi(t) = \zeta(x_t)$ P_x-a.s., where $\zeta(x) = \mathbb{E}_x \psi(0)$. Now for any $t \geq 0$,

$$\mathbb{E}_x \zeta(x_t) = \mathbb{E}_x[\mathbb{E}_x(\psi(t) \mid \mathfrak{F}_t)] = \mathbb{E}_x \psi(t) \leq \mathbb{E}_x \psi(0) = \zeta(x).$$

For any stopping time $\tau \in \Sigma_f$ it is clear that $\mathbb{E}_x b(x_\tau) \leq \mathbb{E}_x[\sup_{s \geq 0} b(x_s)] = \zeta(x)$, and therefore

$$v(x) = \sup_{\tau \in \Sigma_f} \mathbb{E}_x b(x_\tau) \leq \zeta(x).$$

Now fix a time $s \geq 0$ and take $t \geq s$. Then $\psi(s) \geq \psi(t)$ a.s., so

$$\mathbb{E}_x[\psi(s) \mid \mathfrak{F}_t] \geq \mathbb{E}_x[\psi(t) \mid \mathfrak{F}_t] = \zeta(x_t) \geq v(x_t) \quad \text{a.s.}$$

and

$$\limsup_{t \to \infty} \mathbb{E}_x[\psi(s) \mid \mathfrak{F}_t] \geq \limsup_{t \to \infty} v(x_t). \tag{51.22}$$

Now $y_t := \mathbb{E}_x[\psi(s) \mid \mathfrak{F}_t]$ is a bounded martingale; by the martingale convergence Theorem (13.5) the limit exists as $t \to \infty$ and

$$\limsup_{t \to \infty} \mathbb{E}_x[\psi(s) \mid \mathfrak{F}_t] \leq \mathbb{E}_x[\psi(s) \mid \mathfrak{F}_\infty] = \psi(s) \quad \text{a.s.,}$$

so that from (5.14), $\psi(s) \geq \limsup_{t \to \infty} v(x_t)$. Hence

$$\limsup_{t \to \infty} v(x_t) \leq \inf_{s \geq 0} \psi(s) = \limsup_{t \to \infty} b(x_t).$$

This completes the proof of (51.21).

Proof that $\mathbb{E}_v(x_{U_\varepsilon}) = v(x)$ *when* $U_\varepsilon \in \Sigma_f$. We will show that for any n

and $\varepsilon > 0$

$$\mathbb{E}_x v(x_{U_\varepsilon \wedge T_n}) = v(x). \tag{51.23}$$

If $U_\varepsilon \in \Sigma_f$, $U_\varepsilon \wedge T_n = U_\varepsilon$ for large n and the conclusion follows from (51.23) by bounded convergence. To show (51.23), start with $n = 1$. From the representation of $\mathfrak{F}_t$-stopping times, we know that $U_\varepsilon \wedge T_1 = u \wedge T_1$ for some $u \geqslant 0$. Now, with $s_* := s_*(x)$

$$\begin{aligned} \mathbb{E}_x v(x_{U_\varepsilon \wedge T_1}) &= \mathbb{E}_x v(x_{u \wedge T_1}) \\ &= J(v, v)(u, x) I_{(u < s*)} + Kv(x) I_{(u \geqslant s*)} \\ &= (J_u(v, b)(x) \vee Kv(x)) I_{(u < s*)} + Kv(x) I_{(u \geqslant s*)}. \end{aligned} \tag{51.24}$$

The last equality (51.24) follows from Lemma (51.7) and the fact that $v = L(v, b)$. Indeed, from (51.8) and with $u < s_*$ we have

$$\begin{aligned} J_u(v, b)(x) \vee Kv(x) &= L(v, b)(\phi(u, x)) \mathrm{e}^{-\Lambda(u,x)} + Iv(u, x) \\ &= v(\phi(u, x)) \mathrm{e}^{-\Lambda(u,x)} + Iv(u, x) \\ &= J(v, v)(u, x). \end{aligned}$$

If $Kv(x) \geqslant J_0(v, b)(x)$, then $v = L(v, b) = Kv$ and from (51.24)

$$\mathbb{E}_x v(x_{U_\varepsilon \wedge T_1}) = Kv(x) = v(x).$$

Now suppose $Kv(x) < J_0(v, b)(x)$. We show first that this implies $u < s_*$. If the supremum in the definition of J_0 is attained at some $s_0 \in [0, s_*[$ then

$$J(v, b)(s_0, x) = J_{s_0}(v, b)(x) \vee Kv(x).$$

Using this together with Lemma (51.7) and the equality $v = L(v, b)$ gives

$$\begin{aligned} v(\phi(s_0, x)) \mathrm{e}^{-\Lambda(s_0,x)} &= J(v, b)(s_0, x) - Iv(s_0, x) \\ &= Hb(s_0, x) \\ &= b(\phi(s_0, x)) \mathrm{e}^{-\Lambda(s_0,x)} \end{aligned}$$

(since $s_0 < s_*$). Hence $v(\phi(s_0, x)) = b(\phi(s_0, x))$ and therefore $u < s_0 < s_*$. If the supremum in J_0 is *not* attained at any s_0 then it must be the case that $s_* = t_* < \infty$, $\mathrm{e}^{-\Lambda(t_*,x)} > 0$ and

$$J(v, b)(s_*, x) = J_0(v, b)(x). \tag{51.25}$$

Indeed, recall that

$$J(v,b)(t,x) = \int_0^t Qv(\phi(s,x))f(s,x)\mathrm{d}s + I_{(t<s*)}b(\phi(t,x))\mathrm{e}^{-\Lambda(t,x)}$$
$$+ I_{(t\leqslant s*)}b(\phi(t_*,x))\mathrm{e}^{-\Lambda(t*,x)}.$$

If $s_* = \infty$ then $J(v,b)(t,x) \to Kv(x)$ as $t \to \infty$, and since $\sup_t J(v,b)(t,x) > Kv(x)$, the supremum must be attained at some finite t, contrary to our assumption; hence $s_* < \infty$. A similar contradiction is obtained if $s_* < \infty$ and $\mathrm{e}^{-\Lambda(t*,x)} = 0$; (5.13) follows. From (51.25), and using an argument similar to that used earlier, we conclude that

$$\lim_{s\to s*} v(\phi(s,x)) = \lim_{s\to s*} b(\phi(s,x))$$

and hence that $u < s_*$. We have now shown that $u < s_*$ holds whenever $Kv(x) < J_0(v,b)(x)$.

Continuing to suppose that $Kv(x) < J_0(v,b)(x)$ we now have, for $t < u$,

$$\begin{aligned} J(v,b)(t,x) &= Iv(t,x) + b(\phi(t,x))\mathrm{e}^{-\Lambda(t,x)} \\ &\leqslant Iv(t,x) + v(\phi(t,x))\mathrm{e}^{-\Lambda(t,x)} - \varepsilon\mathrm{e}^{-\Lambda(t,x)} \quad \text{(since } t<u) \\ &= J_t(v,b)(x) \vee Kv(x) - \varepsilon\mathrm{e}^{-\Lambda(t,x)} \quad \text{(by Lemma (51.7))} \\ &\leqslant J_0(v,b)(x) \vee Kv(x) - \varepsilon\mathrm{e}^{-\Lambda(u,x)} \\ &= J_0(v,b)(x) - \varepsilon\mathrm{e}^{-\Lambda(u,x)}. \end{aligned}$$

Hence $J_0(v,b)(x) = J_u(v,b)(x)$, so that from (51.24), and since $u < s_*$,

$$\begin{aligned} \mathbb{E}_x v(x_{U_\varepsilon \wedge T_1}) &= J_u(v,b) \vee Kv(x) \\ &= J_0(v,b) \vee Kv(x) \\ &= L(v,b)(x) = v(x). \end{aligned}$$

We have shown that $\mathbb{E}_x v(x_{U_\varepsilon \wedge T_1}) = v(x)$ in all cases. Now suppose $\mathbb{E}_x v(x_{U_\varepsilon \wedge T_k}) = v(x)$ for some k; then by the strong Markov property,

$$\begin{aligned} \mathbb{E}_x v(x_{U_\varepsilon \wedge T_{k+1}}) &= \mathbb{E}_x\{I_{(U_\varepsilon < T_1)} v(x_{U_\varepsilon})\} + \mathbb{E}_x\{I_{(U_\varepsilon \geqslant T_1)}(\mathbb{E}_z v(x_{U_\varepsilon \wedge T_k}))|_{z=x_{T_1}}\} \\ &= \mathbb{E}_x\{I_{U_\varepsilon < T_1} v(x_{U_\varepsilon}) + I_{U_\varepsilon \geqslant T_1} v(x_{T_1})\} \\ &= \mathbb{E}_x v(x_{U_\varepsilon \wedge T_1}) = v(x). \end{aligned}$$

Hence $\mathbb{E}_x v(x_{U_\varepsilon \wedge T_n}) = v(x)$ for all n; this is (51.23). □

(51.26) *Remarks*
It is easy to produce examples where U_0 does not exist (Example (51.29) below). It is not necessarily the case that $U_\varepsilon \uparrow U_0$ even when U_0 is finite. To see this, take $E = [0,1[\cup[2,3[$, $\lambda = 0$, $\phi(t,x) = t + x$ for $t < t_*(x)$ and $Q(\mathrm{d}y; 1) = \delta_{\{2\}}(\delta y)$, $Q(\mathrm{d}y; 3) = \delta_{\{0\}}(\mathrm{d}y)$, i.e. the process jumps from 1 to 2 and from 3 to 0, and these are the only jumps. Take $g(x) = 4x(1-x)$ for $x \in [0,1[$ and $g(x) = x - 2$ for $x \in [2,3[$, as depicted in Fig. 5.1. Then $v(x) = 1$, and starting, for example, at $x = 2$ we have $U_\varepsilon = 1 - \varepsilon$ but $U_0 = 3/2$.

Complementing Theorem (51.17) there are two further characterizations of the value function v.

(51.27) *Theorem*

1. The value function v is the smallest function $w \in \mathscr{C}_b$ satisfying $w \geqslant g$ and $w = L(w, w)$.
2. v is the smallest excessive function in $\mathscr{C}_b$ majorizing g.

(51.28) *Remark*
A function $w: \mathbb{E} \to \mathbb{R}$ is *excessive* if it is 0-excessive in the terminology of section 27, i.e. if $w(x) \geqslant \mathbb{E}_x w(x_t)$ for all $t \geqslant 0$, and $\mathbb{E}_x w(x_t) \to w(x)$ as

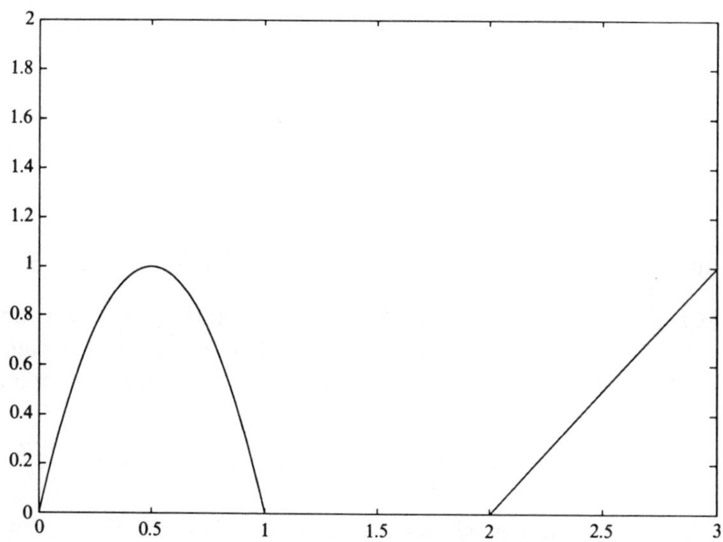

Fig. 5.1. *Payoff.*

$t \downarrow 0$, for each $x \in E$. The first property is equivalent to saying that the process $t \to w(x_t)$ is an $\mathfrak{F}_t$-supermartingale for every P_x.

Proof

1. Consider the function v_n defined by (51.10). We have

$$L(v_n, v_n)(x) = (\sup_{0 \leqslant t < s*} J(v_n, v_n)(t, x)) \vee Kv_n(x)$$

$$\leqslant (\sup_{0 \leqslant t < s*} (Iv_n(t, x) + HL(v_n, g))) \vee Kv_n(x)$$

(since $L(v_n, g) = v_{n+1} \geqslant v_n$). Now for $t < s_*$, and using Lemma (51.7),

$$HL(v_n, b)(t, x) = L(v_n, g)(\phi(t, x))e^{-\Lambda(t,x)}$$
$$= J_t(v_n, b) \vee Kv_n(x) - Iv_n(t, x).$$

Thus

$$L(v_n, v_n)(x) \leqslant J_0(v_n, b)(x) \vee Kv_n(x) = v_{n+1}(x).$$

Conversely, since $b \leqslant v_n$,

$$v_{n+1} = L(v_n, b)(x) \leqslant L(v_n, v_n)(x).$$

Thus $v_{n+1} = L(v_n, v_n)(x)$. The remaining argument is similar to the proof of part 3 of Theorem (51.17).

2. Clearly, for any $t \geqslant 0$,

$$v(x) \geqslant \sup_{\tau \in \Sigma_f^t} \mathbb{E}_x b(x_\tau),$$

where $\Sigma_f^t = \{\tau \in \Sigma_f : \tau \geqslant t \; P_x \text{a.s.}\}$. However, by the Markov property,

$$\sup_{\tau \in \Sigma_f^t} \mathbb{E}_x b(x_\tau) = \mathbb{E}_x v(x_t),$$

and $\lim_{t \downarrow 0} \mathbb{E}_x v(x_t) = v(x)$ since $v \in \mathscr{C}_b$, by construction of the process (x_t). Thus v is an excessive function, $v \in \mathscr{C}_b$ and $v \geqslant b$. If w is any other function with these properties then $t \to w(x_t)$ is a right-continuous bounded supermartingale; hence the optional sampling theorem (13.7) applies and we have

$$v_1(x) = \sup_{t \geqslant 0} \mathbb{E}_x b(x_{t \wedge T_1}) \leqslant \sup_{t \geqslant 0} \mathbb{E}_x w(x_{t \wedge T_1}) = w(x).$$

The same argument shows that if $w \geqslant v_n$ then $w \geqslant v_{n+1}$. Hence $w \geqslant v_n$ for all n and therefore $w \geqslant \lim_n v_n = v$. □

Since $v \in \mathscr{C}_b$, the boundary value $v(z) = \lim_{t \downarrow 0} v(\phi(-t, z))$ exists at every $z \in \Gamma$. Our next task is to investigate these boundary values, and to ask in particular whether the boundary condition $v(z) = Qv(z)$, $z \in \Gamma$ associated with the generator of the process holds for the value function v. In fact this condition does *not* generally hold. To get some insight into this question it is instructive to consider the following family of examples.

(51.29) *Example*
Take $E = [0, 1[$, $\phi(t, x) = t + x$ for $t < t_*(x) = 1 - x$, $\lambda = 0$ and $Q(\mathrm{d}y; 1) = \delta_{\{0\}}(\mathrm{d}y)$. Thus the process moves to the right at unit speed and jumps back to 0 on hitting 1. There are no other jumps. Now fix $\delta > 0$ and consider maximizing $\mathbb{E}_x[\mathrm{e}^{-\delta\tau} b(x_\tau)]$ for some continuous function b. This is actually a special case of the preceding framework: define an extra component $x^0 \in \mathbb{R}_+$ satisfying $\dot{x}_t^0 = -\delta x_t^0$ and $\bar{b}(\bar{x}) = x^0 b(x)$ with $\bar{x} = (x^0, x)$. Then $\mathbb{E}_{\bar{x}} \bar{b}(\bar{x}_\tau) = x^0 \mathbb{E}_x[\mathrm{e}^{-\delta\tau} b(x_\tau)]$ so that $\bar{v}(\bar{x}) = x^0 v(x)$, where $\bar{v}(\bar{x}) = \sup_\tau \mathbb{E}_{\bar{x}} \bar{b}(\bar{x}_\tau)$ and $v(x) = \sup_\tau \mathbb{E}_x[\mathrm{e}^{-\delta\tau} b(x_\tau)]$. Consider first $b(x) = x$. It is clear that $v(x) = v_1(x) = \sup_\tau \mathbb{E}_x[\exp(-\delta(\tau \wedge T_1)) b(x_{\tau \wedge T_1})]$, and a direct calculation based on the equality $v_1 = L(b, b)$ shows the following:

Case 1: $\delta > 1$

$$v(x) = \begin{cases} \dfrac{1}{\delta} \mathrm{e}^{\delta x - 1}, & 0 \leqslant x < \delta^{-1} \\ x, & \delta^{-1} \leqslant x < 1. \end{cases}$$

(Fig. 5.2(a)). Note that:

1. The optimal stopping time is $\tau_0 = \inf\{t : x_t \geqslant \delta^{-1}\}$; the continuation region is $C = [0, \delta^{-1}[$ and the stopping region $S = [\delta^{-1}, 1[$.
2. $\mathrm{d}v(x)/\mathrm{d}x$ is continuous at $x = \delta^{-1}$. This is the so-called 'smooth pasting condition' (Shiryaev, 1978, section 3.8).
3. $Qv(1) = v(0) = (\mathrm{e}\delta)^{-1} \neq v(1) = 1$.

Case 2: $\delta < 1$

$$v(x) = \mathrm{e}^{\delta(x-1)}, \quad x \in C = E = [0, 1[.$$

(Fig. 5.2(b)). In this case no optimal time exists and $v(x) > b(x)$ for

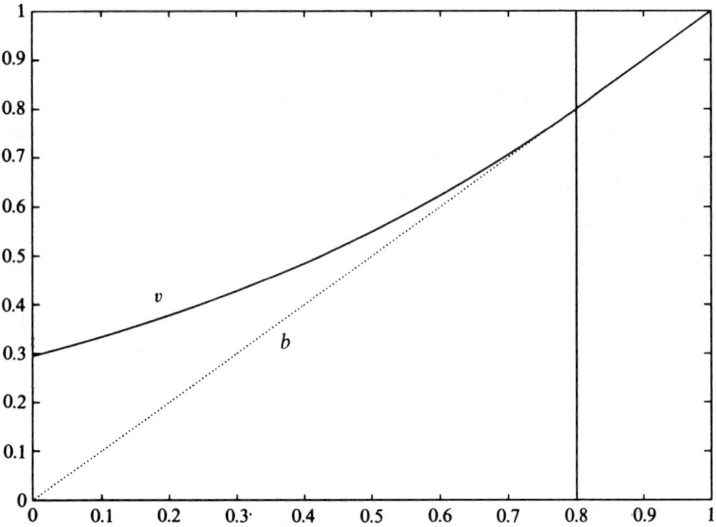

(*a*) $\delta > 1$

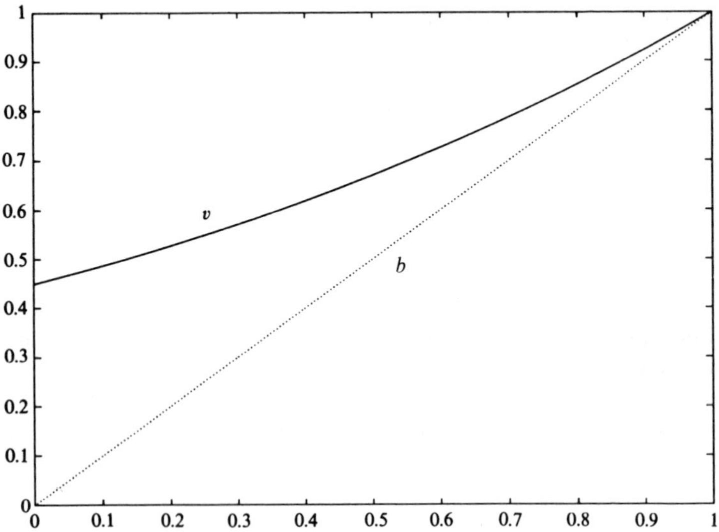

(*b*) $\delta \leqslant 1$

Fig. 5.2. *Payoff and value functions for Example (51.29) with* $g(x) = x$; (*a*) $\delta > 1$, (*b*) $\delta < 1$.

all $x \in E$. An ε-optimal time is equal to τ_θ for some $\theta > 0$, where $\tau_\theta = \inf\{t : x_t > 1 - \theta\}$. It is still true that $v(0) \neq v(1)$.

Now consider $b(x) = 1 - x$. A moment's thought will convince the reader that the only possible candidates for the optimal time are 0 and T_1. Now $\mathbb{E}_x[e^{-\delta T_1} b(x_{T_1})] = e^{\delta(x-1)}$, so that

$$v(x) = v_1(x) = \max\{(1-x), e^{\delta(x-1)}\}.$$

(Fig. 5.3). For any $\delta > 0$ there is a unique number a such that $(1 - a) = e^{\delta(a-1)}$ and the optimal time is $\tau_0 = \inf\{t : x_t \in S = [0, a]\}$ ($= T_1$ if $x > a$, $= 0$ if $x \leqslant a$). In this example therefore:

1. An optimal time always exists.
2. Smooth pasting fails (dv/dx is not continuous at a)
3. The boundary condition $v(0) = Qv(1)$ is satisfied, since $v(0) = v(1) = 1$.

Based on the insight gained from these examples we can now describe the boundary conditions for general PDP optimal stopping problems.

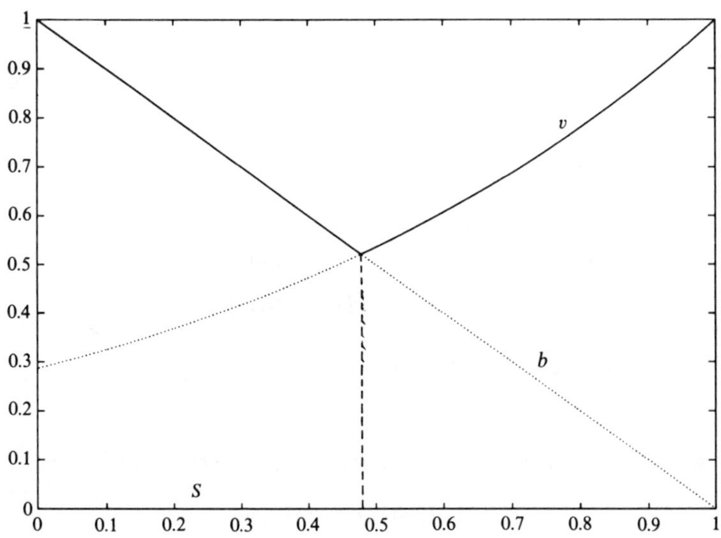

Fig. 5.3. *Payoff and value functions for Example (51.29) with $g(x) = 1 - x$.*

(51.30) *Theorem.* Let $v(x) = \sup_{\tau \in \Sigma_f} \mathbb{E}_x[b(x_\tau)]$ and let $z \in \Gamma$. Then

$$v(z) = b(z) \vee Qv(z).$$

Proof. Fix $z \in \Gamma$ and a sequence of numbers $t_n > 0$, $t_n \downarrow 0$ such that $\phi(-t, z) \in E$ for all $t \leqslant t_1$, and $s_*(x_n) = t_*(x_n) = t_n$ for all n, where $x_n = \phi(-t_n, z)$. Such a sequence exists for any $x \in \Gamma$. Now

$$v(x_n) = L(v, b)(x_n) = \left(\sup_{0 \leqslant t < t_n} J(v, b(t, x_n)) \right) \vee Kv(x_n),$$

where we recall that

$$\begin{aligned} J(v, b)(t, x_n) &= \mathbb{E}_x\{b(\phi(t, x_n))I_{(t < T_1)} + v(x_{T_1})I_{(t < T_1)}\} \\ Kv(x_n) &= \mathbb{E}_{x_n}[v(x_{T_1})I_{(T_1 < \infty)}] \\ &= \int_0^{t_n} \lambda(\phi(t, x_n))\mathrm{e}^{-\Lambda(t, x_n)} Qv(\phi(t, x_n))\mathrm{d}t + \mathrm{e}^{-\Lambda(t_*(x_n), x_n)} Qv(z). \end{aligned}$$

Case 1. There exists n_0 such that $L(v, b)(x_{n_0}) = Kv(x_{n_0})$.

It is then easy to see that $L(v, b)(x_n) = Kv(x_n)$ for all $n \geqslant n_0$. As $n \to \infty$, $v(x_n) \to v(z)$ and $Kv(x_n) \to Qv(z)$, so that $v(z) = Qv(z)$. Also, $v(x_n) \geqslant b(x_n)$, so $v(z) \geqslant b(z)$ and hence $v(z) = b(z) \vee Qv(z)$.

Case 2. For all n, $\sup_{t < t_n} J(v, b(t, x_n)) > Kv(x_n)$. Then for each n there exists $s_n \in [0, t_0[$ such that, with $y_n = \phi(s_n, x_n)$ (so that $y_n \to z$ as $n \to \infty$)

(51.31) $$v(x_n) \geqslant \mathbb{E}_{x_n}\{b(y_n)I_{(s_n < T_1)} + v(x_{T_1})I_{(s_0 \geqslant T_1)}\} > Kv(x_n).$$

Now as $n \to \infty$, $P_{x_n}[T_1 = t_*(x_n)] \to 1$ so $P_{x_n}[s_n \geqslant T_1] \to 0$. Thus taking the limit in (51.31) as $n \to \infty$ we obtain

$$\liminf_{n \to \infty} v(x_n) \geqslant b(z) \geqslant Qv(z).$$

Now suppose the s_n are chosen such that

$$J(v, b)(s_0, x_n) \geqslant L(v, b)(x_n) - \varepsilon = v(x_n) - \varepsilon$$

for some $\varepsilon > 0$. From the above, $J(v, b)(s_n, x_n) \to b(z)$ and hence

$$\limsup_{n \to \infty} v(x_n) \leqslant b(z) + \varepsilon.$$

Thus $\lim_n v(x_n) = v(z) = b(z) \geqslant Qv(z)$, i.e. $v(z) = b(z) \vee Qv(z)$. □

52 Randomized stopping

In the next section, section 53, we will characterize the value function v of the optimal stopping problem as the unique solution of a so-called *variational inequality*. The technique for obtaining this characterization is known as the *penalty method* and involves relating the stopping problem to a family of stochastic control problems, which can be treated by methods similar to those of the last chapter. The interpretation of these control problems is based on the idea of *randomized stopping*, which we explore in this section.

Optimal stopping as formulated so far is a 'yes–no' decision process: at each point in the state space we either continue or stop. One could envisage a more general form of decision rule according to which one stops or continues *with a certain probability*. We will formulate this idea precisely below. The main result of this section is that randomized stopping rules of this kind never increase the value; in particular, if an optimal rule exists it will always be a conventional stopping time as formulated earlier. This strengthens the result we have obtained so far: optimal or ε-optimal times are also (ε-)optimal in the much wider class of randomized decision rules. This result is of some intrinsic interest (it shows the optimal stopping problem has not been formulated in some artificially restricted way) but, primarily, it is required for the developments in section 53.

As before, we are given a function $b \in \mathscr{C}_b$, and obtain a reward $b(x_t)$ if we stop the PDP (x_t) at time t.

(52.1) *Definition.* The class Σ_r of *randomized strategies* is the set of real-valued, $\mathfrak{F}$-adapted stochastic processes F_t such that, for almost every $\omega \in \Omega$, $F_{0-} = 0$, the function $t \to F_t(\omega)$ is increasing and right-continuous, and $\lim_{t \to \infty} F_t(\omega) \leqslant 1$.

Thus Σ_r is the set of *adapted sub-distribution functions*. By stating that $F_{0-} = 0$ we mean that if $F_0 > 0$ then this is interpreted as a discontinuity at $t = 0$, i.e. the distribution function puts weight F_0 on the point $t = 0$. The reward corresponding to $F \in \Sigma_r$ is defined as

$$J_x(F) = \mathbb{E}_x \int_{[0,\infty[} b(x_t) \mathrm{d}F_t, \tag{52.2}$$

where the integral is an ordinary Stieltjes integral, calculated separately for each $\omega \in \Omega$. Since $F_t(\omega)$ is a (sub-) distribution function we have $|J_x(F)| \leqslant \sup_{x \in E} |b(x)| < \infty$, and the *value function* is

$$v_r(x) = \sup_{F \in \Sigma_r} J_x(F).$$

The interpretation of F is simply that $\mathrm{d}F_t = F_{t+\mathrm{d}t} - F_t$ represents the probability of stopping in the interval $[t, t+\mathrm{d}t[$; since F_t is $\mathfrak{F}_t$ adapted, this probability can be chosen in a way that depends on the past evolution of the process (x_t). If $F_\infty = \lim_{t \to \infty} F_t < 1$ then we stop at time $t = \infty$ with probability $1 - F_\infty$; in view of the definition (52.2) the reward when $t = \infty$ is equal to 0. We could remove this possibility by insisting that $F_\infty = 1$, but allowing $F_\infty < 1$ turns out to be more convenient.

A stopping time is a special case of a randomized decision rule. Indeed, if $\tau \in \Sigma$ (Σ is the set of $\mathfrak{F}_t$-stopping times, possibly taking the value $+\infty$), then $F_t^\tau \in \Sigma_r$, where $F_t^\tau := I_{(t \geqslant \tau)}$, and

$$J_x(F^\tau) = \mathbb{E}_x[b(x_\tau)]$$

(where, again, $b(x_\tau) := 0$ on $(\tau = \infty)$). Identifying τ with F^τ we therefore have the inclusions $\Sigma_f \subset \Sigma \subset \Sigma_r$. We will usually write $J_x(\tau)$ rather than $J_x(F^\tau)$. Some further sub-classes of Σ_r will also be required.

(52.3) *Definition.* Let n be a positive integer. Then we define

$$\Sigma_r^n = \left\{ F \in \Sigma_r : F_t = 1 - \exp\left(- \int_0^t u_s \mathrm{d}s \right) \text{ for some adapted process } (u_t) \text{ taking values in } [0, n] \right\};$$

$$\mathscr{S}^n = \{ F \in \Sigma_r : \text{there exist stopping times } \tau_1 < \tau_2 < \cdots < \tau_n \text{ such that } \tau_k \in \Sigma_f \text{ for } k = 1, \ldots, n;\ F_{\tau_1 -} = 0;\ F_t = F_{\tau_k} \text{ for } t \in [\tau_k, \tau_{k+1}[\text{ and } F_\infty = F_{\tau_n} \}.$$

Note that there is a one-to-one correspondence between Σ_r^n and the set of processes appearing in its definition (which would not be true if we had insisted that $F_\infty = 1$). $\mathscr{S}^n$ is the set of 'step function' strategies which stop the process only at times $\tau_1, \ldots, \tau_n$ for some finite number n or at time $+\infty$. The main result of this section is given below.

(52.4) *Theorem*

$$\sup_{F\in\Sigma_r} J_x(F) = \sup_{\tau\in\Sigma} J_x(\tau).$$

To show this, we will first show that for every step function $F\in\mathscr{S}^n$ there is a stopping time τ which is at least as good, and then that an arbitrary $F\in\Sigma_r$ can be approximated by a sequence F^n with $F^n\in\mathscr{S}^n$.

(52.5) *Proposition.* Take $F\in\mathscr{S}^n$. Then there exists a stopping time $\tau\in\Sigma$ such that $J_x(\tau)\geqslant J_x(F)$.

Proof. First, consider a probability space $(\Omega, \mathfrak{F}, P)$ and a discrete-time filtration $(\mathfrak{G}_k, k=0,\ldots,n+1)$, and let Z_k be an E-valued Markov process adapted to $\mathfrak{G}_k$ with a given non-stationary transition function $p_k(A;z)$ such that $P[Z_{k+1}\in A\,|\,\mathfrak{G}_k] = p_k(A;Z_k)$. We are interested in the problem of maximizing $\mathbb{E}[b(\sigma, Z_\sigma)]$ over the class $\Sigma(\mathfrak{G})$ of $\mathfrak{G}_k$-stopping times, i.e. random variables σ taking values in $\{0,1,\ldots, n+1\}$ such that $\{w:\sigma(\omega)=k\}\in\mathfrak{G}_k$, $k=0,\ldots,n+1$, where $b(\cdot,\cdot)$ is some bounded function. In this case a randomized strategy can be described in more intuitive terms as follows. Suppose independent random variables $U_0,\ldots,U_n$, uniformly distributed on $[0,1]$ and independent of $\mathfrak{G}_n$, are defined on $(\Omega, \mathfrak{F}, P)$. A randomized rule is a $\mathfrak{G}_k$-adapted process θ_n such that $\theta_k\in[0,1]$ for $k=0,\ldots,n$ and $\theta_{n+1}=1$ a.s. Denote by Θ the set of such processes. At each time k, assuming we have not stopped up to that point, we stop if $\theta_k > U_k$ (this is equivalent to flipping a coin with heads probability θ_k and stopping on heads). This procedure amounts to stopping at a random time σ whose conditional distribution given $\mathfrak{G}_{n+1}$ is given by

$$R_k := P[\sigma > k\,|\,\mathfrak{G}_{n+1}] = \prod_{j=0}^{k}(1-\theta_j).$$

Then

$$\rho_0 := P[\sigma = 0\,|\,\mathfrak{G}_{n+1}] = \theta_0$$

$$\rho_k := P[\sigma = k\,|\,\mathfrak{G}_{n+1}] = R_{k-1} - R_k = \theta_k\prod_{j=0}^{k}(1-\theta_j), \quad k\geqslant 1 \tag{52.6}$$

so that ρ_k is $\mathfrak{G}_k$-adapted, and the expected reward is

$$\mathbb{E}\{\mathbb{E}(b(\sigma, Z_\sigma)\,|\,\mathfrak{G}_{n+1})\} = \mathbb{E}\left\{\sum_{j=0}^{n+1} b(j, Z_j)\rho_j\right\}.$$

This is the equivalent, in the discrete-time context, of formula (52.2) defining the reward for a randomized strategy F.

Finite-horizon stopping problems of this type are covered in detail by Shiryaev (1978, section 2.2) and we shall just sketch the results here. They are solved by a simple dynamic programming argument. For $\psi \in B(E)$ define $H_k\psi(z) = \int_E \psi(y)p_k(\mathrm{d}y; z)$ so that $\mathbb{E}[\psi(Z_{k+1})|\mathfrak{G}_k] = H_k\psi(Z_k)$ a.s. Now define $V_{n+1}(x) = b(n+1, x)$ and

$$V_k(x) = \max\{b(k,x), H_k V_{k+1}(x)\}, \quad k = n, \ldots, 0.$$

Then

$$\mathbb{E}V_0(Z_0) = \max_{\sigma \in \Sigma(\mathfrak{G})} \mathbb{E}[b(\sigma, z_\sigma)],$$

and the time $\hat{\sigma} = \min\{k: V_k(Z_k) = b(k, Z_k)\}$ is optimal in $\Sigma(\mathfrak{G})$. Now consider the recursion

$$(52.7) \quad \begin{cases} \hat{V}_{n+1}(x) = b(n+1, x) \\ \hat{V}_k(x) = \max\limits_{0 \leqslant \theta \leqslant 1} \{(\theta b(k,x) + (1-\theta)H_k \hat{V}_{k+1}(x)), \quad k = n, \ldots, 0. \end{cases}$$

It is easy to show that, analogously to the above

$$\mathbb{E}\hat{V}_0(Z_0) = \max_{\theta \in \Theta} \mathbb{E}\left\{\sum_{j=0}^{n+1} b(j, Z_j)\rho_j\right\},$$

with ρ_j given by (52.6). However, obviously the maximum in (52.7) is attained at $\theta = 0$ or 1, so that $\hat{V}_k(x) = V_k(x)$ and the time $\hat{\sigma}$ is optimal in Θ (the element of Θ corresponding to $\hat{\sigma}$ being $\hat{\theta}_k = I_{(\sigma = k)}$). Thus randomization does not increase the value.

Let us now return to the original PDP stopping problem and consider a fixed strategy $F \in \mathscr{S}_n$. Thus there are stopping times $\tau_1 < \tau_2 < \cdots < \tau_n$ such that $F_t = \sum_i I_{(T_i \leq t)} \Delta F_{\tau_i}$. Adjoin an isolated point Δ_0 to the state space and define $Z_k := x_{\tau_k}$ and $\mathfrak{G}_k = \mathfrak{F}_{\tau_k}$ for $k = 0, \ldots, n$ and $Z_{n+1} = \Delta_0$, $\mathfrak{G}_{n+1} = \mathfrak{F}_{\tau_n}$. Then by the strong Markov property $(Z_k, \mathfrak{G}_k)$ is a discrete-time Markov process as described above, and F_t corresponds to a stopping strategy

$$\theta_0 = \rho_0, \quad \theta_k = \frac{\rho_k}{\left(1 - \sum_{j=0}^{k-1} \rho_j\right)}, \quad k = 1, \ldots, n, \quad \theta_{n+1} = 1.$$

where $\rho_j = \Delta F_{\tau_j}$. We also define $b(k,x) = b(x)$ for $k = 0, \dots, n$ and $b(n+1, \Delta_0) = 0$. The idea here is that stopping at time $n+1$ in the discrete-time model is equivalent to stopping at $+\infty$ in the PDP. By the above results there is a $\mathfrak{G}_k$-stopping time $\hat{\sigma}$ whose performance is at least as good as that of (θ_k). If we define an $\mathfrak{F}_t$-stopping time $\hat{\tau}$ by

$$\hat{\tau} = \sum_{j=0}^{n} \tau_j I_{(\hat{\sigma}=j)} + \infty I_{(\hat{\sigma}=n+1)},$$

it then follows that

$$\mathbb{E}_x b(x_{\hat{\tau}}) \geqslant \int_{[0,\infty[} b(x_t) \mathrm{d}F_t.$$

This completes the proof. □

(52.8) *Proposition.* Let $F \in \Sigma_r$. Then there exists a sequence $F^n, n = 1, 2, \dots$ such that $F^n \in \mathscr{S}_n$ and for each $x \in E$

(52.9) $$J_x(F^n) \to J_x(F) \quad \text{as } n \to \infty.$$

Proof. If the function $t \to b(x_t)$ were continuous, this result would follow immediately from the standard construction of the Riemann-Stieltjes integral (Protter and Morrey, 1977, pp. 316–17). We need a slight modification to take account of the fact that $t \to b(x_t)$ is discontinuous at the jump times $T_1, T_2, \dots$ (but only at those times, since $b \in \mathscr{C}_b$).

For $n = 1, 2, \dots$ let $t_k^n = k2^{-n}$ for $k = 0, \dots, 4^n$ and let σ_n^k, $k = 0, \dots, n_0$, where $n_0 = 4^n + n$, be the times t_k^n together with the first n jump times $T_1, \dots, T_n$ of the PDP, arranged in increasing order. Clearly, each σ_k^n is a stopping time. Now define

$$\Delta F^n(0) = F(0), \quad \Delta F^n(\sigma_k^n) = F(\sigma_k^n) - F(\sigma_{k-1}^n), \quad k = 1, \dots, n_0$$

and

$$F^n(t) = \sum_{\sigma_k^n \leqslant t} \Delta F^n(\sigma_k^n).$$

Then $F^n \in \mathscr{S}_{n_0+1}$. We claim that (52.9) holds for this choice of F^n.

Fix $\omega \in \Omega$ such that $T_n(\omega) \uparrow \infty$, and recall that the set of such ω has probability 1. Given $\varepsilon > 0$, there exists N such that $F(\infty) - F(N-1) < \varepsilon$, and hence

$$\int_{[N,\infty[} |b(x_s)| \mathrm{d}F(s) < \varepsilon \|b\|, \quad \int_{[N,\infty[} |b(x_s)| \mathrm{d}F^n(s) < \varepsilon \|b\|$$

for all n, where $\|b\| = \sup_x |b(x)|$. We now show that for arbitrary $N > 0$

$$\int_{[0,N[} |b(x_s)|\,\mathrm{d}F^n(s) \to \int_{[0,N[} |b(x_s)|\,\mathrm{d}F(s).$$

Let $T_1(\omega), \ldots, T_j(\omega)$ be the jump times in $[0, N[$ and define $C_k^n =]T_k - 2^{-n}, T_k + 2^{-n}[, k = 1, \ldots, j$ and $C^n = \bigcup_{k=1}^{j} C_k^n$. Now the function $s \to b(x_s(\omega))$ is continuous on $[0, N[\backslash C^n$ and a slight modification of (Protter and Morrey, 1977, pp. 316–17) shows that

$$\int_{[0,N[\backslash C^n} b(x_s)\mathrm{d}F^n(s) \to \int_{[0,N[\backslash C_0} b(x_s)\mathrm{d}F(s),$$

where $C_0 = \bigcup_{k=1}^{j} \{T_k\}$. On the other hand by construction,

$$\int_{C_k^n} b(x_s)\mathrm{d}F^n(s) \to \int_{\{T_k\}} b(x_s)\mathrm{d}F(s)$$

for $k = 1, \ldots, j$. We have therefore shown that

$$\int_{[0,\infty[} b(x_s)\mathrm{d}F^n(s) \to \int_{[0,\infty[} b(x_s)\mathrm{d}F(s) \quad \text{a.s.}$$

The result (52.9) follows by bounded convergence. □

Proof of Theorem (52.4). Denote

$$v_x(\mathscr{A}) = \sup_{F \in \mathscr{A}} J_x(F)$$

for any class of strategies $\mathscr{A}$. It follows from Proposition (52.8) that $v_x(\Sigma_r) = v_x(\mathscr{S})$, where $\mathscr{S} = \bigcup_{n=1}^{\infty} \mathscr{S}_n$. On the other hand, Proposition (52.5) shows that $v_x(\Sigma) \geqslant v_x(\mathscr{S})$. Hence $v_x(\Sigma) = v_x(\Sigma_r)$, as claimed. □

The final result concerns the class Σ_r^n of Definition (52.3). It is not hard to show that the reward for a stopping time τ can be approximated by that of a sequence of elements $F^n \in \Sigma_r^n$.

(52.9) *Proposition.* For any $\tau \in \Sigma$, there exists a sequence F^n with $F^n \in \Sigma_r^n$ such that $J_x(F_n) \to J_x(\tau)$.

Proof. For $\tau \in \Sigma$, define $F^n(t) = 1 - \exp(\int_0^t u_n(s)\,\mathrm{d}s)$, where $u_n(t) = nI_{(t \geqslant \tau)}$. Then

$$\int_{[0,\infty[} b(x_t)\,\mathrm{d}F^n(t) = \int_{\tau}^{\infty} b(x_\tau + t)n\,\mathrm{e}^{-nt}\,\mathrm{d}t \to b(x_\tau) \quad \text{a.s. as } n \to \infty,$$

since x_t has right-continuous paths. Hence $J_x(F^n) \to J_x(\tau)$ by bounded convergence. □

53 Variational inequalities and the penalty method

In this section our aim is to obtain a characterization of the value function v of optimal stopping as the unique solution of a variational inequality. Exactly what this means is explained below, but roughly it is the equivalent for optimal stopping of the Bellman equation of stochastic control theory. We shall use this link explicitly by bringing in the class Σ_r^n of randomized strategies (Definition (52.3)) in which the process (u_t) can be thought of as a control.

In the interests of a simpler exposition we consider here a stopping problem with discounted reward

$$J_x(\tau) = \mathbb{E}_x[e^{-\delta\tau} b(x_\tau)]$$

for $\tau \in \Sigma$, $\delta > 0$, $b \in \mathscr{C}_b$ so that the value function is

$$v(x) = \sup_{\tau \in \Sigma} J_x(\tau). \tag{53.1}$$

As pointed out in Example (51.29), this is actually a special case of the problem without discounting ($\delta = 0$). Note that it is now natural to allow stopping times $\tau \in \Sigma$ which (possibly) take the value $+\infty$ since $\mathbb{E}_x[e^{-\delta t} b(x_t)] \to 0$ as $t \to \infty$. This is consistent with our convention in section 52 above of attaching zero reward to stopping at $\tau = \infty$. We will also assume the following:

(53.2) *Assumptions*

1. λ is bounded.
2. $b \in \mathscr{L}_b$ (i.e. b is bounded and locally Lipschitz along trajectories (see Definition (3.1)).
3. Let $A_\varepsilon = \{x \in E : t_*(x) \geqslant \varepsilon\}$. Then $Q(A_\varepsilon; x) = 1$ for all $x \in \Gamma$, for some $\varepsilon > 0$.

The first of these assumptions implies that $s_*(x) = t_*(x)$ for all $x \in E$. The third is the same as Assumption (31.12) and, as pointed out there, is satisfied in virtually every application. It is similar to Condition (46.16) and is used in the proof of Theorem (53.11) in the same way that Condition (46.16) was used in Chapter 4.

Let us first consider strategies $F \in \Sigma_r^n$ for a fixed value of n.

We define

$$v_n(x) = \sup_{F \in \Sigma_p^n} J_x(F).$$

Note from Definition (52.3) that for $F \in \Sigma_r^n$ we have

$$\frac{\mathrm{d}}{\mathrm{d}t} F(t) = u_t \exp\left(- \int_0^t u_s \, \mathrm{d}s \right)$$

and hence

$$(53.3) \qquad J_x(F) = \mathbb{E}_x \left\{ \int_0^\infty \exp\left(- \int_0^t (\delta + u_s) \mathrm{d}s \right) u_t b(x_t) \mathrm{d}t \right\}.$$

As pointed out in section 52, Σ_r^n is in one-to-one correspondence with the set of adapted processes u taking values in $[0, n]$, and we will write $u \in \Sigma_r^n$ and $J_x(u)$ for the corresponding cost when convenient. It is now a stochastic control problem to maximize $J_x(u)$ over $u \in \Sigma_r^n$. The Bellman equation corresponding to this is as in section 42 (writing $V(x)$ for $v_n(x)$)

$$(53.4) \qquad \max_{u \in [0,n]} \{ \mathfrak{A} V(x) - (\delta + u) V(x) + u b(x) \} = 0, \quad x \in E$$

$$(53.5) \qquad V(x) = Q V(x), \quad x \in \Gamma,$$

where $\mathfrak{A}$ is the generator of (x_t). Note there is no boundary control. (53.4) is equivalent to

$$(53.6) \qquad \mathfrak{A} V(x) - \delta V(x) + n(b(x)) - V(x))^+ = 0,$$

where $h^+ := h \vee 0$. Equation (53.6) is the so-called penalty equation (Bensoussan and Lions, 1975). The intuition behind it is as follows: we know that the value function v defined by (53.1) satisfies $v \geqslant b$, $(b - v)^+ = 0$. Generally, it will not be the case that $v_n(x) \geqslant b(x)$ for all x, and the term $n(b - v_n)^+$ in (53.6) represents a 'penalty' for having $v_n < b$; because of the factor n the penalty gets increasingly severe as $n \to \infty$, enforcing $v \geqslant b$ in the limit.

The above control problem is of a rather simple kind, in that the control u_t does not affect the system trajectory, but only the payoff. As can be seen from (53.3), if we make u large initially we obtain a large initial reward but condemn ourselves to small rewards later because of the discounting. The optimization problem is therefore one of balancing immediate versus future rewards. As in Chapter 4, we will study the problem by considering controls between successive jump times. For $u \in \Sigma_r^n$, define the operator $G_u : B(E) \to B(E)$ by

$$(53.7)\quad \begin{aligned} G_u\psi(x) &= \mathbb{E}_x\left\{\int_0^{T_1} \exp\left(-\int_0^t (\delta + u_s)\,\mathrm{d}s\right) u_t b(x_t)\,\mathrm{d}t \right. \\ &\qquad \left. + \exp\left(-\int_0^{T_1} (\delta + u_s)\,\mathrm{d}s\right)\psi(x_{T_1})\right\} \\ &= \int_0^{t_*} \exp\left(-\int_0^t (\delta + \lambda(s) + u_s)\,\mathrm{d}s\right)(u_t b(t) + \lambda(t) Q\psi(t))\,\mathrm{d}t \\ &\qquad + \exp\left(-\int_0^{t_*} (\delta + \lambda(s) + u_s)\,\mathrm{d}s\right) Q\psi(t_*). \end{aligned}$$

Here $t_* = t_*(x)$ and we use the abbreviated notation $\lambda(s) = \lambda(\phi(s, x))$ and similarly for b, $Q\psi$. Now let $\mathscr{U}_n$ denote the set of measurable functions $u : \mathbb{R}_+ \to [0, n]$ and define

$$(53.8)\quad G\psi(x) = \sup_{u \in \mathscr{U}_n} G_u\psi(x).$$

Calculating this is a deterministic optimal control problem.

(53.9) *Proposition.* There exists an optimal control $\hat{u} \in \mathscr{U}_n$ for the control problem (53.8).

Proof. We use relaxed controls, as in section 43, the set of relaxed controls $\mathscr{R}_n$ being the set of measurable functions $r : \mathbb{R}_+ \to \mathscr{P}([0, n])$. For $x \in E, r \in \mathscr{R}_n$, define $y_0(t), y_1(t) \in \mathbb{R}$ by

$$(53.10)\quad \left.\begin{aligned} \frac{\mathrm{d}}{\mathrm{d}t} y_1(t) &= -\int_{[0,n]} y_1(t)(\delta + \lambda(t) + u) r_t(\mathrm{d}u), & y_1(0) &= 1, \\ \frac{\mathrm{d}}{\mathrm{d}t} y_0(t) &= \int_{[0,n]} y_1(t)(u b(t) + \lambda(t) Q\psi(t)) r_t(\mathrm{d}u), & y_0(0) &= 0. \end{aligned}\right\}$$

Then $y_1(t) \leqslant \mathrm{e}^{-\delta t}$ and from (53.7) the cost corresponding to $r \in \mathscr{R}_n$ is

$$G_r\psi(x) = y_0(t_*) + y_1(t_*) Q\psi(t_*).$$

Note that if r is an ordinary control, i.e. $r_t(\mathrm{d}u) = \delta_{u(t)}(\mathrm{d}u)$ for some $u \in \Sigma_r^n$ then this coincides with $G_u\psi$ as previously defined. If $t_* < \infty$ then it follows directly from Theorem (43.5) that the map $r \to G_r\psi(x)$ is continuous in the Young topology. If $t_* = \infty$ the same conclusion follows using the uniform bound $y_1(t) \leqslant e^{-\delta t}$. Since $\mathscr{R}_n$ is compact there is therefore a maximizing element, i.e. an element $\hat{r} \in \mathscr{R}_n$ such that $G_{\hat{r}}\psi(x) \geqslant G_r\psi(x)$ for all $r \in \mathscr{R}_n$. Now define the ordinary control $\hat{u}(t) = \int_{[0,n]} u\hat{r}_t(\mathrm{d}u)$. Because the right-hand sides of (53.10) are linear in u, we see that $G_{\hat{u}}\psi(x) = G_r\psi(x)$. Hence $\hat{u}$ is optimal in $\mathscr{U}_n$. □

(53.11) *Theorem.* There exists a control process $u_t^0 \in \Sigma_r^n$ which maximizes $J_x(u)$ given by (53.3).

Proof. The argument here is exactly the same as that of Chapter 4. Assumption (53.2) (3) implies that the operators G_u are two-stage contraction operators, uniformly in $u \in \mathscr{U}_n$ and hence that G is a two-stage contraction. It follows that (i) G has a unique fixed point $V \in B(E)$ which coincides with the value function for the control problem (5.28), and (ii) if $GV(x) = G_{\hat{u}(x)}V(x)$, i.e. $\hat{u}(x)$ is the optimal control in $\mathscr{U}_n$ for problem (53.8) with $\psi = V$, then $\hat{u}(x)$ is an optimal piecewise open loop control for problem (53.3). Thus if we define $u_t^0 = \hat{u}(x_{T_k})(t - T_k)$ for $t \in]T_k, T_{k+1}]$ then $u_t^0 \in \Sigma_r^n$ and u_t^0 is optimal in Σ_r^n. □

We now want to understand the connection between the optimal control thus obtained and the penalty equation (53.6). To do so, we use the Pontryagin maximum principle of deterministic optimal control theory (Fleming and Rishel, 1975, Theorem II.5.1). In view of (53.10), we can formulate the deterministic optimal control problem (53.9) with $x \in E$ fixed and $\psi = V$ as follows: maximize

(53.12)
$$\int_0^{t_*} y(t)(u(t)b(t) + \lambda(t)QV(t))\,\mathrm{d}t + y(t_*)QV(t_*)$$

over pairs (y, u) where $u \in \mathscr{U}_n$ and

(53.13)
$$\frac{\mathrm{d}}{\mathrm{d}t}y(t) = -(\delta + \lambda(t) + u(t))y(t), \quad y(0) = 1.$$

Since $\hat{u} = \hat{u}(x)$ is optimal for this problem, $\hat{u}$ must satisfy the maximum princple. For $(t, y, p, u) \in \mathbb{R}_+ \times \mathbb{R}_+ \times \mathbb{R} \times [0, n]$ define

$$H(t, y, p, u) = -p(\delta + \lambda(t) + u)y + (ub(t) + \lambda(t)QV(t))y.$$

Then for almost every t,

(53.14)
$$H(t, \hat{y}(t), \hat{p}(t), \hat{u}(t)) = \max_{u \in [0,n]} H(t, \hat{y}(t), \hat{p}(t), u),$$

where $\hat{y}(t)$ is given by (53.13) with $u(t) = \hat{u}(t)$ and where the *adjoint variable* $\hat{p}(t)$ is the unique solution of the differential equation

$$(53.15) \quad -\frac{\mathrm{d}}{\mathrm{d}t}\hat{p}(t) = -(\delta + \lambda(t) + \hat{u}(t))\hat{p}(t) + \hat{u}(t)b(t) + \lambda(t)QV(t)$$

with boundary condition

$$\hat{p}(t_*) = QV(t_*), \quad t_* < \infty,$$

$$\lim_{t\to\infty} \hat{p}(t) = 0, \qquad t_* = \infty.$$

Now (53.15) is a linear equation, whose solution is

$$\hat{p}(t) = \int_t^{t_*} \exp\left(-\int_t^s (\delta + \lambda(\theta) + \hat{u}(\theta))\,\mathrm{d}\theta\right)(\hat{u}(s)b(s) + \lambda(s)QV(s))\,\mathrm{d}s$$
$$+ \exp\left(-\int_t^{t_*} (\delta + \lambda(\theta) + \hat{u}(\theta))\,\mathrm{d}\theta\right)QV(t_*),$$

the last term vanishing if $t_* = \infty$. Comparing this with (53.7) and noting that $G_{\hat{u}}V(x) = V(x)$ we have the interesting result that for this problem $\hat{p}(t) = V(\phi(t,x))$. Thus, using our abbreviated notation $V(t) = V(\phi(t,x))$ we have from (53.14),

$$\begin{aligned} H(t, \hat{y}(t), \hat{p}(t), \hat{u}(t)) &= \max_{u\in[0,n]} (-V(t)(\delta + \lambda(t) + u) + ub(t) + \lambda(t)QV(t))\hat{y}(t) \\ &= \max_{u\in[0,n]} (\lambda(t)(QV(t) - V(t)) - \delta V(t) \\ &\quad + u(b(t) - V(t)))\hat{y}(t) \\ &= (\lambda(t)(QV(t) - V(\dot{t})) - \delta V(t) + n(b(t) - V(t))^+)\hat{y}(t). \end{aligned}$$

Application of the maximum principle thus tells us that the optimal control $\hat{u}(t)$ satisfies

$$(53.16) \qquad \hat{u}(t)(b(t) - V(t)) = n(b(t) - V(t))^+.$$

Recall from Definition (31.1) that $\mathscr{C}^1$ is the set of bounded measurable functions $W: E \to \mathbb{R}$ such that for each $x \in E$ the map $t \to W(\phi(t,x))$ is in $C^1([0, t_*(x)[)$.

(53.17) *Theorem.* Let $v_n(x) = \sup_{u\in\Sigma_r^n} J_x(u)$. Then $v_n \in \mathscr{C}^1$ and v_n is the unique solution in $\mathscr{C}^1$ of the Bellman (penalty) equation

$$\mathfrak{A}V(x) - \delta V(x) + n(b(x) - V(x))^+ = 0, \qquad x \in E \tag{53.18}$$

$$V(x) = QV(x), \quad x \in \Gamma. \tag{53.19}$$

Proof. Let $\hat{u} \in \mathscr{U}_n$ be as above. Then from (53.7) we have for any $x \in E$ (using the notation $\lambda(t) = \lambda(\phi(t, x))$ etc., as before)

$$\begin{aligned} v_n(x) - v_n(\phi(t,x)) &= \int_0^t \exp\left(- \int_0^s (\delta + \lambda(\theta) + \hat{u}(\theta))\mathrm{d}\theta \right)(\hat{u}(s)b(s) + \lambda(s)Qv_n(s))\mathrm{d}s \\ &\quad + \left(\exp\left(- \int_0^s (\delta + \lambda(\theta) + \hat{u}(\theta))\mathrm{d}\theta \right) - 1 \right) v_n(t). \end{aligned} \tag{53.20}$$

Since v_n is bounded, this shows that $v_n(\phi(t,x)) \to v_n(x)$ as $t \to 0$, i.e. $v_n \in \mathscr{C}_b$. From (53.20) we have, in compressed notation,

$$\begin{aligned} \frac{1}{t}(v_n(x) - v_n(\phi(t,x))) &= \lambda(Qv_n - v_n) - \delta v_n + \hat{u}(b - v_n) + o(1) \\ &= \lambda(Qv_n - v_n) - \delta v_n + n(b - v_r)^+ + o(1) \end{aligned}$$

(from (53.16)). It follows on taking the limit as $t \downarrow 0$ that v_n satisfies (53.19) and this also implies, since $v_n \in \mathscr{C}_b$, that $v_n \in \mathscr{C}^1$. The boundary condition $v_n(x) = Qv_n(x), x \in \Gamma$ is a direct consequence of the representation (53.7) and the fact that v_n satisfies $v_n = Gv_n$.

Conversely, suppose $V \in C^1$ and V satisfies (53.18) and (53.19). Then, from (53.4), for any $u \in [0, n]$,

$$\mathfrak{A}V(x) - (\delta + u)V(x) + ub(x) \leqslant 0. \tag{53.21}$$

Let $u_t \in \Sigma_r^n$ be any control and define

$$e_t = \exp\left(- \int_0^t (\delta + u_s)\mathrm{d}s \right).$$

Then by the differential formula (31.4) we have

$$\mathrm{d}(e_t V(x_t)) = -(\delta + u_t)e_t V(x_t)\mathrm{d}t + e_t \mathfrak{A}V(x_t)\mathrm{d}t + \mathrm{d}M_t$$

where M_t is a martingale. Thus, in view of (53.21),

$$\mathrm{d}(e_r V(x_t)) \leqslant e_t u_t b(x_t)\,\mathrm{d}t + \mathrm{d}M_t,$$

so that

$$V(x) \geqslant \mathbb{E}_x\left\{\int_0^t e_s u_s b(x_s)\,\mathrm{d}s + e_t V(x_t)\right\}.$$

Since V is bounded and $e_t \leqslant \mathrm{e}^{-\delta t}$ the last term vanishes as $t \to \infty$ and we obtain $V(x) \geqslant J_x(u)$. If we define $\check{u}(x) := nI_{(b(x) > v(x))}$ and $u_t^0 := \check{u}(x_t)$ then $u^0 \in \Sigma_r^n$ and taking $u_t = u_t^0$ the above argument goes through with equality replacing inequality, so that $V(x) = J_x(u^0)$. Thus u^0 is optimal, and $V(x) = v_n(x)$, as claimed. □

This completes our study of the optimization problem over Σ_r^n. We now want to consider what happens when the bound on the controls is relaxed, i.e. $n \to \infty$. Now $v_{n+1}(x) \geqslant v_n(x)$ since $\Sigma_r^{n+1} \supset \Sigma_r^n$, and the v_n are uniformly bounded; hence $\hat{v}(x) = \lim_{n\to\infty} v_n(x)$ exists. It is clear that

$$\hat{v}(x) = v(x) = \sup_{\tau \in \Sigma} J_x(\tau).$$

Indeed, by Theorem (52.4), $v_n(x) \leqslant v(x)$, and hence $\hat{v}(x) \leqslant v(x)$. Conversely, from Proposition (52.9) for any $\tau \in \Sigma$ and $\varepsilon > 0$ there exists n and $u \in \Sigma_t^n$ such that $|J_x(\tau) - J_x(u)| < \varepsilon$, and hence $\hat{v}(x) \geqslant v(x)$.

We can now establish the main properties of the value function v.

(53.22) *Theorem.* Suppose Assumptions (53.2) hold and define $v(x) = \sup_{\tau \in \Sigma} J_x(\tau)$. Then:

1. $v \in \mathscr{L}_b$;
2. $v(x) \geqslant b(x)$ for all $x \in E$;
3. $\mathfrak{A}v(x) - \delta v(x) \leqslant 0$ for all $x \in E$;
4. for any $x \in E$, either $v(x) = b(x)$ or $\mathfrak{A}v(x) - \delta v(x) = 0$.

Proof. Part 2 is true since v is the value function, and also follows from the representation $v = \lim v_n$ since (taking a sequence of controls $v_n(t) \equiv n$)

$$v(x) \geqslant v_n(x) \geqslant \mathbb{E}_x \int_0^\infty n\,\mathrm{e}^{-nt} b(x_t)\mathrm{d}t \xrightarrow{n\to\infty} b(x).$$

Now fix $x_0 \in E$ and, as before, write $b(t) = b(\phi(t, x_0))$, etc. Since $v(t)$ is the limit of an increasing sequence $v_n(t)$ of continuous functions, $v(t)$ is lower semicontinuous (actually we know from the results in section 51 that $v(t)$ is continuous). Define $t^*(x) = \sup\{t : \phi(-t, x_0) \in E\}$

and $D_n = \{t \in]-t^*(x_0), t_*(x_0)[: v_n(t) > b(t)\}$. Then D_n is open and $D_n \uparrow D := \{t : v(t) > b(t)\} = \bigcup_n D_n$. Hence D is open. Thus for any $t \in D$, there exists n and $\varepsilon > 0$ such that $]t - \varepsilon, t + \varepsilon[\subset D_n$. It now follows from (53.16) that $\hat{u}_n(\phi(t + s, x_0)) = 0$ for $|s| < \varepsilon$, where $\hat{u}_r$ is the optimal control in Σ_r^n. Applying (53.20) with $x = \phi(t, x_0)$, we obtain for $s < \varepsilon$

$$v_n(x) - v_n(\phi(t,x)) = \int_0^t \exp\left(- \int_0^\zeta (\delta + \lambda(\theta))\mathrm{d}\theta \right)\lambda(\zeta)Qv_n(\zeta)\mathrm{d}\zeta$$
$$+ \left(\exp\left(- \int_0^s (\delta + \lambda(\theta))\mathrm{d}\theta \right) - 1 \right) v_n(\phi(s,x)).$$

Letting $n \to \infty$ and using bounded convergence, we obtain the same relation for v. It now follows exactly as in the proof of Theorem (53.17) that $s \to v(\phi(s,x))$ is differentiable at $s = 0$ and $\mathfrak{A}v(x) - \delta v(x) = 0$. Thus part 3 holds with equality on $C := \{x : v(x) > b(x)\}$. Now take $x \in S = \{x : v(x) = b(x)\}$. In this case $\tau = 0$ is an optimal time, so for any $t > 0$,

$$\text{(52.23)} \qquad v(x) = b(x) \geqslant \mathbb{E}_x[\mathrm{e}^{-\delta(t \wedge T_1)} v(x_{t \wedge T_1})]$$
$$= \int_0^t \exp\left(- \int_0^s (\delta + \lambda(\theta))\mathrm{d}\theta \right)\lambda(s)Qv(s)\mathrm{d}s$$
$$+ \exp\left(- \int_0^t (\delta + \lambda(\theta))\mathrm{d}\theta \right) v(t),$$

and it follows that $\limsup_{t \downarrow 0} v(\phi(t,x)) \leqslant v(x)$. On the other hand $v(t)$ is lower semicontinuous, i.e. $\liminf_{t \downarrow 0} v(\phi(t,x)) \geqslant v(x)$, and thus $v(\phi(t,x))$ is continuous at $t = 0$. (Again, we know this independently from section 51.) From (53.23) we have

$$0 \geqslant \frac{1}{t}(v(\phi(t,x)) - v(x)) + \frac{1}{t}\int_0^t \exp\left(- \int_0^s (\delta + \lambda(\theta))\mathrm{d}\theta \right)\lambda(s)Qv(s)\mathrm{d}s$$
$$+ \frac{1}{t}\left(\exp\left(- \int_0^t (\delta + \lambda(\theta))\mathrm{d}\theta \right) - 1 \right) v(\phi(t,x))$$

and hence

$$\text{(53.24)} \quad 0 \geqslant \liminf_{t \downarrow 0} \frac{1}{t}(v(\phi(t,x)) - v(x)) + \lambda(x)(Qv(x) - v(x)) - \delta v(x).$$

We now show that $v(x)$ is locally Lipschitz continuous along trajectories at points $x \in S$. Indeed, take x such that $v(x) = b(x)$; then $v_n(x) \leqslant b(x)$ for all n, and $J_x(u) \leqslant u_n(x)$, for any $u \in \Sigma_r^n$. Hence

$$|b(x) - v_n(x)| \leqslant b(x) - J_x(u).$$

Now take $u_t \equiv n$ and write $J(x) = J_x(u)$. For any $t_1 < t_*(x)$, $J(x)$ satisfies

$$J(x) = \int_0^{t_1} \exp\left(-\int_0^t (\delta + \lambda(s) + n)\mathrm{d}s\right)(nb(t) + \lambda(t)Q\,J(t))\mathrm{d}t$$
$$+ \exp\left(-\int_0^{t_1} (\delta + \lambda(s) + n)\mathrm{d}s\right)Q\,J(t_1).$$

Subtract from this the following identity:

$$b(x) = \int_0^{t_1} \exp\left(-\int_0^t (\delta + \lambda(s) + n)\mathrm{d}s\right)(\delta + \lambda(t) + n)\mathrm{d}t\, b(x)$$
$$+ \exp\left(-\int_0^{t_1} (\delta + \lambda(s) + n)\mathrm{d}s\right)b(x).$$

Using the fact that $(\delta + \lambda(s)) > 0$ we obtain

$$\text{(53.25)} \quad |J(x) - b(x)| \leqslant \int_0^{t_1} \mathrm{e}^{-nt} n |b(\phi(t,x)) - b(x)|\mathrm{d}t$$
$$+ \int_0^{t_1} \mathrm{e}^{-nt}(\lambda(t)Q\,J(t) + (\delta + \lambda(t))b(x))\mathrm{d}t$$
$$+ \mathrm{e}^{-nt_1}|Q\,J(t_1) + b(x)|.$$

Now by Assumption (53.2) (2) the function $t \to b(\phi(t,x))$ is locally Lipschitz, so we can choose $t_1 > 0$ and $L < \infty$ such that $|b(\phi(t,x)) - b(x)| \leqslant Lt$ for $t \in [0, t_1]$. Also the functions appearing in the last two terms on the right of (53.25) are all bounded, so for some constants K_1, K_2

$$|J(x) - b(x)| \leqslant L\int_0^{t_1} \mathrm{e}^{-nt} nt\,\mathrm{d}t + K_1 \int_0^{t_1} \mathrm{e}^{-nt}\mathrm{d}t + K_2\mathrm{e}^{-nt_1}$$
$$\leqslant (L + K_1)\frac{1}{n} + K_2\mathrm{e}^{-nt_1} \leqslant (L + K_1 + K_2 n\mathrm{e}^{-nt_1})\frac{1}{n}$$
$$\leqslant K_3\frac{1}{n},$$

where $K_3 = L + K_1 + K_2(t_1 e)^{-1}$. Hence

$$|b(x) - v_n(x)| \leqslant K_3 \frac{1}{n}. \tag{53.26}$$

Now v_n satisfies the Bellman equation

$$\mathfrak{X} v_n(x) + \lambda(x)(Qv_n(x) - v_n(x)) - \delta v_n(x) + n(b(x) - v_n(x))^+ = 0, \tag{53.27}$$

where we recall that

$$\mathfrak{X} v_n(x) = \frac{\mathrm{d}}{\mathrm{d}t} v_n(\phi(t, x))|_{t=0}.$$

In view of (53.26) the last three terms on the right of (53.27) are bounded, uniformly in n. Hence $\mathfrak{X} v_n(x)$ is bounded uniformly in n, which implies that $t \to v_n(\phi(t, x))$ is locally Lipschitz uniformly in n, and hence that $t \to v(\phi(t, x)) = \lim_n v_n(\phi(t, x))$ is locally Lipschitz.

We have shown that $t \to v(\phi(t, x))$ is locally Lipschitz at $x \in S$. For $x \in C$ we know $t \to v(\phi(t, x))$ is C^1 and *a fortiori* locally Lipschitz. This proves part 1 of the theorem.

A locally Lipschitz function is differentiable almost everywhere. Hence the limit in (53.24) exists for almost every t, and (53.24) becomes

$$\mathfrak{A} v(x) - \delta v(x) \leqslant 0, \quad x \in S.$$

For $x \in C$ we already know that $\mathfrak{A} v(x) - \delta v(x) = 0$, and statements 3 and 4 in the theorem follow. □

We can rephrase the statement of Theorem (53.22) as a so-called *variational inequality*. Consider the following system of equations for a function $W \in \mathscr{L}_b$

$$\left.\begin{aligned} b - W &\leqslant 0 \\ \mathfrak{A} W - \delta W &\leqslant 0 \\ (b - W)(\mathfrak{A} W - \delta W) &= 0 \end{aligned}\right\} \quad \text{in } E \tag{53.28}$$

with the boundary condition $W(x) = b(x) \vee QW(x), x \in \Gamma$, which can be expressed in a manner similar to (53.28) as

$$\left.\begin{aligned} b - W &\leqslant 0 \\ QW - W &\leqslant 0 \\ (b - W)(QW - W) &= 0 \end{aligned}\right\} \quad \text{on } \Gamma. \tag{53.29}$$

From (53.21) and Theorem (51.30), we know that $v \in \mathscr{L}_b$ and v satisfies

(53.28) and (53.29). In fact, (53.28) and (53.29) provide a sufficient condition for optimality as well.

(53.30) *Theorem.* Under Assumptions (53.2), the value function v is the unique solution in the class $\mathscr{L}_b$ of (53.28) and (53.29).

Proof. We have already shown that $v \in \mathscr{L}_b$ and v satisfies (53.28) and (53.29). Conversely, suppose $W \in \mathscr{L}_b$ and W satisfy (53.28) and (53.29). Then by the differential formula (31.4), for any $x \in E$

$$(53.31)\quad e^{-\delta t}W(x_t) = \int_0^t e^{-\delta s}(\mathfrak{A}W - \delta W)\mathrm{d}s + \int_0^t e^{-\delta s}(QW - \delta W)\mathrm{d}p_s^* + \int_0^t \int_E e^{-\delta s}\mathfrak{B}W \mathrm{d}q.$$

The last term on the right is a martingale and by (53.28) and (53.29) the integrands in the other two terms are non-positive. Hence

$$W(x) \geqslant \mathbb{E}_x[e^{-\delta t}W(x_t)].$$

It follows from the Markov property that the process $t \to e^{-\delta t}W(x_t)$ is a supermartingale, and hence by the optional sampling theorem (31.7) that for any stopping time τ,

$$W(x) \geqslant \mathbb{E}_x[e^{-\delta\tau}W(x_\tau)].$$

Since, from (53.28), $W \geqslant b$, this implies that $W(x) \geqslant \mathbb{E}_x[e^{-\delta\tau}b(x_\tau)]$ and hence that $W(x) \geqslant v(x)$. It remains to show that $W(x) \leqslant v(x)$.

For $\varepsilon > 0$ let

$$U_\varepsilon = \inf\{t: e^{-\delta t}b(x_t) \geqslant e^{-\delta t}W(x_t) - \varepsilon\}$$

and note that $U_\varepsilon < \infty$ a.s. since b and W are bounded. Now consider (53.31) with $t = U_\varepsilon$. From (53.28), $\mathfrak{A}W(x_s) - \delta W(x_s) = 0$ for $s \leqslant U_\varepsilon$, so the first integral on the right of (53.31) vanishes for $s \leqslant U_\varepsilon$. If T_k is a boundary hitting time less than or equal to U_ε then $W(x_t) > b(x_t) + \varepsilon e^{\delta t}$ for $t < T_k$ and hence $W(x_{T_k -}) \geqslant b(x_{T_k -}) + \varepsilon e^{\delta T_k}$. Now from (53.29) $W(z) > b(z)$ implies that $W(z) = QW(z)$ for $z \in \Gamma$; thus the second integral on the right of (53.31) also vanishes for $t \leqslant U_\varepsilon$ and by optional sampling we obtain

$$W(x) = \mathbb{E}_x[e^{-\delta U_\varepsilon}W(x_{U_\varepsilon})].$$

By right-continuity of the sample path of x_t, $e^{-\delta U_\varepsilon}W(x_{U_\varepsilon}) \leqslant$

$e^{-\delta U_\varepsilon} b(x_{U_\varepsilon}) + \varepsilon$ and hence

$$W(x) \leqslant \mathbb{E}_x[e^{-\delta U_\varepsilon} b(x_{U_\varepsilon})] + \varepsilon \leqslant \sup_{\tau \in \Sigma} \mathbb{E}_x[e^{-\delta \tau} b(x_\tau)] + \varepsilon = v(x) + \varepsilon.$$

Thus $W(x) \leqslant v(x)$. This completes the proof. □

Finding a solution of the variational inequality is often the easiest way of solving an optimal stopping problem given that, by Theorem (53.30), *any* solution is guaranteed to be the value function.

(53.32) *Example*
This is Example (51.29) revisited, with $b(x) = x$. One can guess that the stopping region must be an interval of the form $[a, 1]$; in the continuation region we have $\mathfrak{A}v(x) - \delta v(x) = 0$, i.e. $\dot{v}(x) - \delta v(x) = 0$, so that $v(x) = Ae^{\delta x}$ for some constant A, and application of the smooth pasting condition gives $a = \delta^{-1}$ and $A = (e\delta)^{-1}$ when $\delta > 1$. The function $W(x) = \delta^{-1} e^{\delta x - 1}, x \in [0, \delta^{-1}]$, $W(x) = x, x \in [\delta^{-1}, 1[$ certainly satisfies (53.29) (i.e. $W(1) = b(1) \vee QW(1)$), and $\mathfrak{A}W(x) - \delta W(x) = 0$ for $x \in [0, \delta^{-1}]$, $\mathfrak{A}W(x) - \delta W(x) = 1 - \delta x < 0$ for $x \in]\delta^{-1}, 1[$, so that W satisfies (53.28), i.e. $W = v$. Any barrier $a < \delta^{-1}$ is ruled out for this reason: if we suppose that $S = [a, 1[$ with $a < \delta^{-1}$, then $W(x) = x$ for $x \in [a, 1[$, but then $\mathfrak{A}W(x) - \delta W(x) > 0$ for $x \in [a, \delta^{-1}[$, so that (53.28) fails. If $S = [a, 1[$ with $a > \delta^{-1}$, then $\dot{W}(a-) > 1$, so that $W(a \dot{-} \varepsilon) < a - \varepsilon = b(a - \varepsilon)$ for small ε, contradicting (53.28). The smooth pasting condition thus identifies the only value of a for which (53.28) holds. On the other hand, smooth pasting does not hold in every problem. Indeed, we showed in Example (51.29) that smooth pasting fails when $b(x) = 1 - x$, but it is easy to check that the function $W(x) = \max\{(1 - x), e^{\delta(x-1)}\}$, which we claimed was the value function, satisfies (53.28) and (53.29). In this case the boundary condition $W(1) = QW(1)$ is satisfied, and this replaces smooth pasting as the extra condition needed to pin down uniquely the solution.

This example is important in another respect: it shows that our result in Theorem (53.22), stating that $v \in \mathscr{L}_b$, cannot generally be improved. One might think that v is actually continuously differentiable along trajectories and that one should be able to prove this, given adequate technique. However, this is not so: the value function $v(x) = \max\{(1 - x), e^{\delta(x-1)}\}$ for Example (51.29) with $b(x) = 1 - x$ is Lipschitz continuous but *not* continuously differentiable, even though there is no lack of smoothness in the functions defining the problem.

In the next section we shall have occasion to study the optimal stopping problem of *minimizing* over stopping times $\tau \in \Sigma$ a cost function taking the form

(53.33) $$I_x(\tau) = \mathbb{E}_x\left\{\int_0^\tau e^{-\delta t} l(x_t) dt + e^{-\delta\tau}\psi(x_\tau)\right\},$$

with bounded l, ψ. This is in fact (almost) a special case of the discounted reward $\mathbb{E}_x[e^{-\delta\tau}b(x_\tau)]$ considered in this section, and we now show what the correspondence is and what form the variational inequality (53.28) and (53.29) takes in this case.

The integral cost term in (53.33) is accommodated by the usual procedure of state augmentation. We define a new process $\bar{x}_t = (x_t^0, x_t)$ in $\mathbb{R} \times E$ by taking (x_t) as the original PDP and x_t^0 to satisfy

$$\frac{d}{dt}x_t^0 = \delta x_t^0 - l(x_t).$$

Then

$$\frac{d}{dt}e^{-\delta t}x_t^0 = -\delta e^{-\delta t}x_t^0 + e^{-\delta t}\frac{d}{dt}x_t^0 = -e^{-\delta t}l(x_t),$$

so that

(53.34) $$e^{-\delta t}x_t^0 = x_0^0 - \int_0^t e^{-\delta t} l(x_s) ds.$$

Thus, if we define $b(\bar{x}) := x^0 - \psi(x)$, then $I_x(\tau) = -\mathbb{E}_{(0,x)}[e^{-\delta\tau}b(\bar{x}_\tau)]$ and

$$V(x) := \inf_{\tau\in\Sigma} I_x(\tau) = -\bar{W}(0, x),$$

where

$$\bar{W}(x) := \sup_{\tau\in\Sigma} \mathbb{E}_{\bar{x}}[e^{-\delta\tau}b(\bar{x}_\tau)].$$

(53.35) *Proposition.* Suppose that Assumptions (53.2) (1) and (3) are satisfied and that $\psi \in \mathscr{L}_b$ and $l \in B(E)$. Then $V(x) = \inf_{\tau\in\Sigma} I_x(\tau)$ is the unique solution in $\mathscr{L}_b$ of the variational inequality

(53.36) $$\left.\begin{aligned} Qw - \delta w + l \geqslant 0 \\ \psi - w \geqslant 0 \\ (Qw - \delta w + l)(\psi - w) = 0 \end{aligned}\right\} \text{ in } E$$

with the boundary condition $w(x) = \psi(x) \wedge Qw(x), x \in \Gamma$.

Proof. Under the stated assumption the augmented model described above satisfies all the conditions of Theorem (53.30) except that $b(\bar{x}) = x^0 - \psi(x)$ is not bounded for $x^0 \in \mathbb{R}$. However, the initial point x^0 plays no role in the optimization process and b is bounded if x^0 is confined to a compact set; the conclusions of the theorem remain valid. It is clear from (54.34) and the definition of b that $\bar{W}(\bar{x}) = x^0 + W(x)$, where $W(x) := \bar{W}(0, x)$. The generator of the augmented process is

$$\bar{\mathfrak{A}} f(\bar{x}) = (\delta x^0 - l(x)) \frac{\partial f}{\partial x^0}(\bar{x})$$

$$+ \mathfrak{X} f(\bar{x}) + \lambda(x) \int_E (f(x^0, y) - f(x^0, x)) Q(\mathrm{d}y; x).$$

Now Theorem (53.30) asserts that $\bar{W} \in \mathscr{L}_b$ and applying the above formula we see that

$$\begin{aligned}\bar{\mathfrak{A}} \bar{W}(\bar{x}) - \delta W(\bar{x}) &= (\delta x^0 - l)1 + \mathfrak{X} W + \lambda(QW - W) - \delta(x^0 + W) \\ &= \mathfrak{A} W - l - \delta W.\end{aligned}$$

Recalling that $V(x) := \inf_{\tau \in \Sigma} I_x(\tau) = -W(x)$ we see that the variational inequality (53.28) and (53.29) is equivalent to (54.36) when expressed in terms of V. □

Our argument so far shows that the value function V is in the class $\mathscr{L}_b$, so that the map $t \to V(\phi(t, x))$ is continuous but V is not necessarily continuous on E. Under additional smoothness conditions on the coefficients we can show that in fact $V \in C_b(E)$, and this will be important in connection with the impulse control problem in the next section. The additional conditions required are as follows.

(53.37) *Assumptions*

1. $\lambda, l, \psi \in C_b(E \cup \Gamma)$.
2. $Qw \in C_b(E \cup \Gamma)$ for $w \in C_b(E \cup \Gamma)$.
3. The map $x \to t_*(x)$ is continuous on E.

First, we need a technical lemma.

(53.38) *Lemma.* Let $S: E \to \mathbb{R}_+ \cup \{\infty\}$ be a continuous function and define $F_0 = \{(x, t): x \in E, 0 \leqslant t \leqslant S(x)\}$. Now let $f, c \in C_b(F_0)$ and define

$w: F_0 \to \mathbb{R}$ by

$$w(x,t) = \inf_{t \leqslant s \leqslant S(x)} e^{\delta t}\left\{\int_t^s e^{-\delta u} f(x,u)\,du + e^{-\delta s} c(x,s)\right\},$$

where $\delta > 0$. Then $w \in C_b(F_0)$.

Proof. We will just show that w is continuous in x. Joint continuity in (x, t) follows by similar arguments. Fix $t \in [0, S(x)[$, define $\tilde{w}(x) = e^{-\delta t} w(x,t)$, take $x, x' \in E$ and $\varepsilon > 0$ and let s be an ε-infimal choice of minimizing time at x. Then for all $s' \in [t, S(x')]$,

$$\begin{aligned}
\tilde{w}(x') - \tilde{w}(x) &\leqslant \int_t^{s'} e^{-\delta u} f(x',u)\,du + e^{-\delta s'} c(x', s) \\
&\quad - \int_t^s e^{-\delta u} f(x,u)\,du - e^{-\delta s} c(x,s) + \varepsilon \\
&\leqslant \left| \int_t^{s'} e^{-\delta u} f(x',u)\,du - \int_t^s e^{-\delta u} f(x,u)\,du \right| \\
&\quad + |e^{-\delta s'} c(x', s) - e^{-\delta s} c(x,s)| + \varepsilon.
\end{aligned}$$

Now choose s' to be the closest point in $[t, S(x')]$ to s. As $x' \to x, s' \to s$ and eventually the right-hand side of the inequality is less than, say, 2ε. A similar inequality can be obtained with x and x' interchanged, showing that $\tilde{w}(x') \to \tilde{w}(x)$ as $x' = x$. □

Next, define

$$V_n(x) = \inf_{\tau \in \Sigma_n} \mathbb{E}_x \left\{\int_0^\tau e^{-\delta t} l(x_t)\,dt + e^{-\delta \tau} \psi(x_\tau)\right\}, \quad n = 1, 2 \ldots.$$

From Theorem (51.11) and (51.17) we know that $V_n(x) \downarrow V(x)$ and that $V_n(x) = L(V_{n-1}, \psi)(x)$, $V_0(x) = \psi(x)$ where the operator L, in modified form as appropriate to minimizing $I_x(\tau)$ of (53.38), is given as follows

$$\begin{aligned}
J(w,h)(t,x) := \mathbb{E}_x \Bigg\{ &\int_0^{T_1 \wedge t} e^{-\delta t} l(\phi(s,x))\,ds + e^{-\delta T_1} Qw(\phi(T_1, x)) I_{(t \geqslant T_1)} \\
&+ e^{-\delta t} h(\phi(t,x)) I_{(t < T_1)} \Bigg\}
\end{aligned}$$

$$\begin{aligned}
Kw(x) &:= \mathbb{E}_x \left\{ \int_0^{T_1} e^{-\delta s} l(\phi(s,x))\,ds + e^{-\delta T_1} Qw(\phi(T_1, x)) \right\} \\
&= J(w,h)(\infty, x)
\end{aligned}$$

(53.39) $$L(w,h)(x) := \left(\inf_{0 \leqslant t < t_*(x)} J(w,h)(t,x) \right) \wedge Kw(x).$$

(53.40) *Theorem.* Suppose Assumptions (53.37) hold, in addition to the standing assumptions in section (24.8). Then $V \in C_b(E)$, where $V(x) = \inf_{\tau \in \Sigma} I_x(\tau)$.

Proof. L can be expressed in explicit form as in section 51, and it then follows from Lemma (53.38) that $L(w,\psi) \in C_b(E)$ for $w, \psi \in C_b(E)$, and hence by induction that $V_n \in C_b(E)$ for $n = 1, 2, \ldots$. Now

$$|I_x(\tau) - I_x(\tau \wedge T_n)|$$
$$= \left| I_x \left\{ I_{(\tau > T_n)} \left(\int_{T_n}^{\tau} e^{-\delta t} l(x_t) dt + e^{-\delta\tau} \psi(x_\tau) - e^{-\delta T_n} \psi(x_{T_n}) \right) \right\} \right|$$
$$\leqslant \left| \mathbb{E}_x \left\{ \frac{1}{\delta} \| l \| e^{-\delta T_n} + e^{-\delta T_n} (e^{-\delta(\tau - T_n)} \psi(x_\tau) - \psi(x_{T_n})) \right\} \right|$$
$$\leqslant \left(\frac{1}{\delta} \| l \| + 2 \| \psi \| \right) \mathbb{E}_x [e^{-\delta T_n}].$$

Since $V_n(x) = \inf_{\tau \in \Sigma} I_x(\tau \wedge T_n)$, this implies that

$$0 \leqslant V(x) - V_n(x) \leqslant \left(\frac{1}{\delta} \| l \| + 2 \| \psi \| \right) \chi_n(x),$$

where $\chi_n(x) := \mathbb{E}_x[e^{-\delta T_n}]$. Now $\chi_n(x) \downarrow 0$ as $n \to \infty$ by bounded convergence, and it is easily shown that χ_n is continuous, using the recursive representation $\chi_n(x) = \mathbb{E}_x[e^{-\delta T_1} \chi_{n-1}(x_{T_1})]$ which follows from the strong Markov property of (x_t). Hence by Dini's Theorem (Dudley, 1989, Theorem 2.4.10) $\chi_n \to 0$ uniformly on compact sets. Thus $V_n(x) \to V(x)$ uniformly on compact sets, which implies that V is continuous. □

54 Impulse control

Impulse control concerns a situation in which a process runs uncontrolled until the decision maker decides to intervene, moving the process instantaneously to some new point in the state space, from which its motion continues until a second intervention is made, and so on. The repair shop model of section (21.15) is a case in point. Generally, for any point x in the state space E there will be a set $K(x) \subset E$ of points to which the process may be moved, and a cost $c(x, y)$ is paid when the process is moved from x to $y \in K(x)$. Cost accumulates at rate $l(x)$ in between interventions. It will be assumed throughout this section that K, c and l satisfy at least the following conditions. Further conditions will be added later, as required.

(54.1) *Assumptions*

1. There exist a compact set $Y \subset E$ and a closed set $\mathscr{K} \subset E \times Y$ such that, for all $x \in E, K(x) = \{y \in Y : (x, y) \in \mathscr{K}\}$, and if $y \in K(x)$, then $K(y) \subset K(x)$.
2. c is a continuous function on $\mathscr{K}$.
3. $c(x, y) + c(y, z) \geqslant c(x, z)$ for $x \in E, y \in K(x), z \in K(y)$.
4. $0 < c_0 \leqslant \inf_{y \in K(x)} c(x, y) \leqslant c_1 < \infty$.
5. $l : E \to [0, c_2]$ is continuous ($c_2 < \infty$).

Assumption 1 is used below in the application of a *selection theorem*, while Assumption 2 is mainly a matter of convention: if there exists $z \in K(y)$ such that $z \notin K(x)$, we can adjoin z to $K(x)$ with transfer cost $c(x, z) = c(x, y) + c(y, z)$. Assumption 3 says that it is never positively advantageous to move from x to z via a third point y, while 4 ensures that there is a minimal, but bounded, cost for any intervention. The effect of these assumptions is that it will never be optimal to do anything other than make a sequence of isolated interventions. If the intervention times are denoted $\tau_1, \tau_2, \ldots$, then the average cost paid will take the form

$$J_x = \mathbb{E}_x \left\{ \int_0^\infty e^{-\delta t} l(X_t) dt + \sum_{k=1}^\infty e^{-\delta \tau_k} c(X_{\tau_k}, X_{\tau_k +}) \right\},$$

where $\delta > 0$ and X_t is the controlled process (i.e. including the interventions) and $X_{\tau_k +}$ is the position to which the process is moved at τ_k. Note that if l is bounded there is no loss of generality in assuming that $l \geqslant 0$, because if l is replaced by $l + \kappa$, where κ is a constant, then the average cost becomes simply $J_x + \kappa/\delta$.

The formal probabilistic apparatus necessary to describe the above situation precisely, which we give next, is unfortunately rather cumbersome. This seems inevitable, but – fortunately – the details of the construction are only occasionally used explicitly in the ensuing developments, as will be seen below. Readers can therefore skip over the details if they so desire.

Throughout, (x_t, P_x) is a piecewise-deterministic Markov process as defined in section 24. As in the preceding sections, its state space will be a set $E \subset \mathbb{R}^d$ rather than a countable union of such sets. The sample functions of x_t then take values in the set $\Omega = D_E[0, \infty[$ and P_x is a measure on Ω for each $x \in E$. Now let Δ be an isolated point and $\hat{\Omega}$ be the set of functions $\hat{\omega}: \mathbb{R}_+ \to E \cup \{\Delta\}$ such that one of 1, 2 or 3 below holds:

1. $\hat{\omega} \in \Omega$.
2. For some $a \in \mathbb{R}_+$ $\hat{\omega}(\cdot)$ is right continuous with left-hand limits on $]0, a[$, is right continuous at $t = 0$ and has a left-hand limit at $t = a$, and $\hat{\omega}(t) = \Delta$ for $t > a$.
3. $\hat{\omega}(t) = \Delta$ for all $t \in \mathbb{R}_+$.

Define a function $\eta: \hat{\Omega} \to \mathbb{R}_+ \cup \{\infty\}$ as follows: $\eta(\hat{\omega}) = \infty, a, 0$ in cases 1, 2 and 3, respectively. Now define $\hat{x}_t(\tilde{\omega}) = \hat{\omega}(t)$ and the following σ-fields in $\hat{\Omega}$: $\hat{\mathfrak{F}}_t^0 = \sigma\{\hat{x}_s, s \leqslant t\}$, $\hat{\mathfrak{F}}$ is the universal completion of $\hat{\mathfrak{F}}_\infty^0$, and, for each $t \in \mathbb{R}_+$, $\hat{\mathfrak{F}}_t$ is the $\hat{\mathfrak{F}}$-completion of $\hat{\mathfrak{F}}_t^0$. Now let

$$\mathscr{W}_k = \prod_{i=1}^{k} \hat{\Omega}^i, \quad \mathfrak{G}_k^0 = \sigma\left\{\prod_{i=1}^{k} \hat{\mathfrak{F}}^i\right\},$$

$$\mathscr{W} = \prod_{i=1}^{\infty} \hat{\Omega}^i, \quad \mathfrak{G}^0 = \sigma\left\{\prod_{i=1}^{\infty} \hat{\mathfrak{F}}^i\right\},$$

where $(\hat{\Omega}^i, \hat{\mathfrak{F}}^i)$ is a copy of $(\hat{\Omega}, \hat{\mathfrak{F}})$, $i = 1, 2, \ldots$. An element w of $\mathscr{W}$ is then a sequence $w = (\hat{\omega}_1, \hat{\omega}^2, \ldots)$, where $\hat{\omega}_i \in \hat{\Omega}^i$. We also define

$$w_k(w) = (\hat{\omega}_1, \ldots, \hat{\omega}_k) \in \mathscr{W}_k$$

$$\tau_k(w) = \sum_{i=1}^{k} \eta_i(\hat{\omega}_i) \quad (\text{with } a + \infty := \infty, a \in \mathbb{R}_+)$$

$$\tau_\infty(w) = \lim_{k \to \infty} \tau_k(w).$$

Let Δ_∞ be a further isolated point. An $E \cup \{\Delta_\infty\}$-valued function X on $\mathbb{R}_+ \times \mathscr{W}$ is now defined by

$$X_t(w) = \begin{cases} \hat{x}_0(\hat{\omega}_1) = \hat{\omega}_1(0), & \text{if } t = 0 \\ \hat{x}_{t - \tau_i(\omega)}(\hat{\omega}_{i+1}), & \text{if } \tau_i(\omega) < t \leqslant \tau_{i+1}(w) \\ \Delta_\infty, & \text{if } t > \tau_\infty(\omega). \end{cases} \tag{54.2}$$

Note that the function $t \to X_t(\omega)$ is right continuous with left-hand limits except when $t = \tau_i(w)$ for some i, at which points it is generally neither left nor right continuous, although both left and right limits exist.

An *intervention strategy* is a sequence $\mathscr{S} = \{(R_n, S_n),\ n = 1, 2, \ldots\}$, where the following conditions are satisfied.

(54.3) $R_n: \mathscr{W}_n \to E \cup \{\Delta\}$ is a universally measurable function such that

$$R_n(w_n)\begin{cases} = \Delta, & \text{if } \tau_n(w_n) = \infty \\ \in K(\hat{x}_{\eta_n(\hat{\omega}_n)}(\hat{\omega}_n)), & \text{otherwise.} \end{cases}$$

(54.4) $S_1: \Omega \to \mathbb{R}_+ \cup \{\infty\}$ is an $\mathfrak{F}_t$-stopping time such that $S_1(\omega) > 0$ for all $\omega \in \Omega$. For $m = 2, 3, \ldots, S_n: \mathscr{W}_{n-1} \times \Omega \to \mathbb{R}_+ \cup \{\infty\}$ is a measurable function such that $S_n(w_{n-1}, \cdot)$ is an $\mathfrak{F}_t$-stopping time for all $w_{n-1} \in \mathscr{W}_{n-1}$. If $\tau_{n-1}(w_{n-1}) = \infty$ then $S_n(w_{n-1}, \omega) = 0$ for all $w \in \Omega$; otherwise, $S_n(w_{n-1}, \omega) > 0$ for all $\omega \in \Omega$.

A strategy $\mathscr{S} = (S_n, R_n)$ can be used to construct sample paths of the controlled process in the following way: S_1 defines a map $\chi_1: \Omega \to \hat{\Omega}_1$ by

$$\chi_1\omega(t) = \begin{cases} \omega(t), & t \leqslant S_1(\omega) \\ \Delta, & t > S_1(\omega). \end{cases}$$

For $n \geqslant 2$, given $w_n \in \mathscr{W}_n$, $\mathscr{S}$ determines $\chi_n: \Omega \to \hat{\Omega}$ by the following rule:

If $R_n(w_n) = \Delta$, then $\chi_{n+1}\omega(t) = \Delta$ for all $t \in \mathbb{R}_+$; if $R_n(w_n) \neq \Delta$, then

$$\text{if } S_{n+1}(w_n, \omega) = \infty, \quad \text{then } \chi_{n+1}\omega = \omega;$$

otherwise

$$\chi_{n+1}\omega(t) = \begin{cases} \omega(t) & t \leqslant S_{n+1}(w_n, \omega) \\ \Delta, & t > S_{n+1}(w_n, \omega). \end{cases}$$

Recall that P_x is a measure on Ω for each $x \in E$. Given a strategy $\{(R_n, S_n)\} \in \mathscr{S}$ and $x \in E$ define a measure μ_x^1 on $\hat{\Omega}_1$ by

$$\mu_x^1(F) = (P_x \chi_1^{-1})(F) = P_x\{\omega \in \Omega: \chi_1(\omega) \in F\}, \quad F \in \hat{\mathfrak{F}}^1$$

and for $n = 2, 3\ldots$ and each $w_{n-1} \in \mathscr{W}_{n-1}$ a conditional measure on

$\hat{\Omega}_n$ by

$$\mu_x^n(F; w_{n-1}) = P_{R_{n-1}(w_{n-1})}\chi_{n-1}^{-1}.$$

There is then a unique probability measure $\mathbb{P}_x^{\mathscr{S}}$ on $\mathscr{W}$ which is consistent with this family of conditional measures, i.e. such that if $F \in \mathfrak{G}_n$ with $F = F_1 \times F_2 \times \cdots \times F_n, F_k \in \hat{\mathfrak{F}}_k$, then

$$\mathbb{P}_x^{\mathscr{S}}(F) = \int_{F_1} \cdots \int_{F_{n-2}} \int_{F_{n-1}} \mu_x^n(F_n; (\hat{\omega}_1, \ldots, \hat{\omega}_{n-1}))$$
$$\times \mu_x^{n-1}(\mathrm{d}\hat{\omega}_{n-1}; (\hat{\omega}_1, \ldots, \hat{\omega}_{n-2})) \cdots \mu_x^1(\mathrm{d}\hat{\omega}_1).$$

(54.5) *Definition.* A sequence $\mathscr{S} = \{(S_n, R_n): n = 1, 2, \ldots\}$ is an *admissible strategy* if (S_n, R_n) satisfy conditions (54.3) and (54.4) and $\mathbb{P}_x^{\mathscr{S}}\{w: \tau_\infty(w) = \infty\} = 1$ for all $x \in E$. We denote by $\mathbb{S}$ the set of admissible strategies, and by $\mathbb{S}_n$ the set of admissible strategies such that $\tau_{n+1} = \infty$ a.s., i.e. those for which *at most n* interventions are taken.

From the above construction we can see that S_n determines the nth intervention time, and R_n the starting point for the $(n+1)$st 'piece' of the process, i.e. the intervention taken at the nth intervention time. Indeed, if X_t is the sample path defined by (54.2) then $X_0(w) = x, X_t(w) = \hat{x}_t(\hat{\omega}_1)$ for $t \leqslant \tau_1(\hat{\omega}_1)$ and $\tau_1(\hat{\omega}_1) = S_1(\hat{\omega}_1)$. If $\tau_1 = \infty$ then necessarily $R_1(\omega_1) = \Delta$ and hence $\chi_n \omega \equiv \Delta$ for all $n \geqslant 2$. If $\tau_1 < \infty$, then $R_1(\hat{\omega}_1) \in K(x_{\tau_1}(\hat{\omega}_1))$ and $P_x^{\mathscr{S}}[\hat{\omega}_2(0) = R_1(\hat{\omega}_1)] = 1$. Now $X_{\tau_1 + t}(w) = \hat{\omega}_2(t)$ for $t \in]0, \tau_2]$, so that

$$X_{\tau_1 +}(w) = \lim_{t \downarrow 0} X_{\tau_1 + t}(w) = R_1(\hat{\omega}_1).$$

The construction continues in a similar manner.

Fix $\delta > 0$ and suppose c, K and I satisfy Assumptions (54.1). The *cost* of a strategy $S \in \mathbb{S}$ starting at $x \in E$ is

$$J_x(\mathscr{S}) = \mathbb{E}_x^{\mathscr{S}} \left\{ \int_0^\infty \mathrm{e}^{-\delta t} l(X_t) \mathrm{d}t + \sum_{k=1}^\infty \mathrm{e}^{-\delta \tau_k} c(X_{\tau_k}, X_{\tau_k +}) \right\}, \tag{54.6}$$

where $\mathbb{E}_x^{\mathscr{S}}$ denotes expectation with respect to the measure $\mathbb{P}_x^{\mathscr{S}}$ on W, and $X_t(w)$ is defined as above. A strategy $\mathscr{S}^0 \in \mathbb{S}$ is *optimal* if $J_x(\mathscr{S}^0) = v(x)$ for all $x \in E$, where

$$v(x) = \inf_{\mathscr{S} \in \mathbb{S}} J_x(\mathscr{S}). \tag{54.7}$$

Note that $J_x(\mathscr{S})$ is not necessarily finite, even if c is bounded, because there is no guarantee that the sum on the right of (54.6) converges. Let us define $\mathbb{S}^* = \{\mathscr{S}\in\mathbb{S}: J_x(\mathscr{S}) < \infty$ for all $x\in E\}$. There is always at least one strategy with finite cost, namely the 'no-intervention' strategy $\mathscr{S}_{\mathrm{NI}}$ with $\tau_1 \equiv \infty$. This has cost

$$J_x(\mathscr{S}_{\mathrm{NI}}) = \mathbb{E}_x^{\mathscr{S}_{\mathrm{NI}}}\left\{\int_0^\infty \mathrm{e}^{-\delta t} l(X_t)\mathrm{d}t\right\},$$

which, by Assumption (54.1) (5), is bounded by c_2/δ and coincides with

(54.8)
$$v_0(x) := \mathbb{E}_x\left\{\int_0^\infty \mathrm{e}^{-\delta t} l(x_t)\mathrm{d}t\right\},$$

where, in the latter expression, (x_t) is the ordinary PDP defined on (Ω, P_x) and, as usual, $\mathbb{E}_x$ denotes integration with respect to P_x. Thus $v(x)$ is bounded since $v(x) \leqslant v_0(x)$. By the strong Markov property and standard calculations, for $t \geqslant 0$,

(54.9)
$$v_0(x) = \mathbb{E}_x\left\{\int_0^{T_1\wedge t} \mathrm{e}^{-\delta s} l(x_s)\mathrm{d}s + \mathrm{e}^{-\delta(T_1\wedge t)} v_0(T_1\wedge t)\right\}$$
$$= \mathrm{e}^{-(\Lambda(t,x)+\delta t)} v_0(\phi(t,x)) + \int_0^t \mathrm{e}^{-(\Lambda(s,x)+\delta s)}(l(s) + \lambda(s)Qv_0(s))\mathrm{d}s,$$

where, as in the previous section, we write $l(s) = l(\phi(s,x))$, etc. It follows from (54.9) that $v_0 \in \mathscr{C}_b$.

Analysis of optimal intervention strategies hinges on the properties of the operator M defined for $\psi\in B(E)$ by

$$M\psi(s) = \inf_{y\in K(x)} [c(x,y) + \psi(y)].$$

We will assume that M maps $B(E)$ into itself. Assumption (54.1) (4) assures that $|M\psi(x)| \leqslant c_1 + \|\psi\|$, so that $M\psi$ is bounded, but it is not obvious that $M\psi$ is a measurable function of x. The following lemma gives some sufficient conditions for this. In many applications it can be checked directly without recourse to these conditions.

(54.10) *Proposition.* Suppose c and K satisfy Assumptions (54.1). Then for $\psi\in B(E)$:

1. $M\psi$ is measurable if $K(x) = \{y_1(x), y_2(x), \ldots\}$ for some sequence of measurable functions $y_n: E\cup\Gamma \to E$;

2. $M\psi$ is *continuous* if for each $y \in E$ the set $I(y) = \{x : y \in K(x)\}$ is relatively open in E.

Proof

1. In this case

$$M\psi(x) = \inf_{n \in \mathbb{N}} c(x, y_n(x)) + \psi(y_n(x)).$$

The map $x \to c(x, y_n(x)) + \psi(y_n(x))$ is measurable, as the composition of two measurable maps, and the countable infimum preserves measurability.

2. It suffices to show that for each $a \in \mathbb{R}$ the set $\{x : M\psi(x) < a\}$ is open. But

$$\{x : M\psi(x) < a\} = \bigcup_{y \in E} \{x : c(x, y) + \psi(y) < a\} \cap I(y),$$

and each set in the intersection is open under the stated conditions. The result follows. □

Condition 1 is what is needed in applications such as the example in (21.15) where the intervention set is 'small' for each x (in the example, one or two points). As to condition 2, it is worth remarking that this is satisfied if c is continuous and $K(x) = K$ does not depend on x, because then $I(y) = E$ or $\varnothing$. Thus M has a 'smoothing' effect $M : B(E) \to C(E)$ in this case.

Further properties of M can be obtained if extra assumptions on K and c are introduced. Denote by $L(E)$ the set of Lipschitz continuous functions, on E, i.e. real valued functions f such that $|f(x) - f(x')| \leqslant \kappa_f |x - x'|$ for some constant κ_f. A *set-valued* map such as $K(x)$ is continuous if $d_{\mathrm{H}}(K(x), K(x')) \to 0$ as $x' \to x$ and Lipschitz continuous if $d_{\mathrm{H}}(K(x), K(x')) \leqslant \kappa |x - x'|$, where d_{H} denotes the Hausdorff distance

$$d_{\mathrm{H}}(A, B) := \max \left\{ \sup_{a \in A} d(a, B), \sup_{b \in B} d(b, A) \right\}.$$

Lipschitz continuity implies that for any $y' \in K(x')$ there exists $y \in K(x)$ such that $|y - y'| \leqslant \kappa |x - x'|$.

(54.11) *Proposition.* Suppose that Assumptions (54.1) hold. Then

1. if $K(x)$ is continuous then M maps $C_b(E)$ into $C_b(E)$;

2. if $c(x, y)$ and $K(x)$ are Lipschitz continuous then M maps $L(E)$ into $L(E)$;
3. if $K(x) \equiv K$ does not depend on x and c is Lipschitz continuous, then M maps $B(E)$ into $L(E)$.

Proof. We will prove 2, since the proof of 1 is similar. For $x, x' \in E$

$$M\psi(x) - M\psi(x') = \inf_{y \in K(x)} [c(x, y) + \psi(y)] - \inf_{y' \in K(x')} [c(x', y') + \psi(y')].$$

Choose $\varepsilon > 0$ and let y' be ε-optimal in the second infimum. Then for any $y \in K(x)$

$$M\psi(x) - M\psi(x') \leqslant c(x, y) + \psi(y) - c(x', y') - \psi(y') + \varepsilon. \tag{54.12}$$

Choosing y such that $|y - y'| \leqslant \kappa |x - x'|$ and using the Lipschitz continuity of c we obtain

$$\begin{aligned} M\psi(x) - M\psi(x') &\leqslant \kappa_c(|x - x'| + |y - y'|) + \kappa_\psi |y - y'| + \varepsilon \\ &\leqslant ((\kappa_c + \kappa_\psi)\kappa + \kappa_c)|x - x'| + \varepsilon. \end{aligned}$$

The same inequality holds with x and x' interchanged, establishing 2 since ε is arbitrary. For 3, note that if $K(x) = K$ for all x then it is permissible to choose $y = y'$ in (54.12), giving

$$\begin{aligned} M\psi(x) - M\psi(x') &\leqslant c(x, y') - c(x', y') + \varepsilon \\ &\leqslant \kappa_c |x - x'| + \varepsilon, \end{aligned}$$

from which the result follows as before. □

The following corollary is proved in the same way. Recall the definition of the class of functions $\mathscr{L}$ from Definition (31.1).

(54.13) *Corollary.* Suppose that $K(x) = K$ is constant and that for each $x \in E$ there exist $C, \varepsilon > 0$ such that for all $y \in K$

$$|c(\phi(t, x), y) - c(x, y)| \leqslant Ct, \quad 0 \leqslant t < \varepsilon.$$

Then M maps $B(E)$ into $\mathscr{L}$.

The quantity $M\psi(x)$ represents the minimal cost of starting at x, moving immediately to $y \in K(x)$ and paying $\psi(y)$. Part 3 of Proposition (54.11) implies in particular that if K is constant then M maps $\mathscr{C}_b$ into itself; however, if $K(x)$ is not constant then in general M does *not* map $\mathscr{C}_b$ into itself, and this is why we have to consider global continuity, i.e. the function spaces $C_b(E), L(E)$.

We now introduce a further operator $\mathscr{G}$ on $B(E)$ giving the value of the following optimal stopping problem for the PDP (x_t):

$$\mathscr{G}\psi(x) := \inf_{\tau\in\Sigma} \mathbb{E}_x\left(\int_0^\tau e^{-\delta t} l(x_t)dt + e^{-\delta\tau} M\psi(x_\tau) \right). \tag{54.14}$$

(54.15) *Proposition*

1. Suppose that Assumptions (54.1) hold and that K is constant. Then $\mathscr{G}$ maps $B(E)$ into $\mathscr{C}_b$.
2. Suppose in addition that the following conditions hold:

(54.16) (a) $\lambda, l \in C_b(E\cup\Gamma)$,
(b) $Qw \in C_b(E\cup\Gamma)$ for $w \in C_b(E\cup\Gamma)$,
(c) the map $x \to t_*(x)$ is continuous on E,
(d) the map $x \to K(x)$ is continuous.

Then $\mathscr{G}$ maps $C_b(E)$ into $C_b(E)$.

Proof

1. $M\psi \in L(E) \subset \mathscr{C}_b$ for $\psi \in B(E)$, from part 3 of Proposition (54.11). The result now follows from part 4 of Theorem (51.17).
2. Conditions (54.16) (a)–(c) coincide with Assumptions (53.37) and imply from Theorem (53.40) that the optimal stopping value function is in $C_b(E)$ if the terminal cost function is in $C_b(E)$. But, under (54.16(d)), $M\psi \in C_b(E)$ for $\psi \in C_b(E)$. □

(54.17) *Proposition.* Under Assumptions (54.1) and (54.16) the no-intervention cost v_0 given by (54.8) is in $C_b(E)$.

Proof. We know that v_0 is bounded by c_2/δ, and continuity follows by the same argument as in the proof of Theorem (53.40). □

In the following discussion we will suppose that Assumptions (54.1) and (54.16) hold. Take $\psi = v_0$ in (54.14), where v_0 is given by (54.8). Then it is clear that

$$v_1(x) := \mathscr{G}v_0(x) = \inf_{\mathscr{S}\in\mathbb{S}_1} J_x(\mathscr{S}),$$

i.e. $v_1(x)$ is the minimal cost if at most one intervention is made. Indeed, if we intervene at τ then the process can be moved to any point $y \in K(x_\tau)$ at a cost of $c(x_\tau, y)$; from y, since no further interventions are allowed, the remaining cost is $e^{-\delta t}v_0(y)$, by the strong Markov property. Thus the choice of y dictated by M is optimal given that the intervention is made at τ, and the remaining problem is simply to choose the optimal intervention *time*. One possible choice is $\tau = \infty$, i.e. not to intervene at all, in which case we recover the cost $v_0(x)$. Hence $v_1(x) \leqslant v_0(x)$.

It follows directly from the definitions that both M and $\mathcal{G}$ are *monotone operators*: if $\psi_1 \leqslant \psi_2$ (i.e. $\psi_1(x) \leqslant \psi_2(x)$ for all $x \in E$) then $M\psi_1 \leqslant M\psi_2$ and $\mathcal{G}\psi_1 \leqslant \mathcal{G}\psi_2$. The latter property implies that if we define functions v_n recursively by $v_n := \mathcal{G}v_{n-1}, n = 1, 2, \ldots$, so that $v_n = \mathcal{G}^n v_0$, then v_n is a decreasing sequence. We have already seen that $v_1 \leqslant v_0$, and if $v_n \leqslant v_{n-1}$ for some n then $v_{n+1}(x) - v_n(x) = \mathcal{G}v_n(x) - \mathcal{G}v_{n-1}(x) \leqslant 0$ by monotonicity. A straightforward induction argument shows that

$$v_n(x) = \inf_{\mathcal{S} \in \mathbb{S}_n} J_x(\mathcal{S}),$$

the minimal cost over strategies involving at most n interventions.

(54.18) *Proposition*

$$\lim_{n \to \infty} v_n(x) = v(x) = \inf_{\mathcal{S} \in \mathbb{S}} J_x(\mathcal{S}).$$

Proof. The limit $w(x) := \lim_{n \to \infty} v_n(x)$ certainly exists, since v_n is monotone and bounded. Since $\mathbb{S}_n \subset \mathbb{S}$, $v_n(x) \geqslant v(x)$ and hence $w(x) \geqslant v(x)$. To get the reverse inequality we have only to show that the cost of any strategy $\mathcal{S} \in \mathbb{S}^*$ can be approximated by that of some strategy $\mathcal{S}_n \in \mathbb{S}_n$ (we can of course discard strategies in $\mathbb{S} \backslash \mathbb{S}^*$ for which the cost is infinite.) For any $\mathcal{S} \in \mathbb{S}^*$ the intervention times τ_n satisfy $\tau_n \to \infty$ a.s.; this follows from Assumption (54.1) (4) positing the existence of a minimum intervention cost c_0. Let $\mathcal{S}_n$ be the strategy that follows $\mathcal{S}$ up to and including time τ_n and then takes no further interventions. Then

$$J_x(\mathcal{S}) - J_x(\mathcal{S}_n) = \mathbb{E}_x^{\mathcal{S}}\left(\int_{\tau_n}^{\infty} \mathrm{e}^{-\delta t}(l(X_t) - l(\hat{x}_{t-\tau_n}(\hat{\omega}_{n+1})))\mathrm{d}t \right.$$
$$\left. + \sum_{k \geqslant n+1} \mathrm{e}^{-\delta \tau_k} c(X_{\tau_k}, X_{\tau_{k+}}) \right),$$

so that

$$|J_x(\mathcal{S}) - J_x(\mathcal{S}_n)| \leqslant \mathbb{E}_x^{\mathcal{S}} \left\{ \frac{2\|l\|}{\delta} \mathrm{e}^{-\delta \tau_n} + \sum_{k \geqslant n+1} \mathrm{e}^{-\delta \tau_k} c(X_{\tau_k}, X_{\tau_{k+}}) \right\}$$

and the right-hand side converges to 0 as $n \to \infty$ by dominated convergence (the sum is convergent for $n = 0$ with finite expectation since $\mathcal{S} \in \mathbb{S}^*$). We have shown that

$$v(x) = \inf_{\mathcal{S} \in \mathbb{S}^*} J_x(\mathcal{S}) = \inf_{\mathcal{S} \in \bigcup_n \mathbb{S}_n} J_x(\mathcal{S})$$

which implies that $w(x) \leqslant v(x)$. □

We can now state the first important characterization of the value function v of the impulse control problem.

(54.19) *Theorem.* Suppose Assumptions (54.1) are satisfied. Then the value function v is the unique fixed point in $B(E)$ of the operator $\mathcal{G}$ defined at (54.14), i.e. the unique solution if the equation $w = \mathcal{G}w$.

Proof. We know that $v_n \downarrow v$, where v_0 is given by (54.8) and

$$v_{n+1}(x) = \inf_{\tau \in \Sigma} \mathbb{E}_x \left\{ \int_0^{\tau} \mathrm{e}^{-\delta t} l(x_t) \mathrm{d}t + \mathrm{e}^{-\delta \tau} M v_n(x_\tau) \right\}. \tag{54.20}$$

Now

$$\begin{aligned} Mv(x) &= \inf_{y \in K(x)} [c(x, y) + v(y)] \\ &= \inf_{y \in K(x)} \inf_{n \in \mathbb{N}} [c(x, y) + v_n(y)] \\ &= \inf_{n \in \mathbb{N}} \inf_{y \in K(x)} [c(x, y) + v_n(y)] \\ &= \inf_{n \in \mathbb{N}} M v_n(x). \end{aligned}$$

The same argument using the bounded convergence theorem shows that if ψ_n is a decreasing sequence of bounded functions and $\psi_n \downarrow \psi_\infty$ then $w_n \downarrow w_\infty$, where

$$w_n(x) := \inf_{\tau \in \Sigma_1} \mathbb{E}_x \left\{ \int_0^{\tau} \mathrm{e}^{-\delta t} l(x_t) \mathrm{d}t + \mathrm{e}^{-\delta \tau} \psi_n(x_\tau) \right\}$$

for $0 \leqslant n \leqslant \infty$. Taking $\psi_n = M v_n$ and taking the limit as $n \to \infty$ on both sides of (54.20) we conclude that v satisfies $v = \mathcal{G} v$.

To obtain uniqueness we use a general result on concave operators due to Krasnoselskii (1964) and Zabczyk (1983). For completeness, a proof of this result is given at the end of this section. It is easy to check that both M and $\mathscr{G}$ *are* concave operators: for $\theta \in [0, 1]$ we have

$$M(\theta\psi_1 + (1-\theta)\psi_2) = \inf_{y \in K(x)} c(x, y) + \theta\psi_1(y) + (1-\theta)\psi_2(y)$$
$$= \inf_{y \in K(x)} \theta(c(x, y) + \psi_1(y)) + (1-\theta)(c(x, y) + \psi_2(y))$$
$$\geqslant \theta M\psi_1(x) + (1-\theta)M\psi_2(x).$$

An exactly similar argument applies to $\mathscr{G}$.

(54.21) *Lemma.* Suppose that H is a non-decreasing and concave operator on $B(E)_+ = \{f \in B(E) : f \geqslant 0\}$, and define $\mathscr{P}_H = \{f \in B(E)_+ : f \leqslant \kappa HO \text{ for some } k \in \mathbb{R}_+\}$. Then there exists at most one solution in $\mathscr{P}_H$ of the equation $Hw = w$.

To apply the lemma, take $H = \mathscr{G}$ and note that if $w \in B(E)_+$ and $w = \mathscr{G}w$, then $w \leqslant v_0$. Now for any $\tau \in \Sigma$ we have

$$v_0(x) = \mathbb{E}_x\left\{\int_0^\tau e^{-\delta t} l(x_t)\mathrm{d}t + \int_\tau^\infty e^{-\delta t} l(x_t)\mathrm{d}t\right\}$$
$$\leqslant \mathbb{E}_x\left\{\int_0^\tau e^{-\delta t} l(x_t)\mathrm{d}t + \frac{\|l\|}{\delta} e^{-\delta\tau}\right\}$$
$$\leqslant \left(\frac{\|l\|}{c_0\delta} \vee 1\right)\mathbb{E}_x\left\{\int_0^\tau e^{-\delta t} l(x_t)\mathrm{d}t + c_0 e^{-\delta\tau}\right\}$$
$$\leqslant \left(\frac{\|l\|}{c_0\delta} \vee 1\right)\mathscr{G}0(x)$$

from Assumption (54.1) (4).

Thus if $w \in B(E)_+$ and $w = \mathscr{G}w$ then $w \in \mathscr{P}_{\mathscr{G}}$; it therefore follows from the lemma that v is the unique fixed point of $\mathscr{G}$ in $B(E)_+$. □

The 'uniqueness' statement of Lemma (54.21) is complemented in Zabczyk (1983) by an 'existence' result which in fact implies that under the conditions of Proposition (54.15) convergence of v_n to v is uniform and hence $v \in C_b(E)$ (Gątarek, 1992).

Theorem (54.19) shows that the value function v is, in principle, determined by the equation

$$v(x) = \inf_{\tau \in \Sigma} \mathbb{E}_x\left\{\int_0^\tau e^{-\delta t} l(x_t)\mathrm{d}t + e^{-\delta\tau} Mv(x_\tau)\right\}. \tag{54.22}$$

If Mv were replaced by a known function ψ then (54.22) would be an optimal stopping problem similar to those we have already studied. However, determining Mv is of course part of the problem, and for this reason (54.22) is sometimes called an *implicit optimal stopping problem.* In fact we already have – again in principle – a way of solving it, namely by computing the sequence v_n, using (54.20) which is an *explicit* optimal stopping problem determining v_{n+1} in terms of v_n, and then recognizing that $v = \lim_{n\to\infty} v_n$. Once v is determined, it is easy to see what optimal or ε-optimal strategies consist of: the state space divided into a continuation region C on which $v < Mv$ and an intervention region S on which $v = Mv$. For $x \in S$ the optimal intervention is to move to a point $y \in K(x)$ (assuming it exists) satisfying $c(x, y) + v(y) = Mv(x)$, and this y will be in C. The PDP then restarts from y and the next intervention time is the first time the restarted process enters S; and so on.

Formalizing the above discussion is complicated by the fact that (as in the optimal stopping problem) in many cases no optimal strategy exists; Example (54.29) below indicates why. However, ε-optimal strategies can be constructed using Theorem (51.20). Proposition 7.33 of Bertsekas and Shreve (1978) shows that if v is continuous and Assumptions (54.1) (1) and (2) hold then there is a Borel selector for Mv, i.e. a Borel measurable function $\eta: E \to Y$ such that, for all $x \in E, \eta(x) \in K(x)$ and $Mv(x) = c(x, \eta(x)) + v(\eta(x))$. In any case there is for each $\varepsilon > 0$ a Borel function $\eta_\varepsilon: E \to E$ such that $\eta_\varepsilon(x) \in K(x)$ and $c(x, \eta_\varepsilon(x)) + v(\eta_\varepsilon(x)) \leqslant Mv(x) + \varepsilon$. From Theorem (51.20) we know that the time $\tau_1 := U_\varepsilon(x) := \inf\{t: Mv(x_t) - v(x_t) < \varepsilon\}$ is ε-optimal for the stopping problem (5.59), i.e. satisfies

$$\mathbb{E}_x\left\{\int_0^{\tau_1} e^{-\delta t} l(x_t)dt + e^{-\delta\tau_1} Mv(x_{\tau_1})\right\} \leqslant v(x) + \varepsilon.$$

Thus

$$\mathbb{E}_x\left\{\int_0^{\tau_1} e^{-\delta t} l(x_t)dt + e^{-\delta\tau_1} c(x_{\tau_1}, \eta_\varepsilon(x_{\tau_1})) + e^{-\delta\tau_1} v(\eta_\varepsilon(x_{\tau_1}))\right\}$$
$$\leqslant v(x) + \varepsilon + \varepsilon e^{-\delta\tau_1}.$$

In the notation of (54.3) and (54.4) we define a strategy $\mathcal{S}_\varepsilon$ by taking $S_1 = \tau_1$ and $R_1(\omega) = \eta_\varepsilon(x_{\tau_1}(\omega))$ and, for $n \geqslant 2, S_n = U_{\varepsilon 2^{-n+1}}(X_{\tau_{n-1+}}(w_{n-1}))(\omega)$, $R_n(w_n) = \eta_{\varepsilon 2^{-n+1}}(X_{\tau_n}(w_n))$. Applying the above inequality recursively we find that

$$\mathbb{E}_x\left\{\int_0^{\tau_n} e^{-\delta t} l(X_t)dt + e^{-\delta\tau_1}c(X_{\tau_1}, X_{\tau_1+}) + \cdots + e^{-\tau_n}Mv(X_{\tau_n})\right\}$$

$$\leqslant v(x) + \varepsilon + \mathbb{E}_x\left\{\frac{3}{2}\varepsilon e^{-\delta\tau_1} + \cdots + \frac{3}{2^n}\varepsilon e^{-\delta\tau_n}\right\}$$

$$\leqslant v(x) + \varepsilon\left(1 + \frac{3}{2} + \cdots + \frac{3}{2^n}\right) \leqslant v(x) + 4\varepsilon.$$

Thus $J_x(\mathscr{S}_\varepsilon) \leqslant v(x) + 4\varepsilon$, i.e. $\mathscr{S}_\varepsilon$ is 4ε-optimal.

In the remainder of this section we will apply the methods already developed for the optimal stopping problem to obtain more explicit characterizations of the value function v of the impulse control problem, with a view to obtaining insight into the problem as well as an avenue to computational techniques. We first examine the approach taken in section 51, i.e. analysis of the 'single-jump' operator L, and then go on to consider a variational inequality formulation along the lines of section 53. In this application L is defined by (53.39) and, from Corollary (51.9), L maps $B(E) \times \mathscr{C}_b$ into $\mathscr{C}_b$. Now define

$$\hat{L}w(x) := L(w, Mw)(x).$$

This corresponds to the minimal cost if the process stops at the first jump or intervention time with terminal cost w, an intervention being taken at a chosen time t if $t < T_1$ (or no intervention being taken at all). The following result extends Theorem (51.17) to the impulse control setting.

(54.23) *Theorem*

1. The value function v is the biggest solution of the functional equation

(54.24)
$$\left.\begin{aligned} w &= \hat{L}w \\ w &\leqslant v_0. \end{aligned}\right\}$$

2. Define $w_0 = v_0$ and $w_{n+1} = \hat{L}w_n$ for $n = 0, 1, \ldots$. Then $w_n \downarrow v$ as $n \to \infty$.

Proof. $\hat{L}$ is a monotone operator and it is easy to see from the definition that $w_1 \leqslant w_0$. If $w_n \leqslant w_{n-1}$ for some n then $w_{n+1} - w_n =$

$Lw_n - \hat{L}w_{n-1} \leqslant 0$; hence $w_{n+1} \leqslant w_n$ for all n. Let $w := \lim_{n\to\infty} \downarrow w_n$. The argument given in the proof of Theorem (51.17) shows that w satisfies (54.24), while if (54.24) is satisfied by some function w' then $w' = \hat{L}w' \leqslant \hat{L}v_0 = w_1$ by monotonicity. The same argument shows that if $w' \leqslant w_n$ then $w' \leqslant w_{n+1}$; hence $w' \leqslant w$ and w is therefore the biggest solution of (54.24). It remains to show that $w = v$. We will show that w satisfies $w = \mathcal{G}w$, which implies that $w = v$ and hence that $w_n \downarrow v$ since v is the unique fixed point of $\mathcal{G}$.

1. $w \leqslant \mathcal{G}w$. Consider the optimal stopping problem with terminal cost Mw. Its value is $\mathcal{G}w$ and from Theorem (52.27) we know that $u_n \downarrow \mathcal{G}w$, where $u_0 = Mw, u_{n+1} = L(u_n, Mw)$. We claim that $w \leqslant u_n$ for all n; then $w \leqslant \mathcal{G}w$. Indeed, we know that $w = L(w, Mw)$ and therefore $w \leqslant Mw = u_0$, while if $w \leqslant u_n$ for some n then by monotonicity $w = L(w, Mw) \leqslant L(u_n, Mu_n) = \hat{L}u_n = u_{n+1}$. Hence $w \leqslant u_n$ for all n, as claimed.
2. $w \geqslant \mathcal{G}w$. The same argument as used in the proof of Theorem (54.19) shows that

$$\begin{aligned}\mathcal{G}w(x) &= \inf_{\tau\in\Sigma} \mathbb{E}_x \left\{ \int_0^\tau \mathrm{e}^{-\delta t} l(x_t)\,\mathrm{d}t + \mathrm{e}^{-\delta\tau} M\left(\inf_{n\in\mathbb{N}} w_n \right)(x_\tau) \right\} \\ &= \inf_{n\in\mathbb{N}} \mathcal{G}w_n(x).\end{aligned}$$

Now $\mathcal{G}w_n(x) \leqslant \hat{L}w_n(x)$ since in computing $\mathcal{G}w_n$ and $\hat{L}w_n$ we are taking the infimum over stopping times τ in Σ and $\Sigma_1 \subset \Sigma$, respectively. Hence $\mathcal{G}w_n(x) \leqslant w_{n+1} (= \hat{L}w_n)$ and therefore

$$\mathcal{G}w = \inf_n \mathcal{G}w_n(x) \leqslant \inf_n w_{n+1} = w.$$

This completes the proof. □

The main point of Theorem (54.23) lies in part 2 which provides a computational scheme for computing v, namely recursive application of the operator $\hat{L}$. (Some details as to exactly how this is carried out are provided in section 55). However, in order to get started we apparently need to know v_0, and computing this is usually a non-trivial problem in its own right. The next result shows that in fact any function that majorizes v_0 can be used as a starting point. A possible candidate is the constant function $v_0'(x) = \| l \| / \delta$, although more efficient choices might be found in specific cases.

(24.25) *Proposition.* Take $v'_0(x) \geqslant v_0, v'_0 \in C_b(E)$ and define $w'_0(x) = \hat{L}^n v'_0$. Then $w'_0 \downarrow v$.

Proof. Since v_0 and v'_0 are bounded there is a constant a such that $v_0 \leqslant v'_0 \leqslant v_0 + a$. Let us define $v''_0 = v_0 + a$ and $v''_n(x) = \hat{L}^n v''_0$. We claim:

1. $w_n \leqslant w'_n \leqslant w''_n$;
2. $w''_n(x) \leqslant v_0(x) + a\mathbb{E}_x(e^{-\delta T_n})$;
3. $w''_n \downarrow w''$ for some $w'' \in B(E)$; and
4. w'' satisfies $w'' = \hat{L}w''$.

Given these, 1 and 2 imply that $v \leqslant w''$. Since $T_n \to \infty$ a.s., 3 implies that $w'' \leqslant v_0$ which, together with 4 and part 1 of Theorem (54.23), shows that $w'' \leqslant v$. (v is the biggest solution of Lemma (54.21).) Hence $w'' = v$ and it follows from 1 that $\lim_{n\to\infty} w'_0 = v$.

Of the four properties above, 1 and 3 are immediate, while 4 uses by now familiar arguments. For 2, we use an induction argument. It is certainly true when $n = 0$, and if it holds for $n = k$ then, denoting $\chi_n(x) = \mathbb{E}_x e^{-\delta T_n}$, we have

$$\begin{aligned} w''_{k+1}(t) &= \hat{L}w''_k(x) \\ &\leqslant Kw''_k(x) \\ &= \mathbb{E}_x\left\{\int_0^{T_1} e^{-\delta t} l(x_t)dt + e^{-\delta T_1} w''_k(x_{T_1})\right\} \\ &\leqslant \mathbb{E}_x\left\{\int_0^{T_1} e^{-\delta t} l(x_t)dt + e^{-\delta T_1}(v_0(x_{T_1}) + a\chi_k(x_{T_1}))\right\} \\ &= v_0(x) + a\mathbb{E}_x[e^{-\delta T_1}\chi_k(x_{T_1})] \\ &= v_0(x) + a\chi_{k+1}(x), \end{aligned}$$

where the second inequality uses the induction hypothesis and the last equality follows from time homogeneity and the strong Markov property. This completes the proof. □

We now turn to the characterization of the value function v as the solution of a *quasi-variational inequality* (QVI). The idea is straightforward: referring to Proposition (53.35) we see that the variational inequality corresponding to the optimal stopping problem $w = \mathcal{G}\psi$

of (54.14) is

$$\left.\begin{aligned} \mathfrak{A}w - \delta w + l \geqslant 0 \\ M\psi - w \geqslant 0 \\ (\mathfrak{A}w - \delta w + l)(M\psi - w) = 0 \end{aligned}\right\} \quad \text{in } E$$

with boundary condition $w(x) = Qw(x) \wedge M\psi(x), x \in \Gamma$. Since v is a fixed point of $\mathscr{G}$, i.e. satisfies $v = \mathscr{G}v$, it should satisfy the set of equations

(54.26)
$$\left.\begin{aligned} \mathfrak{U}w - \delta w + l \geqslant 0 \\ Mw - w \geqslant 0 \\ (\mathfrak{A}w - \delta w + l)(Mw - w) = 0 \end{aligned}\right\} \quad \text{in } E$$

with boundary condition

(54.27)
$$w(x) = Qw(x) \wedge Mw(x), \quad x \in \Gamma.$$

This is known as a quasi-variational inequality (QVI)[1]. The main problem in justifying (54.26) and (54.27) is to decide in what class of functions w to look for a solution. There is one case in which the appropriate result follows directly from the optimal stopping results were already have.

(54.28) *Theorem.* Suppose that c is Lipschitz continuous, that Assumptions (54.1) hold and that the intervention set K does not depend on the intervention point x. Then the value function v is in $\mathscr{L}_b$ and is the unique solution in $\mathscr{L}_b$ of the QVI (54.26) and (54.27).

Proof. We know that v is bounded, i.e. $v \in B(E)$, and hence, by part 3 of Proposition (54.11), $Mv \in L(E)$ so that in particular $Mv \in \mathscr{L}_b$. Now v satisfies $v = \mathscr{G}v$, i.e. v is the value function of the optimal stopping problem with terminal cost $b = Mv$, so that v satisfies (54.26) and (54.27) by Proposition (54.35). Conversely, if $w \in \mathscr{L}_b$ satisfies (54.26) and (54.27) then, by the uniqueness part of Proposition (53.35), $w = \mathscr{G}w$. But v is the unique fixed point of $\mathscr{G}$ in $B(E)$ and, *a fortiori*, in $\mathscr{L}_b$. □

The following example shows that the degree of smoothness predicted by Theorem (54.28) cannot generally be improved on, and that optimal

[1] The terminology comes from the analogous problem in partial differential equations which arises in impulse control of diffusion processes, and its aptness here is questionable. It is however conventional.

strategies may not exist, although ε-optimal strategies can be derived as in the optimal stopping problem.

(54.29) *Example*
The PDP is again that of Examples (51.29) and (53.32): $E = [0, 1[$, $\phi(t, x) = t + x$ for $t < t_*(x) = 1 - x$, $\lambda = 0$, $Q(\mathrm{d}y; 1) = \delta_{\{0\}}(\mathrm{d}y)$. The intervention set is $K = E$, the intervention cost a constant $c_0 > 0$ and the running cost $l(x) = 1 - x$. Since cost is accumulated at a lower rate when x is large, it is natural to conjecture that the continuation set C must take the form $[a, 1[$ for some $a \in E$. In C the value function v satisfies $\mathfrak{A}v - \delta v + l = 0$, i.e.

$$\frac{\mathrm{d}}{\mathrm{d}x} v(x) - \delta v(x) + (1 - x) = 0,$$

so that

$$v(x) = \left(v(1) + \frac{1}{\delta^2} \right) \mathrm{e}^{-\delta(1-x)} - \frac{1}{\delta^2}(1 - \delta(1 - x)). \tag{54.30}$$

This function has a minimum at the point $x_1 = (1/\delta)\ln(1 + \delta^2 v(1))$, and the value of $v(1)$ is fixed by the requirement that $v(1) - v(x_1) = c_0$. Then a is fixed by requiring that $v(a) = v(1)$.

Figure 5.4 shows the solution with $\delta = 0.5$, $c_0 = 0.08$, giving $v(1) = 0.8542$, $x_1 = 0.6129$ and $a = 0.1991$. In $S = [0, a[$ the optimal intervention is to transfer the process to x_1, so that $v(x) = v(x_1) + c_0$ for $x \in S$. We note that v is in $\mathscr{L}_b$ but is not continuously differentiable. The no-intervention cost v_0 is also shown in Fig. 5.4; it is given by (54.30) together with the boundary condition $v_0(1) = v_0(0)$ $(= 1.083)$. It will be optimal not to intervene at all in cases where $v_0(1) - v_0(x_1^0) < c_0$, where x_1^0 is the minimum point $x_1^0 = (a/\delta)\ln(1 + \delta^2 v_0(1))$, since v_0 then satisfies the QVI.

Existence of an optimal strategy can fail if c and/or K is not constant. Suppose for example that $c(x, y) = c_0 + \alpha(1 - x)$ for some $\alpha > 0$. The intervention set is then $[0, a_\alpha[$ for some $a_\alpha < a$ and the value function is unchanged on $[a, 1]$. Figure 5.5 shows the value function with δ, c_0 as above and $\alpha = 0.05$, giving $a_\alpha = 0.0910$.

Note that now $v(1) = Mv(1) < Qv(1) = v(0)$, and it is ε-optimal to intervene in some small neighbourhood $[\theta, 1]$ of 1, transferring the process to x_1, since this transfer is cheaper than if done from 0. Note that the intervention time $\tau = \inf\{t : v(x_t) = Mv(x_t)\}$ is *not optimal*, but

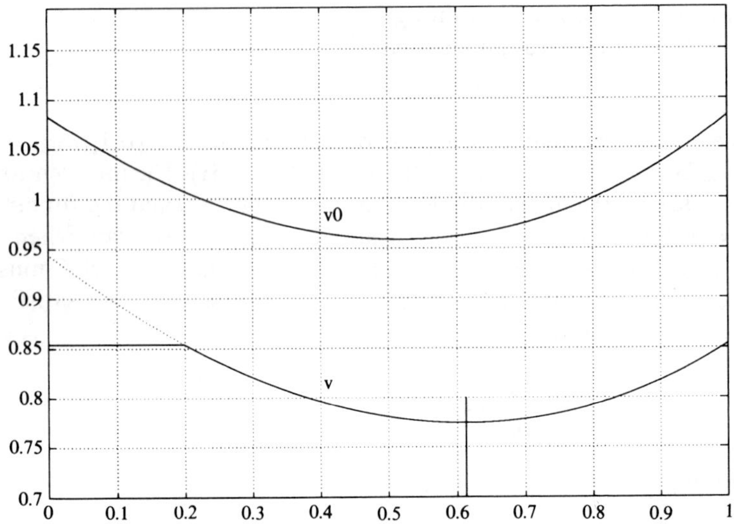

Fig. 5.4. *Value function v and no-intervention cost v_0 for Example (54.29).*

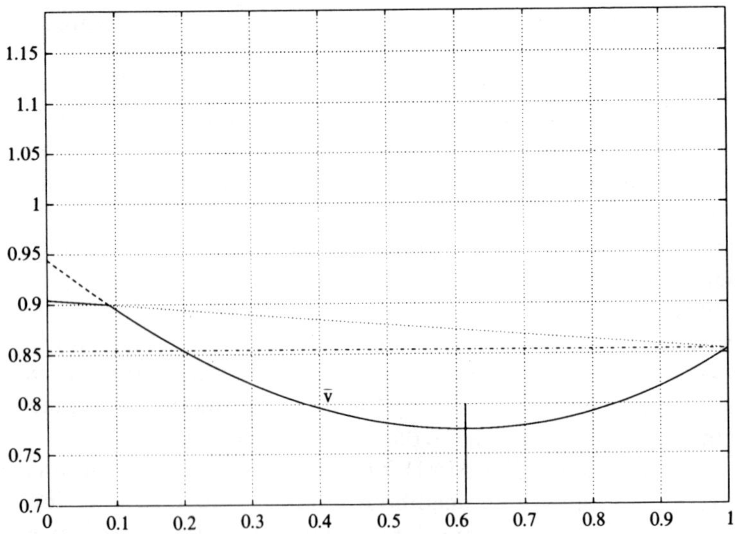

Fig. 5.5. *Value function for modified intervention cost, Example (54.29).*

times of the form $\tau = \inf\{t : Mv(x_t) - v(x_t) < \varepsilon'\}$ are ε-optimal. Similar examples can be constructed with non-constant $K(x)$.

Things become very much more difficult if the intervention set $K(x)$ is not independent of x. The proof of Theorem (54.28) uses the strong smoothing property of part 2 of Proposition (54.11) in an essential way. When $K(x)$ depends on x, no such result is available and in order to show that $v \in \mathscr{L}_b$ (so that the QVI (54.26) makes sense) we have to compare values of v *across* integral curves of $\mathfrak{X}$ as well as along them, i.e. use a 'globally Lipschitz' framework with functions in $L(E)$ instead of the 'Lipschitz along integral curves' setting $\mathscr{L}_b$. The situation is complicated, because even the no-intervention cost $v_0(x)$ is not in $L(E)$ unless a stability condition is imposed, as was demonstrated in Example (46.7). Here we shall just give some partial results. The following extra conditions relating to the data of the problem will be required.

(54.31) *Assumptions*

1. The functions g_i, λ, l, t_* belong to $C_b^1(E)$ (the space of bounded continuously differentiable functions on E). Here g_i is the ith component of the vector field $\mathfrak{X}f = \sum_i g_i(x)(\partial f/\partial x^i)$.
2. Let $\lambda_0 = \inf_{x \in F} \lambda(x) (\geqslant 0)$. Then for all $x \in E$

$$\delta + \lambda_0 - \sum_{i,j=1}^{d} \left| \frac{\partial g_i}{\partial x^j}(x) \right| \geqslant \alpha > 0.$$

Assumption 2 is a stability condition. The $d \times d$ matrix-valued function $\Phi(t, x)$, where $\Phi_{ij}(t, x) = \partial \phi_i(t, x)/\partial x^j$, satisfies the linear equation

$$\frac{\partial}{\partial t} \Phi(t, x) = A(t)\Phi(t, x), \quad \Phi(0, x) = I,$$

where I denotes the $d \times d$ identity matrix and $A_{ij}(x) = (\partial g_i/\partial x^j)(\phi(t, x))$. It follows from the Gronwall lemma that

(54.32) $$\left\| \frac{\partial \phi}{\partial x^j}(t, x) \right\| \leqslant \exp\left(\int_0^t a(s) \mathrm{d}s \right),$$

where $a(s) = \sum_{i,j} |\partial g_i (\partial x^j)(\phi(s, x))|$, and Assumption 2 guarantees that

(54.33) $$(\delta + \lambda_0)t - \int_0^t a(s) \mathrm{d}s \geqslant \alpha > 0.$$

Let us accept for a moment the following additional assumptions.

Some conditions under which they are satisfied will be given later in the section.

(54.34) *Assumptions*

1. $$v_0(x) = \mathbb{E}_x\left\{\int_0^\infty e^{-\delta t} l(x_t)\mathrm{d}t\right\} \quad \text{is in } L(E).$$

2. $\mathscr{G}\psi(x) \in L(E)$ for $\psi \in L(E)$, where $\mathscr{G}$ is the optimal stopping operator (54.14).

(54.35) *Theorem.* Suppose Assumptions (54.1), (54.31) and (54.34) hold. Suppose also that $c(x, y) \equiv c_0$ is constant and that $K(x)$ is Lipschitz continuous with constant $\kappa \leqslant 1$. Then the value function v of the impulse control problem is in $\mathscr{L}_b$ and is the unique solution in $\mathscr{L}_b$ of the QVI (54.26) and (54.27).

Proof. Let $w = \mathscr{G}\psi$, where $\psi \in L(E)$. We know that $w = L(w, \psi)$, which we can write as

$$w(x) = \min\left\{\inf_{0 \leqslant s < t_*(x)} \int_0^s e^{-(\delta t + \Lambda(t,x))}(l(t) + \lambda(t)Qw(t))\mathrm{d}t\right.$$

$$+ e^{-(\delta s + \Lambda(s,x))}\psi(s), \tag{54.36}$$

$$\left. e^{-(\delta t_* + \Lambda(t_*,x))}Qw(t_*) + \int_0^{t_*} e^{-(\delta t + \Lambda(t,x))}(l(t)\lambda(t)Qw(t))\mathrm{d}t\right\},$$

where, as before, we have written $t_* = t_*(x)$ and $l(t) = l(\phi(t, x))$, etc. Then under Assumptions (54.31) we can calculate for any $s \geqslant 0$ the derivative

$$\frac{\partial}{\partial x^i}\left\{\int_0^s e^{-(\delta t + \Lambda(t,x))}(l(t) + \lambda(t)Qw(t))\mathrm{d}t + e^{-(\delta s + \Lambda(s,x))}\psi(s)\right\}$$

$$= \int_0^s e^{-(\delta t + \Lambda(t,x))}\left(-\int_0^t \lambda'(u)\psi'(u)\mathrm{d}u(l(t) + \lambda(t)Qw(t))\right.$$

$$\left. + (l'(t) + \lambda'(t)Qw(t) + \lambda(t)(Qw)'(t))\phi'(t)\right)\mathrm{d}t$$

$$+ e^{-(\delta s + \Lambda(s,x))}\left(-\int_0^s \lambda'(u)\phi'(u)\mathrm{d}u\psi(s) + \psi'(s)\phi'(s)\right),$$

where we use the notation $l'(t) = l'(\phi(t, x))$ with $l'(y) = (\partial l/\partial y^i)(y)$, etc.

Now using the assumptions that all functions and their derivatives are bounded, together with conditions (54.32) and (54.33), we can bound the various terms as follows (writing $|f| = \sup_{x\in E} |f(x)|$)

$$\left|\int_0^s e^{-(\delta t + \Lambda(t,x))} \int_0^t \lambda'(u)\phi'(u)\mathrm{d}u(l(t) + \lambda(t)Qw(t))\mathrm{d}t\right|$$

$$\leqslant \left|\int_0^s \lambda'(u)\phi'(u) \int_u^s e^{-(\delta+\lambda_0)t}(l + \lambda Qw)\mathrm{d}t\mathrm{d}u\right|$$

$$\leqslant \left|\int_0^s \lambda'(u)e^{\int_0^u a(r)\mathrm{d}r} \frac{1}{\delta+\lambda_0} e^{-(\delta+\lambda_0)u}(|l| + |\lambda||Qw|)\mathrm{d}u\right|$$

$$\leqslant \frac{|\lambda'|}{\alpha(\delta+\lambda_0)}(|l| + |\lambda||Qw|) =: c_1$$

$$\left|e^{-(\delta s + \Lambda(s,x))} \int_0^s \lambda'(u)\phi'(u)\mathrm{d}u\psi(s)\right| \leqslant e^{-(\delta+\lambda_0)s} s|\lambda'|e^{\int_0^s a(r)\mathrm{d}r}|\psi|$$

$$\leqslant \frac{1}{e\alpha}|\lambda'||\psi| =: c_2,$$

using the fact that $se^{-\alpha s} \leqslant (e\alpha)^{-1}$. Thus

$$\left|\frac{\partial}{\partial x^i}\{\cdots\}\right| \leqslant c_1 + \frac{1}{\alpha}(|l'| + |\lambda'||Qw| + |\lambda||(Qw)'|) + c_2 + |\psi'|,$$

so that all partial derivatives are bounded by a constant that is independent of s. Note in particular that the coefficient of $|\psi'|$ in this bound is 1. A similar calculation applies to the second term in the 'min' in (54.36). We conclude that $w(x)$ is the infimum of a collection of Lipschitz functions with uniformly bounded Lipschitz constant, and hence w is Lipschitz with a constant satisfying the same bound. Again, note that this bound takes the form

(54.37) $$\left|\frac{\partial w}{\partial x^i}(x)\right| \leqslant a + \kappa,$$

where a is some constant and κ is the Lipschitz constant for ψ.

Now fix a number $\rho \in]0, 1[$ and define $v_0^\rho = v_0$ and for $n = 0, 1, \ldots$.

(54.38) $$v_{n+1}^\rho(x) = \inf_{\tau\in\Sigma} \mathbb{E}_x\left\{\int_0^\tau e^{-\delta t}l(x_t)\mathrm{d}t + e^{-\delta\tau}\rho M v_n^\rho(x_\tau)\right\}.$$

By the same argument as in Proposition (54.18) the sequence v_n^ρ is decreasing and converges to a limit v^ρ as $n \to \infty$. If we denote by l_n the Lipschitz constant of v_n^ρ then, from (54.37), $l_{n+1} \leqslant a + \rho l_n$, where a is a constant independent of n, and hence by induction

$$l_n \leqslant a + \rho a + \cdots + \rho^{n-1} a + \rho^n l_0 \leqslant \frac{a}{1-\rho} + l_0.$$

Since the functions v_n^ρ have uniformly bounded Lipschitz constant, the limit u^ρ is also Lipschitz, and it now follows as before that v^ρ is a fixed point of the operator defined by (54.38) and satisfies the QVI

(54.39)
$$\left.\begin{aligned} Av^\rho - \delta v^\rho + l \geqslant 0 \\ \rho M v^\rho - v^\rho \geqslant 0 \\ (Av^\rho - \delta v^\rho + l)(\rho M v^\rho - v^\rho) = 0. \end{aligned}\right\}$$

We now let $\rho \uparrow 1$; then $v^\rho \uparrow v$ and hence $v^\rho - \rho M v^\rho \to v - Mv$. Let $C = \{x : v(x) < Mv(x)\}$; then for $x \in C$ there exists ρ^0 such that $v^\rho(x) < \rho M v^\rho(x)$ for all $\rho > \rho^0$. Hence, from (5.69), $\mathfrak{A} v^\rho(x) - \delta v^\rho(x) + l(x) = 0$ for $\rho > \rho^0$, which implies that $\mathfrak{X} v^\rho(x) = \delta v^\rho(x) - l(x) - Qv^\rho(x) + v^\rho(x)$ is uniformly bounded for $\rho > \rho^0$. Thus the map $t \to v^\rho(t, x)$ is uniformly locally Lipschitz continuous, which implies that $v = \lim_{\rho \uparrow 1} v^\rho$ is in $\mathscr{L}_b$, and taking the limit in (54.39) we conclude that v satisfies the QVI (54.26) and (54.27). Uniqueness follows as in Theorem (54.28). □

It remains to give conditions under which Assumptions (54.34) (Lipschitz continuity of v_0 and $\mathscr{G}\psi$) hold. We have the following result.

(54.40) *Proposition.* Suppose that Assumptions (54.31) hold. Then for δ sufficiently large $v_0 \in L(E)$ and $\mathscr{G}$ maps $L(E)$ into $L(E)$, i.e. Assumptions (54.34) are satisfied.

Sketch of proof. The proof involves lengthy computations (which incidentally allow us to quantify what 'sufficiently large' means) and we will just sketch the argument here. It is based on an interesting contraction mapping idea that has been used by Stettner and Zabczyk (1981) and Bensoussan (1982; p. 282 *et seq.*). Consider first the no-intervention cost

$$v_0(x) = \mathbb{E}_x\left\{\int_0^\infty \mathrm{e}^{-\delta t} l(x_t)\,\mathrm{d}t\right\}.$$

We know that, under part 1 of Assumption (54.31), $v_0 \in \mathscr{L}_b$ and satisfies

$$\left.\begin{aligned} \mathfrak{A}v_0 - \delta v_0 + l = 0 \quad & \text{in} \quad E, \\ v_0 = Qv_0 \quad & \text{on} \quad \Gamma. \end{aligned}\right\} \tag{54.41}$$

Now take a (large) number θ and a function f and consider the equation

$$\begin{aligned} \mathfrak{A}w - (\delta - \theta)w + l + \theta f = 0, \quad & \text{in} \quad E \\ w = Qw, \quad & \text{on} \quad \Gamma. \end{aligned}$$

This has solution

$$w(x) = \mathbb{E}_x\left\{\int_0^\infty e^{-(d+\theta)t}(l(x_t) + \theta f(x_t))dt\right\} \tag{54.42}$$

which we denote $w(x) = Uf(x)$. Note that if w is fixed point of the operator U then w satisfies (54.41) and therefore $w = v_0$. To prove that (54.41) has a unique solution in some Banach space of functions $\mathscr{F}$ it therefore suffices to show that U is a contraction mapping on $\mathscr{F}$ for some θ.

In the present case we take $\mathscr{F} = W^{1,\infty}$, the space of essentially bounded functions on E with essentially bounded first derivatives, with norm

$$\|f\|_{1,\infty} = \|f\| + \sum_{i=1}^{d}\left\|\frac{\partial f}{\partial x^i}\right\|, \quad \text{where}^2 \ \|f\| = \operatorname*{ess\,sup}_{x\in E}|f(x)|.$$

We therefore have to show that for θ large enough (a) U maps $W^{1,\infty}$ into itself, and (b) U is a contraction mapping. For (a), we use the representation $Uf(x) = \lim_{n\to\infty} G^n\psi(x)$, where

$$G\psi(x) = \mathbb{E}_x\left\{\int_0^{T_1} e^{-(\delta+\theta)t}(l(x_t) + \theta f(x_t))dt + e^{-\delta T_1}Q\psi(x_{T_1}-)\right\}.$$

The technique is the same as in the proof of Theorem (54.35): we calculate $\partial G\psi(x)/\partial x^i$ and show that for large θ the sequence $G^n\psi$ is uniformly Lipschitz continuous. This requires Assumption (54.31) (2). Examining how the Lipschitz constant depends on $\|f\|_{1,\infty}$,

[2] $\operatorname*{ess\,sup}_{x\in E}|f(x)| = \inf\{a\in\mathbb{R}_+ : \text{Leb}\{x : |f(x)| > a\} = 0\}$, where Leb denotes Lebesgue measure.

we find that the contraction property (b) holds for θ sufficiently large. Thus $v_0 \in W^{1,\infty}$ and *a fortiori* $v_0 \in L(E)$. Similar arguments show that $\mathscr{G}\psi \in L(E)$ for $\psi \in L(E)$. □

Finally, as promised, we give the proof of Lemma (54.21).

Proof of Lemma (54.21). Suppose w_1 and w_2 are two distinct solutions in $\mathscr{P}_H$ of the equation $w = Hw$. Interchanging w_1 and w_2 if necessary, we can assume that $w_1 - w_2 \notin B(E)_+$. Now let $t_0 = \sup\{t \in \mathbb{R}_+ : w_1 \geqslant tw_2\}$; since $w_1 \geqslant 0$ and $w_1 \ngeqslant w_2$, $t_0 \in [0, 1[$, and clearly $w_1 \geqslant t_0 w_2$. Using the monotonicity and concavity of H we have

$$\begin{aligned} w_1 = Hw_1 &\geqslant H(t_0 w_2) \\ &= H(t_0 w_2 + (1 - t_0)0) \\ &\geqslant t_0 Hw_2 + (1 - t_0)H0 \\ &\geqslant t_0 w_2 + (1 - t_0)H0 \\ &\geqslant \left(t_0 + \frac{(1 - t_0)}{\kappa}\right) w_2, \end{aligned}$$

since $w_2 \in \mathscr{P}_H$. But this inequality contradicts the definition of t_0, so w_1 and w_2 must coincide. □

55 Computational methods

In this section we give a brief introduction to computational methods for optimal stopping and impulse control of PDPs. It is beyond the scope of this book to give a complete treatment of this important subject, but the reader can refer to Costa and Davis (1988, 1989), on which the following discussion is based, for full details.

Let us start with the optimal stopping problem of maximizing $\mathbb{E}_x[e^{-\delta\tau} b(x_\tau)]$ over stopping times $\tau \in \Sigma$, where (x_t) is a standard PDP, $\delta > 0$ and $b \in \mathscr{C}_b$ (the restriction to finite-valued stopping times Σ_f as in Section 51 is unnecessary for the discounted reward problem.) Define the value function v as

$$v(x) = \sup_{\tau \in \Sigma} \mathbb{E}_x[e^{-\delta\tau} b(x_\tau)]$$

and recall from Theorem (51.17) that $v \in \mathscr{C}_b$ and $v_n \uparrow v$ as $n \to \infty$, where $v_0 = b$, $v_{n+1} = L(v_n, b)$ and L is defined by (51.6), *mutatis mutandis*, i.e.

$$(55.1) \quad L(w, b)(x) = \sup_{\tau \in \Sigma_1} \mathbb{E}_x[e^{-\delta\tau} b(x_\tau) I_{(\tau < T_1)} + e^{-\delta T_1} w(x_{T_1}) I_{(\tau = T_1)}].$$

Now consider a special problem in which the PDP (x_t) has the following property: there exist $z_1, z_2, \ldots, z_N \in E$ such that, for all $x \in E \cup \Gamma$, $Q(\{z_1, \ldots, z_N\}; x) = 1$, i.e. the process always jumps to one of a finite number of points. Then the set $E_1 = \{\phi(t, z_i): 0 \leqslant t < t_*(z_i), i = 1, \ldots, N\}$ is an invariant set for (x_t) and the optimal stopping problem can be solved on E_1 independently of its solution on $E \backslash E_1$. The key point to notice is that *computing* $L(v_n, b)$ *involves only* N *one-dimensional maximizations.* Indeed, if we let w_n denote the N-vector with ith component $w_n(i) = v_n(z_i)$ and define $q_i(x) = Q(\{z_i\}; x)$ then from (55.1), for $i = 1, \ldots, N$

$$
\begin{aligned}
w_{n+1}(i) = \sup_{t \in \mathbb{R}_+} \mathbb{E}_x \Bigg[& e^{-\delta t} b(\phi(t, z_i)) I_{(t < T_1)} \\
& + e^{-\delta T_1} I_{(t \geqslant T_1)} \sum_{j=1}^{N} w_n(j) q_j(\phi(T_1, z_i)) \Bigg].
\end{aligned}
\tag{55.2}
$$

Thus L defines a non-linear operator, also denoted L, from $\mathbb{R}^N$ to $\mathbb{R}^N$ such that $w_n = L^n w_0$ and $L^n w_0 \to w$, where $w(i) = v(z_i)$. Computing Lw_n involves the N one-dimensional maximizations indicated in (55.2), and in practice this is easily accomplished using search techniques (Press *et al.*, 1986). Note that once w is computed, the problem is completely solved: since v satisfies $v = L(v, b)$ we have for any $x \in E$ (not just $x \in E_1$)

$$
\begin{aligned}
v(x) = \sup_{t \in \mathbb{R}_+} \mathbb{E}_x \Bigg[& e^{-\delta t} b(\phi(t, x_t)) I_{(t < T_1)} \\
& + e^{-\delta T_1} I_{(t \geqslant T_1)} \sum_{j=1}^{N} w(j) q_j(\phi(T_1, z_i)) \Bigg].
\end{aligned}
$$

Thus one further maximization computes v at any point. The optimal continuation and stopping sets are now defined by $C = \{x : v(x) > b(x)\}$, $S = \{x : v(x) = b(x)\}$ as before, and these define ε-optimal (or optimal) stopping times as shown in Theorem (51.20).

For the general problem, a natural discretization procedure can be used. Let $D := \{z_1, z_2, \ldots\}$ be a countable dense set in E and for $N = 1, 2, \ldots$, define sets A_i^N, $i = 1, \ldots, N$ as follows:

$$
B_i^N := \{x \in E : |x - z_i| \leqslant |x - z_j|, j = 1, \ldots, N\}
$$

$$
A_1^N := B_1^N, \; A_{i+1}^N = B_{i+1}^N \setminus \bigcup_{j=1}^{i} A_j^N, \quad i = 1, \ldots, N-1.
$$

This partitions E into disjoint sets $A_1^N, \dots, A_N^N$, each containing a 'representative point' z_i. We now define a modified PDP (x_t^N) which has local characteristics $\mathfrak{X}$ and λ as before and transition measure

$$Q^N(\{z_i\}; x) := Q(A_i^N, x), \quad i = 1, \dots, N, \quad x \in E \cup \Gamma.$$

Thus in the discretized model all the 'weight' on A_i^N is concentrated on the representative point z_i and the optimal stopping problem of computing

$$v^N(x) := \sup_{\tau \in \Sigma^N} \mathbb{E}_x[e^{-\delta\tau} b^N(x_\tau^N)]$$

can be solved as described above, where Σ^N denotes the set of stopping times for the discretized problem. We take reward functions b^N depending on N because these are needed for the impulse control application below. The following theorem gives conditions under which v^N converges to v as $N \to \infty$.

(55.3) *Theorem.* (Costa and Davis, 1988, Theorem 2 and Corollaries) Suppose that:

1. $t_*: E \to \mathbb{R}_+ \cup \{\infty\}$ is continuous;
2. $\lambda \in C_b(E)$;
3. for every $h \in C(E)$, $v \in C(\Gamma)$, $Qh: E \to \mathbb{R}$ and $Qv: \Gamma = \mathbb{R}$ are continuous;
4. $b, b^N \in C_b(E \cup \Gamma)$ and $b^N \to b$ uniformly on compact sets of $E \cup \Gamma$; and
5. $\Lambda(t_*(x), x) = \infty$ whenever $t_*(x) = \infty$.

Then $v^N, v \in C_b(E \cup \Gamma)$ and $v^N \to v$ uniformly on compact sets of $E \cup \Gamma$.

Of course, in particular applications discretization may not proceed in exactly the way described above. In particular the support of the transition measure may only be some small subset of E, in which case only this subset needs discretization. Also, we have said nothing about discretization of *time*. It may be advantageous to restrict possible stopping times to integer multiples of a step length Δ, reducing the basic calculation step $w_{n+1}^N = L^N w_n^N$ to N finite searches. Convergence as $n \to \infty$, $\Delta \to 0$ holds under the same conditions as before.

Computation of the value function v of the impulse control problem (54.7) is based on the result given in Theorem (54.23), namely that $\hat{L}^n v_0 \downarrow v$, where v_0 is the no-intervention cost (54.8) and $\hat{L}w = L(w, Mw)$ (where L is redefined for the minimization problem, as in

Theorem (54.23). Computation of $\hat{L}w$ reduces to N one-dimensional minimizations only if the intervention set $K(x)$ as well as the support of $Q(\cdot;x)$ are contained in a finite set $\{z_1,\ldots,z_N\}$. In the discretization procedure we therefore discretize $K(x)$ as well as Q. Q is discretized as above using points $z_1,\ldots,z_N$ and sets $A_1^N,\ldots,A_N^N$ and we define

$$K^N(x) = \{z_i : i = 1,\ldots,N \quad \text{and} \quad \Gamma(x) \cap A_i^N \neq \varnothing\}.$$

Note that $K^N(x)$ is not necessarily a subset of $K(x)$. Now define M^N and $\hat{L}^N$ by

$$M^N w(x) = \min_{z_i \in K^N(x)} c(x, z_i) + w(z_i),$$

$$\hat{L}^N w = \hat{L}(w, M^N w).$$

Proposition (54.25) now asserts that $(\hat{L}^N)^n v_0^{N'} \downarrow v^N$, where v^N is the value function for the discretized problem and $v_0^{N'}$ is any upper bound for the no-intervention cost. The following convergence result can be established using Theorem (55.3).

(55.4) *Theorem.* Suppose that the conditions of Theorem (55.3) hold, and that $x \to K(x)$ is continuous in the Hausdorff metric topology, $c \in C_b(\mathscr{K})$ and $l \in C_b(E)$. Then $v^N, v \in C_b(E)$ and $v^N \to v$ uniformly on compact sets of E, where v is the value function (54.7) of the impulse control problem.

(55.5) *Remark*. This is essentially the result of Proposition 13 and Corollary 2 of Costa and Davis (1989). The result is stated slightly differently there in that the PDP is extended for technical convenience to a PDP $(\hat{x}_t)$ on $E \cup \Gamma$ with Γ as the set of 'branch points'.

The methods just described for solving optimal stopping and impulse control problems, based on iteration of the 'single jump' operator L, seem to be the most direct and have been shown to work well in examples, although obviously the amount of computation required is substantial except for very small problems. Another approach would be to attempt direct numerical solution of the variational inequality (53.28) and (53.29) or quasi-variational inequality (54.26) and (54.27). In simple examples it is often easy to arrive at the solution this way, but little appears to have been done in this direction in any general setting. This area merits further investigation.

56 Notes and references

51 There is a huge literature on optimal stopping of Markov processes, tackled at various degrees of generality. The general theory was initiated by Dynkin (1963); a standard reference is Shiryaev (1978); and there is a very informative survey by Zabczyk (1983). Optimal stopping of PDPs was developed by Gugerli (1986), and the presentation here follows Gugerli's paper closely. The proof of part 1 of Theorem (51.20) is adapted from Shiryaev (1978). Costa and Davis (1988) extend the results to cover payoff functions which are not necessarily continuous along trajectories, and developed computational methods. Alternative treatments have been given by Lenhart and Liao (1985) and Gątarek (1990, 1992).
52 Randomized stopping is a standard idea in optimal stopping theory; see Shiryaev (1978), for example.
53 Formulation of optimal stopping in terms of variational inequalities is, again, a much-studied subject, particularly for diffusion processes (Bensoussan and Lions, 1978). The 'penalty method' was introduced by Lions (1969). Variational inequalities for PDPs have been studied by Lenhart and Liao (1985) from a more analytical point of view. The treatment here based on randomized stopping was given by Davis and Lopes (1992); see also Lopes (1992).
54 The theory of impulse control was initiated by Bensoussan and Lions (1975) in the context of diffusion processes; an extensive account is given in Bensoussan and Lions (1982). For PDPs, Costa and Davis (1989) give an account based on optimal stopping theory (Costa and Davis, 1988); Lenhart (1989) studies *viscosity solutions* of the associated quasi-variational inequality; Gątarek (1990) gives an approach based on approximation by Feller processes. A further paper by Gątarek (1992) gives a treatment based on the theory of monotone operators as discussed by Zabczyk (1983); this paper clearly identifies the boundary conditions associated with optimal stopping and impulse control of PDPs and also uses viscosity solutions, which may in fact be the most appropriate approach for PDPs. As we point out in the text, the Lipschitz continuous framework is not entirely natural since even the no-intervention cost v_0 is not Lipschitz without some stability conditions. We have however stuck to this framework in order to conform to the control setting of Chapter 4 and to avoid introducing yet another solution concept. Finally, there is an interesting paper by Dempster and Ye (1992) in

which impulse control is transformed into boundary control (as discussed in Chapter 4), enabling combined problems of 'continuous', 'boundary' and 'impulse' control to be studied in a unified framework. The treatment given here is new but contains elements from many of these papers.

Lepeltier and Marchal (1984) have studied impulse control for right processes, of which the PDP is a special case (section 27). However, more specific results can be obtained by using the special structure of PDPs.

Closely related to impulse control are problems of 'switching control' in which one has a finite number of Markov families $P^1_x, P^2_x, \ldots, P^n_x$ and, at each intervention time, a new starting point $x \in E$ and the index $k \in K_n := \{1, \ldots, n\}$ of the process to be started from that point are chosen. In fact, this is clearly equivalent to impulse control for a new process $\hat{x}_t = (v_t, x_t)$ on the state space $K_n \times E$. There are papers on this topic by Lenhart (1987), Lenhart and Liao (1988) and Lenhart and Yamada (1991). The latter paper concerns game theory, a topic not pursued here.

APPENDIX

Jump processes and their martingales

This Appendix gives an account of the martingale theory of jump processes, which is used in Chapter 2 in showing the strong Markov property of the PDP (section 25) and characterizing the extended generator (section 26). The main result is Theorem (A5.5) which shows that all local martingales of the natural filtration of a jump process can be represented as 'stochastic integrals' of the sort already described in section 26. This result is analogous to the famous result of Ito showing that all martingales of Brownian motion are Ito stochastic integrals (Karatzas and Shreve, 1988, Theorem 3.4.15). The argument follows Davis (1976) closely, but a number of technical improvements make the presentation more self-contained. Specifically, some of the arguments given by Brémaud (1980) for 'right-constant' processes enable us to show rather directly in section A2 that the stopped σ-field $\mathfrak{F}_T$ for a stopping time T is essentially the σ-field generated by the process up to the stopping time. Also, we use systematically a constructive definition of 'predictability' (given in section A3), avoiding the need to introduce 'predictable processes' as defined in the *théorie générale des processus.* This is possible because our filtration is generated in a very particular way. Further comments and references on jump process theory are given in section 28.

The jump process is formally defined in section A1 below. Sections A2 and A3 concern the structure of stopping times and *predictable processes* (our definition). The martingale representation results are stated and proved in sections A4 and A5. As in Chou and Meyer (1975) and Davis (1976), the approach is to study the elementary 'single jump' process first and to use this as a building block for the general case.

A1 Definition of the jump process

A *jump process* is a right-continuous piecewise-constant stochastic process (x_t) taking values in $X \cup \{\Delta_\infty\}$, where X is a Borel space and Δ_∞ is an isolated cemetery state. We assume that the process has discontinuities at an increasing sequence of isolated times (T_k) and is killed at time $T_\infty := \lim_k T_k$. Thus a sample function of the process is specified by giving a constant $z_0 \in X$ and a sequence (S_k, Z_k) for $k = 1, 2, \ldots$ of random variables with $S_k \in \mathbb{R}_+$ and $Z_k \in E$ and defining $T_0 = 0$, $T_k = T_{k-1} + S_k$ and $X_t = Z_k$ for $t \in [T_k, T_{k+1}[$, $X_t = \Delta_\infty$ for $t \geqslant T_\infty$. It is possible that $T_k = \infty$ for some k. We will assume that $P[Z_k = Z_{k-1}] = 0$ for all k, so that the process really does 'jump' at the 'jump times' T_k. We can define the jump process on a canonical space, as follows. Let

$$Y = (\mathbb{R}_+ \times X) \cup \{\Delta\},$$

where Δ is an isolated point and let $\mathfrak{Y}$ denote the Borel sets of Y. Define $\Omega_i = \Pi_{k=1}^{i} \Omega$, $\Omega = \Pi_{k=1}^{i} Y_k$, $\mathfrak{F}^{i,0} = \sigma\{\Pi_{k=1}^{i} \mathfrak{Y}_k\}$ and $\mathfrak{F}^0 = \sigma\{\Pi_{k=1}^{\infty} \mathfrak{Y}_k\}$. Let $\xi_k : \Omega \to Y_k$ denote the coordinate mapping and let $\xi_k(\omega) = (S_k(\omega), Z_k(\omega))$ when $\xi_k(\omega) \in \mathbb{R}_+ \times X$ (otherwise, $\xi_i(\omega) = \Delta$). Let $\omega_k(\omega) = (\xi_1(\omega), \ldots, \xi_k(\omega))$. Now let

$$T_k(\omega) := \begin{cases} \sum_{i=1}^{k} S_i(\omega), & \xi_i(\omega) \neq \Delta, i = 1, \ldots, k \\ \infty, & \text{if } \xi_i(\omega) = \Delta \text{ for some } i = 1, \ldots, k \end{cases}$$

$$T_\infty(\omega) := \lim_k T_k(\omega)$$

and define the sample path $x_t(\omega)$ for $t \in \mathbb{R}_+$ by

$$x_t(\omega) = \begin{cases} z_0 & t < T_1(\omega) \\ Z_k & T_k(\omega) \leqslant t < T_{k+1}(\omega) \\ \Delta_\infty & t \geqslant T_\infty(\omega). \end{cases}$$

Here $z_0 \in Z$ is fixed and Δ_∞ is a point isolated from X. The natural filtration of the process (x_t) in Ω is

$$\mathfrak{F}_t^0 := \sigma\{x_s(\cdot), s \leqslant t\}.$$

A probability measure on Ω is defined by giving the following family of conditional distribution functions: μ^1 is a probability measure on

$(Y, \mathfrak{Y})$ such that

$$\text{(A1.1)} \qquad \mu^1((\{0\} \times X) \cup (\mathbb{R}_+ \times \{z_0\})) = 0,$$

and for $k = 2, 3, \ldots$, $\mu^k : \Omega_{k+1} \times \mathcal{Y} \to [0, 1]$ is a transition measure satisfying:

1. $\mu^k(\cdot; \Gamma)$ is measurable for each $\Gamma \in \mathfrak{Y}$;
2. $\mu^k(\omega_{k-1}(\omega); \cdot)$ is a probability measure for each $\omega \in \Omega$;
3. $\mu^k(\omega_{k-1}(\omega); (\{0\} \times X) \cup \{\mathbb{R}_+ \times Z_{k-1}(\omega)\})) = 0$ for each $\omega \in \Omega$;
4. $\mu^k(\omega_{k-1}(\omega); \{\Delta\}) = 1$ if $\xi_i(\omega) = \Delta$ for some $i \leqslant k-1$.

Then P is the unique probability measure on $(\Omega, \mathfrak{F}^0)$ such that for each k and bounded measurable function f on Ω_k

$$\int_\Omega f(\xi_1(\omega), \ldots, \xi_k(\omega)) P(\mathrm{d}\omega)$$

$$= \int_{Y_1} \cdots \int_{Y_k} f(\xi_1, \ldots, \xi_k) \mu^k(\xi_1, \ldots, \xi_{k-1}; \mathrm{d}\xi_k)$$

$$\times \mu^{k-1}(\xi_1, \ldots, \xi_{k-2}; \mathrm{d}\xi_{k-1}) \cdots \mu^1(\mathrm{d}\xi_1).$$

Note from 3 that, with probability one, $T_1 > 0$, $T_k > T_{k-1}$ and $Z_k \neq Z_{k-1}$. Also, 4 implies that $\xi_i = \Delta$ for all $i \geqslant k := \min\{j : \xi_j(\omega) = \Delta\}$ and we interpret this as saying that $T_k(\omega) = \infty$. We now define $\mathfrak{F}_t[\mathfrak{F}^k, \mathfrak{F}]$ to be the σ-field $\mathfrak{F}_t^0[\mathfrak{F}^{k,0}, \mathfrak{F}^0]$ completed with all P-null sets of $\mathfrak{F}^0$. Let $\mathbb{F}$ denote the filtration $(\mathfrak{F}_t)_{t \in \mathbb{R}_+}$. This is the completed natural filtration of the jump process x_t. A *stopping time of* $\mathbb{F}$ is an $\mathfrak{F}$-measurable function $T : \Omega \to \mathbb{R}_+^* = \mathbb{R}_+ \cup \{+\infty\}$ such that $(\omega : T(\omega) \leqslant t) \in \mathfrak{F}_t$ for each $t \in \mathbb{R}_+$.

(A1.2) *Lemma*

1. T_k, $k = 1, 2, \ldots$ and T_∞ are $\mathbb{F}$-stopping times;
2. $\mathfrak{F} = \mathfrak{F}_\infty := \bigvee_{t \in \mathbb{R}_+} \mathfrak{F}_t$.

Proof

1. Let $N_t := \sum_i I_{(t \geqslant T_i)}$. In view of condition 3 below (A1.1), (N_t) is clearly an $\mathfrak{F}_t$-adapted process, and $(T_i \leqslant t) = (N_t \geqslant i)$. Also $(T_\infty \leqslant t) = \bigcap_i (T_i \leqslant t)$.
2. By definition $\mathfrak{F}^0 = \sigma\{\xi_i, i = 1, 2, \ldots\}$ so that $\mathfrak{F}_\infty \subset \mathfrak{F}$. For the converse it suffices to show that ξ_i is $\mathfrak{F}_\infty$-measurable for each i. Now

$(\xi_i = \Delta) = \bigcap_n (T_i > n)$, while $(S_i \leqslant t,\ Z_i \in A) = (T_i \leqslant T_{i-1} + t) \cap (T_{i-1} < \infty) \cap (Z_i \in A) \in \mathfrak{F}_\infty$. This completes the proof. □

A2 Structure of stopping times and stopped σ-fields

Recall that for any $\mathfrak{F}_t$-stopping time T, the *stopped σ-field* $\mathfrak{F}_T$ is defined as

$$\mathfrak{F}_T = \{A \in \mathfrak{F} : A \cap (T \leqslant t) \in \mathfrak{F}_t \text{ for all } t \in \mathbb{R}_+\}.$$

We need a more explicit characterization of $\mathfrak{F}_T$, and this is given by the following theorem, which also shows that the filtration $\mathbb{F}$ is right-continuous.

(A2.1) *Theorem*

1. For any $t \in \mathbb{R}_+$ we have $\mathfrak{F}_t = \bigcap_{\varepsilon > 0} \mathfrak{F}_{t+\varepsilon}$.
2. For any stopping time T we have

$$\mathfrak{F}_T = \sigma\{x_{s \wedge T}, s \in \mathbb{R}_+\}.$$

3. For each $k = 1,2,\ldots$

$$\mathfrak{F}_{T_k} = \sigma\{\xi_1, \ldots, \xi_k\} = \mathfrak{F}^k \times \Pi_{i=k+1}^{\infty} Y_i,$$

i.e. $A \in \mathfrak{F}_{T_k}$ if and only if $A = A' \times \Pi_{i=k+1}^{\infty} Y_i$ for some $A' \in \mathfrak{F}^k$.

Parts 1 and 2 of the theorem are true for any right-constant process, and it is expeditious to prove them in this generality. Thus let $(Y_t)_{t \in \mathbb{R}_+}$ be an X-valued stochastic process with right-continuous sample paths defined on some probability space $(B, \mathfrak{A}, m)$ and let $(\mathfrak{Y}_t)$ be the natural filtration of (Y_t) completed as usual with all m-null sets of $\mathfrak{A}$. (Y_t) is *right-constant* if for each $(t, \beta) \in R_+ \times B$ there exists $\varepsilon(t, \beta) > 0$ such that $Y_{t+\delta}(\beta) = Y_t(\beta)$ for $\delta \in [0, \varepsilon(r, \beta)]$.

(A2.2) *Theorem* (Brémaud, 1981, Appendix A2). Suppose (Y_t) is a right-constant process as described above. Then

1. $\mathfrak{Y}_t = \mathfrak{Y}_{t^+} := \bigcap_{\varepsilon > 0} \mathfrak{Y}_{t+\varepsilon}$ for each $t \in \mathbb{R}_+$.
2. $\mathfrak{Y}_S = \sigma\{Y_{s \wedge S}, s \in \mathbb{R}_+\}$ for each $\mathfrak{Y}_t$ stopping time **S**.

Proof

1. It suffices to show that if $A \in \mathfrak{A}$ is a set which is in $\mathfrak{Y}_{t+2^{-k}}$ for all k, then A is in $\mathfrak{Y}_t$. In view of the fact that Y_t has right-continuous

sample paths and the supposition that $A \in \mathfrak{Y}_{t+2^{-k}}$ we can write the indicator function of A in the form

$$I_A = \Phi_k(Y_s, s \in (\mathbb{Q} \cap [0, t+2^{-k}]) \cup \{(t+2^{-k})\})$$

where $\mathbb{Q}$ denotes the set of rational numbers. Now define

$$\eta_k := \Phi'_k(Y_s, s \in \mathbb{Q} \cap [0,t]) := \Phi_k(Y'_s, s \in (\mathbb{Q} \cap [0, t+{}^{2^{-k}}]) \cup \{(t+2^{-k})\}),$$

where

$$Y'_s = \begin{cases} Y_s, & s \in [0, t[\\ Y_t, & s \in [t, t+2^{-k}] \end{cases}$$

and define

$$B_k = \{\beta \in B : Y_{t+\delta}(\beta) = Y_t(\beta), \delta \leqslant [0, 2^{-k}]\}.$$

Then evidently η_k is $\mathfrak{Y}_t$-measurable, $B_k \uparrow B$ and

$$I_{A \cap B_k} = \eta_k I_{B_k}.$$

It follows that $I_A = \liminf_{k \to \infty} \eta_k$, and hence that $A \in \mathfrak{Y}_t$.

2. Let S be a $\mathfrak{Y}_t$ stopping time and

$$\mathfrak{G} = \sigma\{Y_{s \wedge S}, s \geqslant 0\}.$$

If $A := (Y_{s \wedge S} \in G)$ for some $G \in \mathfrak{B}(X)$ then clearly $A \cap (S \leqslant t) \in \mathfrak{Y}_t$. Hence $\mathfrak{G} \subset \mathfrak{Y}_S$. For the converse, suppose first that S takes values $0 \leqslant a_1 < a_2 < \cdots \leqslant \infty$. Then any $A \in \mathfrak{Y}_S$ can be written $A = \bigcup_i A_i$, where $A_i = A \cap (S = a_i) = \mathfrak{Y}_{a_i}$. As above we can write

$$I_{A_i} = \Phi_i(Y_t, t \in \mathbb{Q} \cap [0, a_i]),$$

and since $S = a_i$ on A_i this is the same as

$$I_{A_i} = \Phi_i(Y_{t \wedge S}, t \in \mathbb{Q} \cap [0, a_i]).$$

Thus $A_i \in \mathfrak{G}$ and hence $A \in \mathfrak{G}$.

For the general case, let

$$S_k = \sum_{i=1}^{\infty} \frac{i}{2^k} I_{((i-1)2^{-k} \leqslant S < i2^{-k})} + \infty I_{(S=\infty)};$$

then S_k is countably valued, and $S_k \downarrow S$, so that $\mathfrak{Y}_S \subset \mathfrak{Y}_{S_k}$. From the above $\mathfrak{Y}_{S_k} = \sigma\{Y_{s \wedge S_k}, s \geqslant 0\}$. Let B_k be defined as above but with S replacing t. We then have

$$\mathfrak{Y}_{S_k} \cap B_k = \sigma\{Y_{s \wedge S}, s \geqslant 0\} \cap B_k,$$

where, for a σ-field $\mathfrak{H}$, $\mathfrak{H}\cap B_k = \{H\cap B_k : H\in\mathfrak{H}\}$. Thus if $A\in\mathfrak{Y}_S\subset\mathfrak{Y}_{S_{k+n}}$ then there exists $G_{k+n}\in\mathfrak{G}$ such that $A\cap B_{k+n} = G_{k+n}\cap B_{k+n}$. Intersecting each side with B_k gives $A\cap B_k = G_{k+n}\cap B_k$ and it follows that $A\cap B_k = G\cap B_k$, where $G := \liminf G_n\in\mathfrak{G}$. Since $B_k\uparrow B$ this shows that $A = G\in\mathfrak{G}$ and hence that $\mathfrak{Y}_S\subset\mathfrak{G}$. □

Proof of Theorem (A2.1). Parts 1 and 2 are special cases of Theorem (A2.2). For part 3 denote

$$\mathfrak{H} = \mathfrak{F}^k \times \prod_{i=k+1}^{\infty} Y_i = \sigma\{\xi_1, \xi_2, \ldots, \xi_k\}.$$

From part 2

$$\mathfrak{F}_{T_k} = \sigma\{x_{t\wedge T_k}, t\geqslant 0\} =: \mathfrak{G}.$$

But there is a 1–1 correspondence between $(\xi_1,\ldots,\xi_k)$ and $(x_{t\wedge T_k}, t\geqslant 0)$, and hence $\mathfrak{H} = \mathfrak{G}$. □

We shall also need the following result, giving a very precise description of the class of $\mathfrak{F}_t$ stopping times.

(A2.3) *Theorem.* Let τ be a stopping time of the jump process natural filtration $\mathfrak{F}_t$. Then there exists a constant s_1 and functions $s_k:\Omega_{k-1}\to\mathbb{R}_+$ for $k = 2,3,\ldots$ such that

$$\tau I_{(\tau\leqslant T_1)} = (s_1 \wedge T_1) I_{(\tau\leqslant T_1)}$$

and for $k = 2, 3, \ldots$

$$\text{(A2.4)}\quad \tau I_{(T_{k-1}<\tau\leqslant T_k)} = ((T_{k-1} + s_k(\xi_{k-1}))\wedge T_k) I_{(T_{k-1}<\tau\leqslant T_k)}.$$

(A2.5) *Remark.* An equivalent, and simpler, statement is: there exist $\mathfrak{F}_{T_{k-1}}$ measurable random variables τ_k such that $\tau I_{(\tau<T_k)} = \tau_k I_{(\tau<T_k)}$. However, the more explicit form (A2.4) is what we need in applications.

To prove Theorem (A2.3) we first consider a simple situation in which there is only one jump (the single-jump process is analysed in more detail in section A4). Thus, let $(\Xi, \mathfrak{S})$ be a measurable space and z_0 be a measurable function mapping Ξ into X (where X is as above). Define $\Omega = (\Xi\times((\mathbb{R}_+\times X)\cup\{\Delta\}))\backslash\{(\xi, t, z)\in\Xi\times\mathbb{R}_+\times X : z = z_0(\xi)\}$. For $\omega = (\xi, t, z)\in\Omega$, denote $\xi(\omega) = \xi, T(\omega) = t, Z(\omega) = z$, while for $\omega = (\xi, \Delta)$ denote $\xi(\omega) = \xi, T(\omega) = \infty$. For $t\in\mathbb{R}_+$ define

$$x_t(\omega) = \begin{cases} z_0(\xi), & t < T(\omega) \\ Z(\omega), & t\geqslant T(\omega). \end{cases}$$

Now let $\mathfrak{H}_t$ be the filtration in Ω defined by $\mathfrak{H}_0 = \mathfrak{S} \times \mathbb{R}_+ \times X$, $\mathfrak{H}_t = \mathfrak{H}_0 \vee \sigma\{x_s, s \leqslant t\}$. Then it is easy to see that

(A2.6) $\mathfrak{H}_t = \Omega \cap (\mathfrak{B}(\Xi \times [0,t] \times X) \cup (\mathfrak{S} \times (]t,\infty[\times X) \cup \{\Delta\}).$

and

$$\mathfrak{H}_\infty = \Omega \cap (\mathfrak{B}(\Xi \times (\mathbb{R}_+ \times X) \cup \{\Delta\})).$$

(A2.7) *Lemma.* If τ is an $\mathfrak{H}_t$ stopping time then

(A2.8) $$\tau \wedge T = t_1(\xi) \wedge T$$

for some measurable function $t_1 : \Xi \to \mathbb{R}_+^*$.

Proof. First, suppose that τ takes one of only a countable number of values $0 \leqslant a_1 < a_2 \cdots \leqslant \infty$. Let $A_i = \{\omega : \tau(\omega) = a_i\}$. Then $A_i \in \mathfrak{H}_{a_i}$, and hence for $1 \leqslant i \leqslant \infty$, in view of (A2.6), $A_i = A_i^1 \cup (A_i^2 \times ([a_i, \infty[\times X) \cup \{\Delta\})$, where $A_i^1 \in \Omega \cap \mathfrak{B}(\Xi \times [0, a_i] \times X)$, $A_i^2 \in \mathfrak{S}$. The A_i are disjoint and partition Ω. Now

$$(\tau < T) = \bigcup_{1 \leqslant i \leqslant \infty} (\tau < T) \cap A_i = \bigcup_{1 \leqslant i \leqslant \infty} (a_i < T) \cap A_i$$

$$= \bigcup_{1 \leqslant i \leqslant \infty} A_i^2 \times (]a_i, \infty[\times X) \cup \{\Delta\}).$$

Define

$$h(\xi) = \sum_{1 \leqslant i \leqslant \infty} a_i I_{A_i^2}(\xi) + \infty I_{A_\infty}.$$

Then $a_i = h(\xi) = \tau(\omega)$ on $A_i \cap (T > \tau)$, while on $A_i \cap (T \leqslant \tau)$, $h(\xi) = a_i = \tau(\omega) \geqslant T(\omega)$, so that (A2.8) holds with $t_1(\xi) = h(\xi)$ (it holds trivially on A_∞).

For a general stopping time τ, define for $n = 1, 2, \ldots$,

$$\tau_n = \sum_{k=1}^{\infty} \frac{k}{2^n} I_{((k-1)2^{-n} < \tau \leqslant k2^{-n})} + \infty I_{(\tau = \infty)}.$$

Then τ_n is a countably valued stopping time, $\tau_n \geqslant \tau$ and $\tau_n \downarrow \tau$ as $n \to \infty$. Let $h_n(\xi)$ be the corresponding sequence of functions as above and define $t_1(\xi) = \liminf_{n \to \infty} h_n(\xi)$. Then t_1 is measurable and for $\omega \in (\tau < T)$ there is a number n_0 such that $\tau_n(\omega) < T(\omega)$ for $n > n_0$, so that $\tau_n(\omega) = h_n(\xi)$ for $n > n_0$; thus $\tau(\omega) = h(\xi)$.

Proof of Theorem (A2.3). If τ is a stopping time of $\mathfrak{F}_t$ then $\tau' :=$

$(\tau \wedge T_k - T_{k-1}) \vee 0$ is a stopping time of $\mathfrak{H}_t := \mathfrak{F}_{(T_{k-1}+t)\wedge T_k}$. Indeed,

$$(\tau' \leqslant t) = (\tau \leqslant T_{k-1}) \cup ((\tau > T_{k-1}) \cap (\tau \wedge T_k \leqslant (t + T_{k-1}) \wedge T_k))$$

and this is an $\mathfrak{H}_t$-set since $(\tau \leqslant T_{k-1}) \in \mathfrak{F}_{T_{k-1}} \subset \mathfrak{H}_t$. But from parts 2 and 3 of Theorem (A2.1) we know that $\mathfrak{H}_t = \mathfrak{F}_{T_{k-1}} \vee \sigma\{X_{(s+T_{k-1})\wedge T_k}, s \in [0,t]\}$ and $\mathfrak{F}_{T_{k-1}} = \sigma\{\xi_1, \ldots, \xi_{k-1}\}$, so that applying Lemma (A2.7) we conclude that

$$\tau' \wedge (T_k - T_{k-1}) = s_k(\xi_1, \ldots, \xi_{k-1}) \wedge (T_k - T_{k-1})$$

for some measurable function $s_k : \Omega_{k-1} \to \mathbb{R}_+$. Hence

$$\begin{aligned} \tau \wedge T_k &= \tau \wedge T_{k-1} I_{(t \leqslant T_{k-1})} + (T_{k-1} + \tau') I_{(\tau > T_{k-1})} \\ &= \tau \wedge T_{k-1} I_{(t \leqslant T_{k-1})} + ((T_{k-1} + s_k) \wedge T_k) I_{(\tau > T_{k-1})}. \end{aligned}$$

This completes the proof. □

A3 Predictability

The concept of predictability was introduced (not originally under that name) by P.A. Meyer to obtain uniqueness in the decomposition of a submartingale into the sum of a martingale and an increasing process. Consider for example a Poisson process $N_t = \sum_i I_{(t \geqslant T_i)}$, where T_0 and $(T_k - T_{k-1})$, $k = 1, 2, \ldots$ are i.i.d. random variables with $P[T_k - T_{k-1} > t] = e^{-\lambda t}$. Then $EN_t = \lambda t$ and it is easy to show that $M_t := N_t - \lambda t$ is a martingale (this is a special case of Proposition (A6.1) below.) We call λt the *compensator* of N_t. Since N_t is an increasing process ($N_t \geqslant N_s$ for $t \geqslant s$) it is certainly a submartingale. We can therefore decompose it into the sum of a martingale and an increasing process in at least two ways, namely $N_t = M_t + \lambda t$ and $N_t = 0 + N_t$. In order to rule out the second, trivial, decomposition we must place some restriction on the class of increasing processes we are prepared to consider as compensators. In this case the process λt is both continuous and deterministic. But it is easy to construct examples where no continuous or deterministic compensator exists (this will be evident in section A5 below) and predictability is just the right requirement to secure both existence and uniqueness in a general context. It is, however, widely regarded as a somewhat arcane concept the intuitive significance of which is not easy to grasp (the reader can consult Elliott (1982), Chapter 5 for a clear account). Fortunately, we do not need it. The only filtrations considered in this book are those associated with stochastic jump processes or, equivalently, piecewise-deterministic processes, and for these filtrations a constructive definition is possible which is equivalent to the general definition when the latter is specialized to the jump process case. We will not demonstrate the equivalence here; a proof can be found in Boel, Varaiya and Wong (1975).

Let $\mathfrak{F}_t$ be the natural filtration of a jump process (x_t) with jump times $T_1, T_2, \ldots$ as defined in section A1.

(A3.1) *Definition.* A stochastic process $\phi(t,\omega)$ is *predictable* if there exist measurable functions $\phi_1 : \mathbb{R}_+ \to \mathbb{R}$, $\phi_k : \mathbb{R}_+ \times \Omega_{k-1} \to \mathbb{R}$, $k=2,3,\ldots$ and $\phi_\infty : \mathbb{R}_+ \times \Omega \to \mathbb{R}$ such that

$$\text{(A3.2)} \quad \phi(t,\omega) = \phi_1(t) I_{(0 \leqslant t \leqslant T_1)} + \sum_{k=2}^{\infty} \phi_k(t, \omega_{k-1}) I_{(T_{k-1} < t \leqslant T_k)} + \phi_\infty(t,\omega) I_{(t \geqslant T_\infty)}.$$

The key point here is that $\phi(t) = \phi_k(t)$ for t up to *and including* T_k.

(A3.3) *Definition.* A stopping time T is *predictable* if the process $I_{(t \geqslant T)}$ is predictable.

(A3.4) *Proposition.* T_∞ is a predictable stopping time.

Proof. $I_{(t \geqslant T_\infty)}$ has the representation (A3.2) with $\phi_k = 0$, $k < \infty$, and $\phi_\infty(t,\omega) = 1$. □

Any deterministic process is predictable so, returning to the Poisson process example, the decomposition $N_t = M_\tau + \lambda t$ does give N_t as the sum of a martingale and a predictable process. However this is not true of the decomposition $N_t = 0 + N_t$, as a consequence of the following theorem, the main result of this section.

(A3.5) *Theorem.* If a process $\phi(t,\omega)$ is $\mathfrak{F}_t$-predictable and is a uniformly integrable martingale with $\phi(0,\omega) = 0$, then $\phi(t,\omega) = 0$ for all t, a.s.

Proof. Apply the optional sampling theorem to the stopping times $s \wedge T_1$, $t \wedge T_1$ with $s \leqslant t$. Then

$$\phi(s \wedge T_1, \omega) = E[\phi(t \wedge T_1, \omega) | \mathfrak{F}_{s \wedge T_1}] \quad \text{a.s.,}$$

and, because ϕ is predictable, $\phi(t,\omega) = \phi_1(t)$ for some non-random function ϕ_1 on the set $(t \leqslant T_1)$. If F denotes the survivor function of T_1, the above conditional expectation is given by

$$I_{(T_1 \leqslant s)}\phi_1(T_1) + I_{(T_1 > s)}\left(\frac{F(t)}{F(s)}\phi_1(t) - \frac{1}{F(s)}\int_{]s,t]}\phi_1(u)\mathrm{d}F(u)\right).$$

Thus on the set $(T_1 > s)$ we have

$$\phi_1(s) = \frac{F(t)}{F(s)}\phi_1(t) - \frac{1}{F(s)}\int_{]s,t]}\phi_1(u)\mathrm{d}F(u). \tag{A3.6}$$

Let $y(t) = \phi_1(t)F(t)$, $\mathrm{d}G(t) = \mathrm{d}F(t)/F(t)$. Then (A3.6) is equivalent to

$$y(t) - y(s) = \int_{]s,t]} y(u)\mathrm{d}G(u), \quad y(0) = 0$$

and $G(u)$ has bounded variation on any interval $[0, t]$ such that $t < c := \inf\{t : F(t) = 0\}$. From Lemma 13.4 of Elliott (1982), the unique (locally bounded) solution to this equation is $y(t) = 0$. Hence $\phi_1(t) = 0$ for $t \in [0, c[$. If $c < \infty$ and $F(c-) = 0$ then can take $\phi_1(t) = 0$ for all $t \geqslant 0$ since $P[T_1 \geqslant c] = 0$. If $c < \infty$ and $P(T_1 = c) = F(c-) > 0$, apply the optional sampling theorem to the stopping times $0, T_1$. Since $\phi_1(t) = 0, t < c$ we see that $0 = E[\phi_1(T_1)] = \phi_1(c)F(c-)$ and hence that $\phi_1(c) = 0$. We can now apply the same argument inductively on stochastic intervals $]T_{k-1}, T_k]$ to show that $\phi_k = 0$, $k = 2, 3, \ldots$. □

(A3.7) *Corollary.* The Poisson process N_t is not predictable, for any $\lambda > 0$.

Proof. We know that $M_t := N_t - \lambda t$ is a martingale, and since N_t is increasing it is clear that $M_{t \wedge n}$ is a uniformly integrable martingale for any $n > 0$. If N_t is predictable then $M_{t \wedge n}$ is a predictable martingale, and hence equal to 0 by Theorem (A3.5). This is a contradiction unless $\lambda = 0$. □

The same reasoning shows that λt is the unique predictable compensator of N_t, since if ϕ_t were another then $M'_t = N_t - \phi_t$ would be a martingale and $M_t - M'_t = \phi_t - \lambda t$ a predictable martingale; hence $\phi_\tau = \lambda t$. Predictable stopping times were defined in Definition (A3.3) above. The following more explicit characterization is easily obtained from the definition of predictability, and complements the description of an *arbitrary* stopping time given by Theorem (A2.3).

(A3.8) *Proposition.* Let T be a predictable $\mathfrak{F}_t$-stopping time. Then there exist a constant s_1, $\mathfrak{F}_{T_{k-1}}$-measurable random variables s_k for $k = 2, 3, \ldots$ and an $\mathfrak{F}$-measurable random variable s_∞, all taking values in $[0, \infty]$ such that, with $T_0 = 0$, $T = T_{p-1} + s_p$, where $p = \inf\{k : T_{k-1} + s_k \leqslant T_k\}$ or $T = T_\infty + s_\infty$ if the set $\{\ldots\}$ is empty.

(A3.9) *Example*
$T := T_7 + 1$ is a predictable time, with $s_k = \infty$, $k \leqslant 7$, $s_8 = 1$, $s_k = (1 - (T_{k-1} - T_7)) \vee 0, 9 \leqslant k \leqslant \infty$.

Finally, it should be emphasized that all the results in this section are entirely special to jump process filtrations. In particular, the main result Theorem (A3.5) is certainly false for general filtrations if 'predictable' is interpreted in the sense of the *théorie générale des processus*. For example, if B_t is a Brownian motion and $\mathfrak{F}_t^B$ its natural filtration, then $B_{t \wedge 1}$ is a non-zero predictable u.i. martingale.

A4 The single-jump process

To analyse jump-process martingales, we begin by studying in detail the single-jump case; the original process can then be treated by decomposing it into a sum of single-jump processes starting at the successive jump times of the original process.

Formally, the single-jump process is the special case of the jump process definition in section A1 in which $\mu^2(\omega_1(\omega); \{\Delta\}) = 1$, but it is more convenient to define it on its own canonical space $(Y, \mathfrak{Y})$ equipped with a probability measure for satisfying (A1.1); we omit the affix '1' throughout and call the coordinate map $\xi = (T, Z)$ for $\xi \in \mathbb{R}_+ \times X$. The process sample path is then

$$x_t = \begin{cases} z_0, & t < T \\ Z, & t \geqslant T. \end{cases}$$

As before $(\mathfrak{F}_t)$ denotes the completed natural filtration of (x_t). It is not hard to see that $\mathfrak{F}_t$ consists of all sets of the form $A \cap ([0, t] \times X)$, where $A \in \mathfrak{F}$ together with $A_0(t) := (]t, \infty[\times X) \cup \{\Delta\}$ as an atom. (By an 'atom' of the completed σ-field, we mean that if $A_0(t)$ is expressed as a disjoint union $A_0(t) = A_1 \cup A_2$ then $PA_1 = 0$ or $PA_2 = 0$.)

(A4.1) *Lemma.* Let τ be an $\mathfrak{F}_t$-stopping time. Then there exists $t_0 \in [0, \infty]$ such that $\tau \wedge T = t_0 \wedge T$.

Proof. This is a special case of Lemma (A2.7). □

For $A \in \mathfrak{B}(X)$, define

$$F^A(t) := \mu(]t, \infty[\times A)$$

and

$$F(t) := F^X(t) + \mu(\{\Delta\}) = P(T > t).$$

These are right-continuous, decreasing functions. Now define

$$c := \inf\{t : F(t) = 0\},$$

so that $P(t \leqslant c) = 1$. We have to distinguish three cases

$$\begin{aligned} &\text{case } 1: c = \infty \\ &\text{case } 2: c < \infty \quad \text{and} \quad F(c-) = 0 \\ &\text{case } 3: c < \infty \quad \text{and} \quad F(c-) > 0. \end{aligned}$$

(Here and throughout $F(c-)$ denotes the left-hand limit: $F(c-) = \lim_{t \uparrow c} F(t)$.)

Any uniformly integrable (u.i.) martingale M_t of $\mathfrak{F}_t$ takes the form $M_t = E[M_\infty | \mathfrak{F}_t]$ for some integrable $\mathfrak{F}_\infty$-measurable random variable M_∞. Here $\mathfrak{F}_\infty = \mathfrak{F}$ and hence all such random variables are of the form $M_\infty = h(T, Z)$ for some measurable function h satisfying

$$E|h(T, Z)| = \int_{]0, \infty] \times X} |h(t, z)| \mu(\mathrm{d}t, \mathrm{d}z) + |h(\Delta)| \mu(\{\Delta\}) < \infty.$$

It will be notationally convenient to write the right-hand side of this expression as

$$\int_{]0, \infty] \times X} |h| \mathrm{d}\mu$$

and a similar convention will apply below to integrals over sets denoted $]t, c] \times X$ when $c = \infty$. One can then check from the definition of conditional expectation that the u.i. martingale M_t has the following explicit expression:

$$M_t = E[h(T, Z) | \mathfrak{F}_t]$$

$$= I_{(t \geqslant T)} h(T, Z) + I_{(t < T)} \frac{1}{F(t)} \int_{]t, c] \times X} h(t, z) \mu(\mathrm{d}t, \mathrm{d}z). \tag{A4.2}$$

A process (M_t) is a *local martingale* if there exists an increasing sequence τ_n of $\mathfrak{F}_t$ stopping times such that $\tau_n \uparrow \infty$ a.s. and $M_t^n := M_{t \wedge \tau_n}$ is a u.i. martingale for each n.

(A4.3) *Theorem.* Let M_τ be a local martingale. Then:

1. M_τ is stopped at T, i.e. $M_\tau = M_{\tau \wedge T}$ a.s.;
2. In cases 1 and 2, M_t is a martingale on $[0, c[$;
3. In case 3, M_t is a u.i. martingale.

Proof.

1. Note from (A4.2) above that any u.i. martingale is stopped at T. Hence if τ_n is a sequence of localizing times, then

$$M_t = \lim_{n \to \infty} M_{t \wedge \tau_n} = \lim_{n \to \infty} M_{t \wedge \tau_n \wedge T} = M_{t \wedge T} \quad \text{a.s.}$$

2. If $\tau_k \geqslant T$ as for some k then using 1 we have

$$M_{t \wedge \tau_k} = M_{t \wedge \tau_k \wedge T} = M_{t \wedge T} = M_t \quad \text{a.s.},$$

so that M_t is a u.i. martingale. Thus suppose $P(\tau_k < T) > 0$ for all k. By Lemma (A4.1) there is a sequence of real numbers t_k such that $\tau_k \wedge T = t_k \wedge T$ and we must have $t_k \uparrow c$ since $\tau_k \uparrow \infty$ a.s. Then

$$M_{t \wedge \tau_k} = M_{t \wedge \tau_k \wedge T} = M_{t \wedge t_k \wedge T} = M_{t \wedge t_k}.$$

Thus M_t is a u.i. martingale on $[0, t_k]$ and hence a martingale on $[0, c[$ since $t_k \uparrow c$.
3. In case $3, 0 < F(c-) = P(T = c)$. Consider the sequence t_k as above. If $t_k < c$ for all k then on the set $(T = c)$ we have $\tau_k = t_k$ for all k, so that $\tau_k \not\to \infty$. Thus there must exist k' such that $t_{k'} = c$. But then $M_t = M_{t \wedge \tau_{k'}}$, so that M_t is a u.i. martingale. □

We now introduce the fundamental family of point processes associated with the jump process (x_t). For $A \in \mathfrak{B}(X)$ and $t \geqslant 0$ define

$$p(t, A) := I_{(t \geqslant T)} I_{(Z \in A)}$$

$$\tilde{p}(t, A) := -\int_{]0, T \wedge t]} \frac{1}{F(s-)} \, \mathrm{d}F^A(s)$$

$$q(t, A) := p(t, A) - \tilde{p}(t, A).$$

Note that the process $t \to p(t, A)$ has sample functions which are either identically zero, or have a unit jump at T if $T < t$ and $Z \in A$. We now show that $\tilde{p}$ is the *compensator* of p.

(A4.4) *Theorem.* For each $A \in \mathfrak{B}(X)$, $\tilde{p}(t, A)$ is the unique predictable process such that the process $t \to p(t, A)$ is an $\mathfrak{F}_t$-martingale.

Proof. $\tilde{p}(t, A)$ is clearly a predictable process in accordance with Definition (A3.1). That it is a compensator follows by direct computation. Take $t > s$; then

$$E[p(t, A) - p(s, A) \mid \mathfrak{F}_s] = I_{(s \leqslant T)} \frac{1}{F(s)} (F^A(s) - F^A(t)).$$

(Note that $p(t, A) - p(s, A) = 0$ if $s \geqslant T$.) On the other hand $\tilde{p}(t, A)$ is a function of T only, and $F(t)$ is the survivor function of T. Hence

$$E[\tilde{p}(t, A) - \tilde{p}(s, A) \mid \mathfrak{F}_s] = I_{(s < T)} \left\{ -\frac{F(t)}{F(s)} \int_{]s,t]} \frac{1}{F(u-)} \mathrm{d}F^A(u) \right.$$
$$\left. + \frac{1}{F(s)} \int_{]s,t]} \int_{]s,r]} \frac{1}{F(u-)} \mathrm{d}F^A(u) \mathrm{d}F(r) \right\}.$$

Interchanging the order of integration, the second term on the right is

$$\frac{1}{F(s)} \int_{]s,t]} \frac{1}{F(u-)} \int_{[u,t]} \mathrm{d}F(r) \mathrm{d}F^A(\mathrm{u})$$
$$= \frac{1}{F(s)} \int_{]s,t]} \frac{1}{F(u-)} (F(t) - F(u-)) \mathrm{d}F^A(u)$$
$$= \frac{F(t)}{F(s)} \int_{]s,t]} \frac{1}{F(u-)} \mathrm{d}F^A(u) + \frac{1}{F(s)} (F^A(t) - F^A(s)).$$

Thus $E[\tilde{p}(t, A) - \tilde{p}(s, A) \mid \mathfrak{F}_s] = E[p(t, A) - p(s, A) \mid \mathfrak{F}_s]$ and hence $q(t, A) = p(t, A) - \tilde{p}(t, A)$ is a martingale, since both p and $\tilde{p}$ are $\mathfrak{F}_t$-adapted processes. The predictable compensator is unique, by Theorem (A3.5). □

We now want to consider *stochastic integrals* with respect to the family of martingales $q(t, A)$. These will simply be differences of ordinary (Stieltjes) integrals with respect to p and $\tilde{p}$, applied to suitable classes of integrands.

For p, the appropriate definition of the integral is clear. Identify $p(t, A)$ with a random set function ν on $\mathbb{R}_+ \times X$ such that

$$\nu([0, t] \times A) = p(t, A).$$

Then it is clear that ν is simply the Dirac measure $\delta_{(T,Z)}$ at (T, Z), since $\nu([0, t] \times A) = 1$ if $(T, Z) \in [0, t] \times A$ and $= 0$ otherwise. We therefore define, for any measurable function $g: Y \to R$

$$\int_Y g\mathrm{d}p = \int_{]0,\infty] \times X} g(t, z)p(\mathrm{d}x, \mathrm{d}z) := g(T, Z).$$

Throughout, we will only consider functions g such that $g(\Delta) = 0$. Then we say that $g \in L_1(p)$ if

$$\| g \|_{L_1(p)} := \mathbb{E} \int_Y |g| \mathrm{d}p = \mathbb{E}|g(T, Z)| < \infty.$$

Thus $L_1(p) = L_1(Y, \mathfrak{Y}, \mu)$. We say that $g \in L_1^{\mathrm{loc}}(p)$ if $gI_{(t < \tau_n)} \in L_1(p)$ for some sequence of stopping times $\tau_n \uparrow \infty$ a.s.

For $\tilde{p}$ we adopt a similar approach. We identify $\tilde{p}$ with the random set function $\tilde{\nu}$ defined by

$$\tilde{\nu}(]0, t] \times A) = \tilde{p}(t, A).$$

It is then easy to see that $\tilde{\nu}$ satisfies

$$\tilde{\nu}([0, t] \times A) = \int_{]0,t] \times A} I_{(s \leqslant T)} \frac{1}{F(s-)} \mu(\mathrm{d}s, \mathrm{d}z)$$

and therefore that $\tilde{\nu}$ coincides with the random measure

$$\tilde{\nu}(F) = \int_F I_{(s \leqslant T)} \frac{1}{F(s-)} \mu(\mathrm{d}s, \mathrm{d}z)$$

for $F \in \mathfrak{Y}$. We thus define

$$\int_Y g\mathrm{d}\tilde{p} := \int_Y g(s, z)\tilde{p}(\mathrm{d}s, \mathrm{d}z) := \int_Y g(s, z)\tilde{\nu}(\mathrm{d}s, \mathrm{d}z)$$

$$= \int_{]0,\infty] \times X} g(s, z) I_{(s \leqslant T)} \frac{1}{F(s-)} \mu(\mathrm{d}s, \mathrm{d}z).$$

Again, we consider only functions g such that $g(\Delta) = 0$. The integral exists when $g \in L_1(\tilde{p})$ defined by

$$L_1(\tilde{p}) = \left\{ g: Y \to R: \| g \|_{L_1(p)} := \mathbb{E} \int_Y |g| \mathrm{d}\tilde{p} < \infty \right\}$$

and

$$L_1^{\mathrm{loc}}(\tilde{p}) = \{ g: gI_{(s<\tau_n)} \in L_1(\tilde{p}) \quad \text{for stopping times } \tau_n \uparrow \infty \text{ a.s.} \}.$$

(A4.5) *Proposition A.3.* $L_1(p) = L_1(\tilde{p})$ and $L_1^{\mathrm{loc}}(p) = L_1^{\mathrm{loc}}(\tilde{p})$. Also

(A4.6) $$\| g \|_{L_1(\tilde{p})} = \| g \|_{L_1(\tilde{p})}.$$

Proof. We need only show (A4.6). Note again that $\int g \mathrm{d}\tilde{p}$ is a function only of T, whose survivor function is F. Hence

$$\begin{aligned} \| g \|_{L_1(\tilde{p})} &= - \int_{]0,\infty]} \int_Y I_{(s \leqslant t)} |g(s,x)| \frac{1}{F(s-)} \mu(\mathrm{d}s, \mathrm{d}x) \mathrm{d}F(t) \\ &= - \int_Y |g(s,x)| \frac{1}{F(s-)} \left(\int_{[s,\infty]} \mathrm{d}F(t) \right) \mu(\mathrm{d}s, \mathrm{d}z) \\ &= \int_Y |g(s,x)| \mu(\mathrm{d}s, \mathrm{d}z) = \| g \|_{L_1(p)}. \end{aligned}$$ □

(A4.7) *Proposition.* $L_1^{\mathrm{loc}}(p) = L_1^{\mathrm{loc}}(\mathrm{d}\mu)$, where

$$L_1^{\mathrm{loc}}(\mathrm{d}\mu) = \{ g: Y \to \mathbb{R}: gI_{(s \leqslant t)} \in L_1(Y, \mathfrak{Y}, \mu) \quad \text{for all} \quad t < c \}.$$

Proof. Suppose $g \in L_1^{\mathrm{loc}}(p)$, let τ_n be a sequence of localizing times and let t_n be the associated sequence of constants such that $\tau_n \wedge T = t_n \wedge T$. Then

$$\int_Y gI_{(t<\tau_k)} \mathrm{d}p = g(T,Z) I_{(T<\tau_k)}$$

and $(T < \tau_k) =]0, t_k[\times X$. Hence $g \in L_1^{\mathrm{loc}}(\mathrm{d}\mu)$, since $t_k \uparrow c$. Conversely, if $g \in L_1^{\mathrm{loc}}(\mathrm{d}\mu)$, take any sequence $t_k \uparrow c$ and introduce the following stopping times:

$$\begin{aligned} &c = \infty: && \tau_k = k \\ &c < \infty, F(c-) = 0: && \tau_k = kI_{(T \leqslant t_k)} + t_k I_{(T > t_k)} \\ &c < \infty, F(c-) > 0: && \tau_k = \infty. \end{aligned}$$

then $\tau_k \uparrow \infty$ a.s. and it is easily shown that $gI_{(t \leqslant \tau_k)} \in L_1(p)$.

For $g \in L_1^{\text{loc}}(p)$ we can now define a process (M_t^g) by

$$\begin{aligned} M_t^g &:= \int_Y I_{(s \leq t)} g(s,z) q(\mathrm{d}s, \mathrm{d}z) \\ &= \int_Y I_{(s \leq t)} g(s,z) p(\mathrm{d}s, \mathrm{d}z) - \int_Y I_{(s \leq t)} g(s,z) \tilde{p}(\mathrm{d}s, \mathrm{d}z). \end{aligned}$$

M_t^g is given more explicitly, from the definition, as follows

(A4.8) $$M_t^g = g(T,Z) I_{(t \geq T)} + \int_{]0, T \wedge t] \times X} g(s,z) \frac{1}{F(s-)} \mu(\mathrm{d}s, \mathrm{d}z).$$

The following result is then proved by direct calculations similar to those in the proof of Theorem (A4.4) above.

(A4.9) *Theorem.* (M_t^g) is a martingale for $g \in L_1(p)$, and a local martingale for $g \in L_1^{\text{loc}}(p)$.

Suppose M_t is a uniformly integrable $\mathfrak{F}_t$-martingale; then M_t takes the form $M_t = \mathbb{E}[M_\infty | \mathfrak{F}_t]$ for some $\mathfrak{F}$-measurable random variable M_∞ such that $\mathbb{E}|M_\infty| < \infty$. However, all such random variables can be written as $M_\infty = h(T,Z)$ for some measurable function h on Y such that $\mathbb{E}|h(T,Z)| < \infty$. Then

$$\begin{aligned} M_t &= \mathbb{E}[h(T,Z) | \mathfrak{F}_t] \\ &= I_{(t \geq T)} h(T,Z) + I_{(t < T)} \frac{1}{F(t)} \int_{]t, \infty] \times X} h \, \mathrm{d}\mu. \end{aligned}$$

If $M_0 = 0$ a.s. then $\mathbb{E}h(T,Z) = 0$, i.e.

$$0 = \int_{]0,t] \times X} h \, \mathrm{d}\mu + \int_{]t, \infty] \times X} h \, \mathrm{d}\mu,$$

so every u.i. martingale such that $M_0 = 0$ takes the form

(A4.10) $$M_t = I_{(t \geq T)} h(T,Z) - I_{(t < T)} \frac{1}{F(t)} \int_{]0,t] \times X} h \, \mathrm{d}\mu.$$

We want to show that $M_t = M_r^g$ for some integrand g. To get an idea what g must be, consider the following example

(A4.11) *Example*
Take $X = \mathbb{R}$ and suppose $\mu(\mathrm{d}s, \mathrm{d}x) = \Psi(s,x) \mathrm{d}s \mathrm{d}x$ for some density

function Ψ. Then from (A4.8), for $q \in L_1(p)$

$$\text{(A4.12)} \quad M_t^g = I_{(t \geqslant T)}\left(g(T,Z) - \int_0^T \int_R \frac{1}{F(s)} g(s,x)\Psi(s,x)\mathrm{d}x\mathrm{d}s \right) + I_{(t<T)} \int_0^t \int_R \frac{1}{F(s)} g(s,x)\Psi(s,x)\mathrm{d}x\mathrm{d}s.$$

If M_t is a martingale with associated function h, then $M_t = M_t^g$ only if the coefficients of $I_{(t \geqslant T)}$ in (A4.10) and (A4.12) agree, i.e.

$$\text{(A4.13)} \qquad h(t,z) = g(t,z) - \int_0^t \int_{\mathbb{R}} \frac{1}{F(s)} g(s,x)\Psi(s,x)\mathrm{d}x\mathrm{d}s.$$

Define $\eta(t) = g(t,z) - h(t,z)$ (noting that it does not depend on z) and

$$f(s) = \int_{\mathbb{R}} \Psi(s,x)\mathrm{d}x, \quad \gamma(s) = \int_{\mathbb{R}} h(s,x)\Psi(s,x)\mathrm{d}x.$$

Then from (A4.13) we have

$$\eta(t) = \int_0^t \int_{\mathbb{R}} \frac{1}{F(s)} (\eta(s) + h(s,x))\Psi(s,x)\mathrm{d}x\mathrm{d}s$$
$$= \int_0^t \frac{f(s)}{F(s)} \eta(s)\mathrm{d}s + \int_0^t \frac{1}{F(s)} \gamma(s)\mathrm{d}s.$$

Thus $\eta(t)$ satisfies the linear ordinary differential equation

$$\frac{\mathrm{d}}{\mathrm{d}t}\eta(t) = \frac{f(t)}{F(t)}\eta(t) + \frac{1}{F(t)}\gamma(t), \quad \eta(0) = 0$$

whose unique solution is

$$\eta(t) = \int_0^t \exp\left(\int_s^t \frac{f(u)}{F(u)} \mathrm{d}u \right) \frac{1}{F(s)} \gamma(s)\mathrm{d}s = \frac{1}{F(t)} \int_0^t \gamma(s)\mathrm{d}s,$$

where the last equality follows from the fact that $f(s) = -\mathrm{d}F(s)/\mathrm{d}s$. This shows that the coefficients of $I_{(t \geqslant T)}$ in (A4.10) and (A4.12) agree if

$$g(t,z) = h(t,z) + \frac{1}{F(t)} \int_0^t \int_{\mathbb{R}} h(s,x)\Psi(s,x)\mathrm{d}x\mathrm{d}s.$$

It is easily shown that with this choice of g the coefficients of $I_{(t \leqslant T)}$ in (A4.10) and (A4.12) agree as well, so $M_t = M_t^g$.

The general result is as follows.

(A4.14) *Theorem.* (M_t) is a local martingale of $\mathfrak{F}_t$ with $M_0 = 0$ if and only if $M_t = M_t^g$ for some $g \in L_1^{\text{loc}}(p)$.

Proof. We have already shown that M_t^g is a local martingale for $g \in L_1^{\text{loc}}(p)$. Thus suppose that M_t is a local martingale with $M_0 = 0$. We have two cases.

Case 1. $c < \infty$, $F(c-) > 0$: It was shown above that M_t is then a u.i. martingale, and hence of the form (A4.10) for some h with $\int |h| \mathrm{d}\mu < \infty$. Consider the function g given by

$$\text{(A4.15)} \qquad g(t,z) = h(t,z) + I_{(t<c)} \frac{1}{F(t)} \int_{]0,t] \times X} h(s,x)\mu(\mathrm{d}s, \mathrm{d}x).$$

We can verify by direct calculation that 1 $M_t = M_t^g$ for $t < c$, and 2 M_t and M_t^g are stopped at c and $M_c = M_c^g$ when $T(\omega) = c$. Now

$$\begin{aligned}
\| g \|_{L_1(p)} &= \int_Y |g| \mathrm{d}\mu \leqslant \int_Y |h| \mathrm{d}\mu \\
&\quad - \int_{]0,c]} \frac{1}{F(t)} \int_{]0,t] \times X} |h(s,x)| \mu(\mathrm{d}s, \mathrm{d}x) \mathrm{d}F(t) \\
&\leqslant \int_Y (h) \mathrm{d}\mu - \frac{1}{F(c-)} \int_{]0,c]} \int_{]0,t] \times X} |h| \mathrm{d}\mu \mathrm{d}F(t) \\
&= \int_Y |h| \mathrm{d}\mu + \frac{1}{F(c-)} \int_Y (F(s) - F(c-)) |h| \mathrm{d}\mu \\
&\leqslant \left(1 + \frac{1}{F(c-)}\right) \| h \|_{L_1(\mathrm{d}\mu)}.
\end{aligned}$$

Thus $g \in L_1(p)$.

Case 2. $c = \infty$ or $c < \infty$, $F(c-) = 0$. Here M_t is a u.i. martingale on $[0, r]$ for any $r < c$, and hence of the form (A4.10) for some h satisfying

$$\int_{]0,r] \times X} |h| \mathrm{d}\mu < \infty, \quad \text{for all} \quad r < c.$$

Calculations as before show that $M_t = M_1^g$ with g given by (A4.15), and for $r < c$

$$\int_{]0,r]\times X} |g|\mathrm{d}\mu \leqslant \int_{]0,r]\times X} |h|\mathrm{d}\mu - \int_{]0,r]} \frac{1}{F(t)} \int_{]0,t]\times X} |h|\mathrm{d}\mu\mathrm{d}F(t).$$

$$\leqslant \left(1 + \frac{1}{F(r)}\right) \int_{]0,r]\times X} |h|\mathrm{d}\mu.$$

This shows that $g \in L_1^{\mathrm{loc}}(p)$ in view of Proposition (A4.7). □

(A4.16) *Remark.* The following extension of the preceding results will be needed in the next section. Suppose that $(\Xi, \mathfrak{S}, m)$ is a probability space and define $\Omega = \Xi \times Y$. Let $\mathfrak{F}_0 = \mathfrak{S} \times Y$ and $\mathfrak{F}_t^0 = \mathfrak{F}_0 \vee \sigma\{x_s, s \leqslant t\}$, where the path x_t is defined as before except that z_0 is now an $\mathfrak{F}_0$-measurable random variable. Let $\mu(\xi; \mathrm{d}t, \mathrm{d}z)$ be the conditional measure of (T, Z) given $\mathfrak{F}_0$, so that a probability measure P on Ω is defined by

$$P(A_1 \times A_2) = \int_{A_1} \mu(\xi; A_2) m(\mathrm{d}\xi), \quad A_1 \in \mathfrak{S}, A_2 \in \mathfrak{Y}.$$

Let $\mathfrak{F}_t$ be the P-completion of $\mathfrak{F}_t^0$. Then the characterization of $\mathfrak{F}_t$-local martingales given in Theorem (A4.14) remains unchanged, except for the obvious modifications to the class of integrands g to allow for ξ-dependence.

A5 Local martingale representation for the general jump process

We now revert to consideration of the general multi-jump process as described in section A1. We define the family of elementary point processes $p(t, A)$ for $t \geqslant 0$, $A \in \mathfrak{B}(X)$ as

$$p(t, A) = \sum_k I_{(t \geqslant T_k)} I_{(Z_k \in A)},$$

and define

$$F^{A,1}(u) = \mu^1(]u, \infty[\times A), \quad F^1(u) = F^{X,1}(u) + \mu^1(\{\Delta\})$$

and for $k = 2, 3, \ldots$

$$F^{A,k}(\omega_{k-1}, h) = \mu^k(\omega^{k-1};]u, \infty[\times A)$$

$$F^k(\omega_{k-1}, u) = F^{X,k}(\omega_{k-1}, u) + \mu^k(\omega_{k-1}; \{\Delta\})$$

$$\Phi_k^A(\omega_{k-1}, s) = -\int_{]0,s]} \frac{1}{F^k(\omega_{k-1}, u-)} \mathrm{d}F^{A,k}(\omega_{k-1}, u).$$

Let

$$\tilde{p}(t, A) = \Phi_1^A(S_1) + \cdots + \Phi_{k-1}^A(\omega_{k-2}, S_{k-1}) + \Phi_k^A(\omega_{k-1}, t - T_{k-1}), \quad \text{for } t \in]T_{k-1}, T_k].$$

This is a predictable process.

(A5.1) *Proposition.* For $A \in \mathfrak{B}(X)$, let

$$q(t, A) = p(t, A) - \tilde{p}(t, A)$$

then for each fixed k, A, the process $q(t \wedge T_k, A)$ is an $\mathfrak{F}_t$-martingale, i.e. $\tilde{p}(t, A)$ is the predictable compensator of $p(t, A)$.

Proof. This is proved by direct calculation. □

An integrand g for stochastic integration is made up from functions g^k in the following way, where for each $k = 1, 2, \ldots, g^k: \Omega_k \to R$ is a measurable function such that $g(\omega_{k-1}; \Delta) = 0$ for all ω_{k-1}.

$$\text{(A5.2)} \qquad g(\omega, t, z) = \begin{cases} g^1(t, z), & t \leqslant T_1(\omega) \\ g^k(\omega_{k-1}; t, z), & T_{k-1}(\omega) < t \leqslant T_k(\omega) \\ 0, & t \geqslant T_\infty(\omega). \end{cases}$$

Equivalently, g is a measurable function such that for each $z \in X$ the map $(t, \omega) \to g(\omega, t, z)$ is a predictable process. The integrals of g with respect to p and $\tilde{p}$ are defined in a way which directly generalizes the definitions in section A4, namely

$$\int_\Omega g \mathrm{d}p = \sum_{k=1}^{\infty} g^k(S_1, Z_1, \ldots, S_k, Z_k)$$

$$\int_\Omega g \mathrm{d}\tilde{p} = \int_{Y_1} I_{(s \leqslant S_1)} g^1(s, z) \frac{1}{F^1(s-)} \mu^1(\mathrm{d}s, \mathrm{d}z)$$

$$+ \sum_{k=2}^{\infty} \int_{Y_k} I_{(s \leqslant S^k)} g^k(\omega_{k-1}, s, z) \frac{1}{F^k(\omega_{k-1}, s)} \mu^k(\omega_{k-1}; \mathrm{d}s, \mathrm{d}z).$$

Note that these are finite sums if $S_k = \infty$ for some k. The definitions of $L_1(p)$, $L_1^{\mathrm{loc}}(p)$, etc., read exactly as before, except that the localizing times τ_n are assumed to converge to T_∞, not ∞.

(A5.3) *Proposition.* Suppose $g \in L_1^{\mathrm{loc}}(p)$ and define

$$M_t^g = \int_{]0,t] \times X} g(w, s, z) q(\mathrm{d}s, \mathrm{d}z),$$

where $q = p - \tilde{p}$. Then there exists a sequence of stopping times τ_n such that $\tau_n < T_\infty, \tau_n \uparrow T_\infty$ and $M^g_{t \wedge \tau_n}$ is a u.i. martingale for each n.

Proof. Take $\tau_n = \sigma_n \wedge T_n$, where σ_n are localizing times for g, i.e. $gI_{t<\sigma_n} \in L_1(p)$ for each n. A direct calculation shows that $M^g_{t \wedge \tau_n}$ is a martingale, and $M^g_{t \wedge \tau_n} = \mathbb{E}[M^g_{\tau_n} | \mathfrak{F}_{t \wedge \tau_n}]$, showing that $M^g_{t \wedge \tau_n}$ is u.i. □

Let M_t be a u.i. $\mathfrak{F}_t$-martingale. Then $M_t = \mathbb{E}[M_\infty | \mathfrak{F}_t]$ for some $M_\infty \in \mathfrak{F}_\infty = \mathfrak{F}$ (Lemma (A1.2)). From Theorem (A2.1) we know that $\mathfrak{F} = \bigvee_n \mathfrak{F}_{T_n}$; hence

(A5.4) $$M_{T_\infty -} = \lim_{n \to \infty} M_{T_n} = \lim_{n \to \infty} \mathbb{E}[M_\infty | \mathfrak{F}_{T_n}] = \mathbb{E}[M_\infty | \mathfrak{F}] = M_\infty.$$

Thus any u.i. martingale is stopped at T_∞ and is left-continuous there. The same therefore applies to local martingales.

We now come to the main result.

(A5.5) *Theorem.* Let M_t be a local $\mathfrak{F}_t$-martingale. Then $M_t = M^g_t$ for some $g \in L^{\text{loc}}_1(p)$.

Proof. First, suppose that M_t is u.i. We can then express M_t in the form

(A5.6) $$M_t = M_{t \wedge T_1} + \sum_{k=2}^{\infty} (M_{t \wedge T_k} - M_{T_{k-1}}) I_{(t \geqslant T_{k-1})}.$$

Indeed, this is an identity if $t < T_\infty$, and the right-hand side is equal to $\lim_k M_{T_k} = M_{T_\infty -}$ if $t \geqslant T_\infty$; from (A5.4), $M_t = M_{T_\infty -}$ in this case. Define

(A5.7) $$X^k_t := M_{(t + T_{k-1}) \wedge T_k} - M_{T_{k-1}}, \quad t \geqslant 0.$$

Then

(A5.8) $$M_{t \wedge T_k} - M_{T_{k-1}} = X_{(t - T_{k-1}) \vee 0}$$

and X^k_t is a u.i. martingale with respect to the filtration $\mathfrak{H}_t = \mathfrak{F}_{(t + T_{k-1}) \wedge T_k}$. From Theorem (A2.1) we know that $\mathfrak{H}_t = \mathfrak{F}_{T_{k-1}} \vee \sigma\{x_{(s + T_{k-1}) \wedge T_k}, s \in [0, t]\}$, and thus X^k_t takes the form $X^k_t = \mathbb{E}[h^k(\omega_{k-1}; S_k, Z_k) | \mathfrak{H}_t]$.

Since $E|X^k_t| < \infty$ we have

$$\int_{\Omega_{k-1}} \int_Y |h^k(\eta; s, z)| \mu^k(\eta; \mathrm{d}s, \mathrm{d}z) \nu^{k-1}(\mathrm{d}\eta) < \infty,$$

where ν^k is the marginal distribution of ω_{k-1}. Thus from the 1-jump result, Theorem (A4.14), and Remark (A4.16), we can represent X_t^k as

$$X_t^k = \int_{]0,t]\times X} g^k(\omega_{k-1}; s, z) q^k(\mathrm{d}s, \mathrm{d}z),$$

where $q^k(t, A) := q((t + T_{k-1}) \wedge T_k, A)$, for some integrand g^k satisfying

$$\text{(A5.9)} \quad \int_{]0,r]\times X} |g^k| \mathrm{d}\mu^k \leqslant \left(1 + \frac{1}{F^k(\omega_{k-1}, r)}\right) \int_{]0,r]\times X} |h^k| d\mu^k$$

for all $r < c^k(\omega_{k-1}) := \inf\{t : F^k(\omega_{k-1}, t) = 0\}$. The collection g^k defines an integrand g such that $M_t = M_t^g$ a.s. for each t; it remains to show that $g \in L_1^{\mathrm{loc}}$.

For $n = 1, 2\ldots$, define $S_n^k(\omega_{k-1})$ as follows (omitting (ω_{k-1})-dependence for convenience): if $c^k = \infty$ or $c^k < \infty$ and $F^k(c^{k-}) \leqslant 1/n^3 : S_n^k :=$ $\inf\{t : F^k(t) \leqslant 1/n^3\}$; if $c^k < \infty$ and $F^k(c^k -) \geqslant 1/n^3 : S_n^k := c^k$.

Then, from (A5.9)

(A5.10)

$$\int_{\Omega_{k-1}} \int_Y I_{(S < S_n^k)} |g^k| \mathrm{d}\mu^k \mathrm{d}\nu^{k-1} \leqslant (1 + n^3) \int_{\Omega_{k-1}} \int_Y |h^k| \mathrm{d}\mu^k \mathrm{d}\nu^{k-1} < \infty.$$

Now define $\tau_n := T_j + S_n^j$, where $j := \min\{k : T_k + S_n^k \leqslant T_{k+1}\}$. Then τ_n is an $\mathfrak{F}_t$-stopping time, and

$$P[\tau_n < T_n] \leqslant P \bigcup_{k=1}^{n} (S_n^k < S_k) \leqslant n \frac{1}{n^3} = \frac{1}{n^2}.$$

Thus $\sum P[\tau_n < T_n] < \infty$, so that by the Borel–Cantelli Lemma (12.2), $P(\liminf(\tau_n \geqslant T_n))] = 1$. It follows that $\tau_n \uparrow T_\infty$ a.s. Now

$$\int_Y |g| I_{(t \leqslant T_n)} \mathrm{d}p = |g(T_1, Z_1)| + \cdots + |g^n(T_1, Z_1, \ldots T_{n-1}, Z_{n-1}; S_n Z_n)|,$$

so, using (A5.10)

$$\mathbb{E} \int_Y I_{(t \wedge \tau_n \wedge T_n)} |g| \mathrm{d}p| \leqslant \sum_{k-1}^{n} \int_{\Omega_{k-1}} \int_Y I_{(S < S_n^k)} |g^k| d\mu^k d\nu^{k-1} < \infty.$$

Since $\tau_n \wedge T_n \uparrow T_\infty$, this shows that $g \in L_{\mathrm{loc}}^1(p)$, as claimed.

(A5.11) *Remark*

1. When $T_\infty = \infty$ a.s., Proposition (A5.3) and Theorem (A5.5) assert

that M_t is a local $\mathfrak{F}_t$-martingale if and only if $M_t = M_t^g$ for some $g \in L_1^{\text{loc}}(p)$.

2. The situation is slightly unsatisfying in that we have shown $g \in L_1(p)$ is a sufficient condition for M_t^g to be a martingale, but we have not shown that this condition is necessary. The same point arises in connection with Ito stochastic integrals: if B_t is a Brownian motion and Ψ_t a nonanticipative integrand then the Ito integral $\int_0^t \Psi_s dB_s$ is defined and is a local martingale when $\int_0^t \Psi_s^2 ds < \infty$ a.s. for all t. The integral is a martingale when $\mathbb{E} \int_0^t \Psi_s^2 ds < \infty$ for all t, but again this is only a sufficient condition.

Bibliography

Asmussen, S. (1987) *Applied Probability and Queues*, Wiley, New York.

Beard, R.E., Pentikäinen, T. and Pesonen, E. (1984) *Risk Theory*, 3rd edn, Chapman & Hall, London.

Bensoussan, A. (1982) *Stochastic Control by Functional Analysis Methods*, North Holland, Amsterdam.

Bensoussan, A. and Lions, J.L. (1975) Nouvelles méthodes en contrôle impulsionnel. *Applied Math. and Optimization*, **1**, 289–312.

Bensoussan, A. and Lions, J.L. (1978) *Applications des Inéquations Variationelles en Contrôle Stochastique*, Dunod, Paris.

Bensoussan, A. and Lions, J.L. (1982) *Contrôle Impulsionnnel et Inéquations Quasivariationelles*, Dunod, Paris.

Bertsekas, D. and Shreve, S.E. (1978) *Stochastic Optimal Control: the Discrete-Time Case*, Academic Press, New York.

Billingsley, P. (1968) *Convergence of Probability Measures*, Wiley, New York.

Billingsley, P. (1986) *Probability and Measure*, 2nd edn, John Wiley, New York.

Boel, R., Varaiya, P. and Wong, E. (1975) Martingales on jump processes I: representation results. *SIAM J. Control and Optimization*, **13**, 999–1021.

Boothby, W. M. (1984) *An Introduction to Differentiable Manifolds and Riemannian Geometry*, 2nd edn, Academic Press, New York.

Brémaud, P. (1981) *Point Processes and Queues: Martingale Dynamics*, Springer-Verlag, Berlin.

Bühlmann, H. (1970) *Mathematical Methods in Risk Theory*, Springer-Verlag, Berlin.

Chou, C.S. and Meyer, P.A. (1975) Sur la répresentation des martingales comme intégrales stochastiques dans les processus ponctuels. *Séminaire de Probabilités IX*, Lecture Notes in Mathematics **465**, Springer-Verlag, Berlin.

Çinlar, E. (1975) *Introduction to Stochastic Processes*, Prentice Hall, Englewood Cliffs, NJ.

Clarke, F.H. (1983) *Nonsmooth Analysis and Optimization*, Wiley, New York.

Coddington, E. A. and Levinson, N. (1955) *Theory of Ordinary Differential Equations*, McGraw-Hill, New York.

Costa, O.L.V. (1990) Stationary distributions for piecewise-deterministic Markov processes. *J. Appl. Prob.*, **27**, 60–73.

Costa, O.L.V. (1991) Impulse control of piecewise-deterministic processes via linear programming. *IEEE Trans. Automatic Control*, **AC-36**, 371–5.

Costa, O.L.V. and Davis, M.H.A. (1988) Approximations for optimal stopping of a piecewise-deterministic process. *Math. Control Signals Systems*, **1**, 123–46.

Costa, O.L.V. and Davis, M.H.A. (1989) Impulse control of piecewise-deterministic processes. *Math. Control Signals Systems*, **2**, 187–206.

Cox, D.R. (1955) The analysis of non-Markovian stochastic processes by the inclusion of supplementary variables. *Proc. Camb. Phil. Soc.*, **51**, 433–41.

Cox, D.R. (1962) *Renewal Theory*, Methuen, London.

Cox, D.R. and Miller, H.D. (1965) *The Theory of Stochastic Processes*, Chapman & Hall, London.

Dassios, A. and Embrechts, P. (1989) Martingales and insurance risk. *Commun. Statist. – Stochastic Models*, **5**(2), 181–217.

Davis, M.H.A. (1976) The representation of martingales of jump processes. *SIAM J. Control and Optimization*, **14**, 623–38.

Davis, M.H.A. (1977) *Linear Estimation and Stochastic Control*, Chapman & Hall, London.

Davis, M.H.A. (1984a) *Lectures on Stochastic Control and Nonlinear Filtering*, Tata Institute of Fundamental Research Series, Springer-Verlag, Berlin.

Davis, M.H.A. (1984b) Piecewise-deterministic Markov processes: a general class of non-diffusion stochastic models (with discussion). *J. Royal Statist. Soc. (B)*, **46**, 353–88.

Davis, M.H.A. (1986) Control of piecewise-deterministic processes via discrete-time dynamic programming, in *Stochastic Differential Systems* (ed. M. Kohlmann), Lecture Notes in Control and Information Sciences **78**, Springer-Verlag, Berlin.

Davis, M.H.A. and Lopes, E.P. (1992) Optimal stopping of Piecewise-deterministic processes via the penalty method. *Applied Math. and Optimization* (to appear).

Davis, M.H.A., Dempster, M.A.H., Sethi, S.P. and Vermes, D. (1987) Optimal capacity expansion under uncertainty. *Adv. Appl. Prob.*, **19**, 156–76.

Dellacherie, C. and Meyer, P.-A. (1975/80) *Probabilités et Potentiel*, Hermann, Paris, Ch. I–IV and V–VIII.

Dempster, M.A.H. (1991) Optimal control of piecewise-deterministic Markov Processes, in *Applied Stochastic Analysis* (ed. M.H.A. Davis and R.J. Elliott), Gordon and Breach, New York, pp. 303–25.

Dempster, M.A.H. and Ye, J.J. (1990) A maximum principle for control of piecewise deterministic Markov processes, in A.G. Law and C. L. Wang (eds.) *Approximation, optimization and Computing: Theory and Applications*, North Holland, Amsterdam, pp. 235–40.

Dempster, M.A.H. and Ye, J.J. (1993a) A generalized Bellman–Hamilton–Jacobi equation for control problems with boundary conditions. *Math. of Control Signals and Systems*, to appear.

Dempster, M.A.H. and Ye, J.J. (1993b) Impulse control of piecewise deterministic Markov processes. *Ann. Appl. Prob.*, to appear.

Dempster, M.A.H. and Ye, J.J. (1992) Necessary and sufficient optimality conditions for control of piecewise-deterministic Markov processes. *Stochastics and Stochastics Reports*, **40**, 125–45.

Doob, J.L. (1953) *Stochastic Processes*, Wiley, New York.

Dudley, R.M. (1989) *Real Analysis and Probability*, Wadsworth & Brooks/Cole, Pacific Grove, Calif.

Dunford, N. and Schwartz, J.T. (1963) *Linear Operators, Part I: General Theory*, Wiley/Interscience, New York.

Dynkin, E.B. (1963) The optimal choice of the instant for stopping a Markov process. *Dokl. Akad. Nauk USSR*, **150**, 238–40.

Dynkin, E.B. (1965) *Markov Processes I, II*, Springer-Verlag, Berlin.

Dynkin, E.B. and Yushkevich, A.A. (1978) *Controlled Markov Processes and their Applications*, Springer-Verlag, Berlin.

Elliott, R.J. (1982) *Stochastic Calculus and Applications*, Springer-Verlag, Berlin.

Erlenkotter, D., Sethi, S. and Okada, N. (1981) Planning for surprise: water resources development under demand and supply uncertainty I: the general model, Working Paper, International Institute for Applied Systems Analysis, Laxenburg, Austria.

Ethier, S.N. and Kurtz, T.G. (1986) *Markov Processes: Characterization and Convergence*, John Wiley, New York.

Fleming, W. H. and Rishel, R. W. (1975) *Deterministic and Stochastic Optimal Control*, Springer-Verlag, New York.

Gątarek, D. (1990) On value functions for impulse control of piecewise-deterministic processes. *Stochastics and Stochastics Reports*, **32**, 27–52.

Gątarek, D. (1991) On first-order quasi-variational inequalities with integral terms. *Appl. Math. Optim.*, **24**, 85–98.

Gątarek, D. (1992) Optimality conditions for impulsive control of piecewise deterministic processes. *Math. Control Signals Systems* **5**, 217–32.

Gerber, H.E. (1979) *An Introduction to Mathematical Risk Theory*, Huebner Foundation Monographs, Wharton School, Philadelphia.

Getoor, R.K. (1975) *Markov Processes: Ray Processes and Right Processes*, Lecture Notes in Mathematics 440, Springer-Verlag, Berlin.

Gnedenko, B.V. and Kovalenko, I.I. (1966) *Introduction to the Theory of Mass Service* [in Russian], Nauka, Moscow [English translation 1968, Jerusalem].

Grandell, J. (1991) *Aspects of Risk Theory*, Springer-Verlag, Berlin.

Gugerli, U.S. (1986) Optimal stopping of piecewise-deterministic Markov processes. *Stochastics*, **19**, 221–36.

Hestenes, M. (1986) *Calculus of Variations and Optimal Control Theory*, Wiley, New York.

Heyman, D.P. and Sobel, M.J. (1990) *Handbooks in Operations Research and Management Science, Vol. 2: Stochastic Models*, North Holland, Amsterdam.

Hirsch, M. and Smale, S. (1974) *Differential Equations, Dynamical Systems and Linear Algebra*, Academic Press, New York.

Jacod, J. (1975) Multivariate point processes: predictable projections, Radon-Nikodym derivatives, representation of martingales. *Z. Wahrscheinlichkeitstheorie ver. Geb.*, **31**, 235–53.

Jacod, J. (1979) *Calcul Stochastique et Problèmes de Martingales*, Lecture Notes in Mathematics **714**, Springer-Verlag, Berlin.

Kalashnikov, V.V. (1975) Some properties of piecewise linear Markov processes. *Teoria Veroyatnostei i ee Primeneniya*, **20**, 560–71.

Kalman, R.E. (1960) Contributions to the theory of optimal control. *Bol. Soc. Math. Mexicana*, **5**, 102–19.

Karatzas, I. and Shreve, S.E. (1988) *Brownian Motion and Stochastic Calculus*, Springer-Verlag, New York.

Kaspi, H. and Mandelbaum, A. (1992) On the ergodicity of a closed queueing network. *Stochastics and Stochastics Reports*, **39**, 239–58.

Kingman, J.F.C. and Taylor, S.J. (1966) *Introduction to Measure and Probability*, Cambridge University Press.

Kolmogorov, A.N. (1933) *Grundbegriffe der Wahrscheinlichkeitsrechnung*, Springer-Verlag, Berlin.

Krasnolselskii, M.A. (1964) *Positive Solutions of Operator Equations*, Noordhoff, Groningen.

Kushner, H.J. (1977) *Probability Methods for Approximations in Stochastic Control and for Elliptic Equations*, Academic Press, New York.

Lenhart, S.M. (1987) Viscosity solutions associated with switching control problems for piecewise-deterministic processes. *Houston J. Math.*, **13**, 405–26.

Lenhart, S.M. (1989) Viscosity solutions associated with impulse control problems for piecewise-deterministic processes. *Int. J. of Math. and Math. Sci.*, **12**, 145–58.

Lenhart, S.M. and Liao, Y.-C. (1985) Integro-differential equations associated with optimal stopping of a piecewise-deterministic process, *Stochastics*, **15**, 183–207.

Lenhart, S.M. and Liao, Y.-C. (1988) Switching control of piecewise-deterministic processes. *J. Optimization Th. and Appls.*, **59**, 99–116.

Lenhart, S.M. and Yamada, N. (1991) Viscosity solutions associated with switching game for piecewise-deterministic processes. *Stochastics and Stochastics Reports*, **38**, 27–47.

Lepeltier, J.P. and Marchal, B. (1984) Théorie générale du contrôle impulsionel markovien. *SIAM J. Control Optim.*, **22**, 645–65.

Lions, J.L. (1969) *Quelques Méthodes de Résolution de Problèmes aux Limits Non-linéaires*, Dunod, Paris.

Lopes, E. Prado (1992) PhD Thesis, University of London.

Merton, R.C. (1991) *Continuous-Time Finance*, Basil Blackwell, Oxford.

Meyer, P.-A. (1966) *Probability and Potentials*, Blaisdell, Waltham Mass.

Meyn, S.P. and Tweedie, R.L. (1992) Stability of Markovian processes. I: Criteria for discrete time chains. *Adv. Appl. Prob.*, **24**, 542–74.

Meyn, S.P. and Tweedie, R.L. (1993) *Markov Chains and Stochastic Stability*, Control and Communication in Engineering, Springer-Verlag, New York.

Michalska, H. and Mayne, D.Q. (1991) Receding horizon control of nonlinear systems without differentiability of the optimal value function, *Systems and Control Letters*, **16**, 123–30.

Pontryagin, L.S., Boltyanskii, V.G., Gamkrelidze, R.V. and Mischenko, E.F. (1962) *The Mathematical Theory of Optimal Processes*, Wiley-Interscience, New York.

Press, W.H., Flannery, B.P., Teukolsky, S.A. and Vetterling, W.T. (1986) *Numerical Recipes: the Art of Scientific Computing*, Cambridge University Press, New York.

Protter, M.H. and Morrey, C.B. (1977) *A First Course in Real Analysis*

Ribeiro Do Val, J.B. (1987) PhD Thesis, Imperial College, University of London.

Ripley, B.D. (1987) *Stochastic Simulation*, Wiley, Chichester.

Rishel, R. (1990) Controlled continuous-time Markov processes (Chapter 9 of D.P. Heyman and M.J. Sobel, eds.), *Handbooks in Operations Research and Management Science, Vol 2: Stochastic Models*, North-Holland, Amsterdam.

Royden, H.L. (1988) *Real Analysis* (3rd edn), Macmillan, New York.

Schornagel, A. (1987) *Optimum Preventive Maintenance for Complex Systems with Independent Equipment Units*, Shell Research Laboratory, Amsterdam.

Sharpe, M. (1988) *General Theory of Markov Processes*, Academic Press, San Diego.

Shiryaev, A.N. (1978) *Optimal Stopping Rules*, Springer-Verlag, Berlin.

Skiadas, C.N. (1986) Undergraduate thesis, Department of Electrical Engineering, Imperial College London.

Soner, H.M. (1986) Optimal control with state-space constraint II. *SIAM J. Control and Optimization*, **24**, 1110–22.

Stettner, Ł. and Zabczyk, J. (1981) Strong envelopes of stochastic processes and a penalty method. *Stochastics*, **4**, 267–80.

Stroock D.W. and Varadhan, S.R.S. (1969) Diffusion processes with continuous coefficients I, II. *Comm. Pure and Appl. Math.*, **22**, 345–400 and 479–530.

Tijms, H.C. (1986) *Stochastic Modelling and Analysis*, John Wiley, Chichester.

Tweedie, R.L. (1975) Sufficient conditions for ergodicity and recurrence of Markov chains on a general state space. *Stoch. Proc. Appls.*, **3**, 385–403.

Ulam, S. (1932) Zum Maßbegriffe in Produkträumen. *Proc. Int. Cong. of Mathematicians*, Zürich, **2**, 118–19.

van der Duyn Schouten, F.A. (1983) *Markov Decision Drift Processes*, CWI Tract, Centre for Mathematics and Computer Science, Amsterdam.

Vermes, D. (1980) Optimal dynamic control of a useful class of randomly jumping processes, *Report PP-80-15*, International Institute for Applied Systems Analysis, Laxenburg, Austria.

Vermes, D. (1985) Optimal control of piecewise-deterministic Markov processes. *Stochastics*, **14**, 165–208.

Vinter, R.B. and Lewis R.M. (1978) A necessary and sufficient condition for optimality of dynamic programming type, making no *a priori* assumptions on the controls. *SIAM J. Control and Optimization*, **16**, 571–83.

Warga, J. (1972) *Optimal Control of Differential and Functional Equations*, Academic Press, New York.

Waters, H. *et al.* (1991) *Permanent Health Insurance*, CMI Report 12, Institute of Actuaries, London.

Wentzell, A.D. (1981) *A Course in the Theory of Stochastic Processes*, McGraw-Hill, New York.

Whittle, P. (1982) *Optimization over Time*, Vol. I, John Wiley, Chichester.

Williams, D. (1979) *Diffusions, Markov Processes and Martingales I*, John Wiley, Chichester.

Williams, D. (1990) *Probability with Martingales*, Cambridge University Press.

Wong, E. and Hajek, B. (1984), *Stochastic Processes in Engineering Systems*, Springer-Verlag, New York.

Ye, J.J. (1990) *Optimal Control of Piecewise-deterministic Markov Processes*, PhD Thesis, Dalhousie University.

Yushkevich, A.A. (1983) Continuous time Markov decision processes with interventions. *Stochastics*, **9**, 235–74.

Yushkevich, A.A. (1987) Bellman inequalities in Markov decision deterministic drift processes. *Stochastics*, **23**, 25–77.

Zabczyk, J. (1983) Stopping problems in stochastic control. *Proc. Int. Cong. Mathematicians, Warsaw*, 1425–37, PWN, Warszawa.

Index of notation

Subject index